Epoch	NALMA	SALMA	ELMA	ALMA
Pleistocene	IRV./RLB	Ensenadan Lujanian	Biharian	Nihewanian
Pliocene	Blancan	Uquian Chapadmalalan	Villafranchian Ruscinian	Youhean Jinglean
Miocene	Hemphillian Clarendonian Barstovian Hemingfordian	Montehermosan Huayquerian Mayoan Laventan Colloncuran Santacrucian Colhuehuapian	Turolian Vallesian Astaracian Orleanian Agenian	Baodean Bahean Tunggurian Shanwangian Tabenbulakian
Oligocene	Arikareean Whitneyan Orellan	Deseadan [hatched gap] Tinguiririca	Arvenian Suevian	Hsandgolian Kekeamuan Houldjinian
Eocene	Chadronian Duchesnean Uintan Bridgerian Wasatchian	Mustersan Casamayoran (Barrancan) [hatched gap] Casamayoran (Vacan)	Ludian Robiacian Geiselthalian Sparnacian	Ergilian Sharamurunian Irdinmanhan Bumbanian
Paleocene	Clarkforkian Tiffanian Torrejonian Puercan	Riochican Itaboraian "Peligran" Tiupampan	Cernaysian Hainin	Gashatan Nongshanian Shanghuan

THE PRINCETON FIELD GUIDE TO PREHISTORIC MAMMALS

Donald R. Prothero
With illustrations by Mary Persis Williams

Princeton University Press
Princeton and Oxford

Published by Princeton University Press, 41 William Street, Princeton, New Jersey 08540
In the United Kingdom: Princeton University Press, 6 Oxford Street, Woodstock, Oxfordshire OX20 1TR

press.princeton.edu

Jacket illustration by Mary Persis Williams

ISBN 978-0-691-15682-8

Library of Congress Control Number: 2016946980

British Library Cataloging-in-Publication Data is available

For details of the mammals depicted in the illustration on the previous page, see page 217.

This book has been composed in Galliard, Goudy and Optima

Printed on acid-free paper. ∞

Designed by D & N Publishing, Baydon, Wiltshire, UK

Printed in China

10 9 8 7 6 5 4 3 2 1

CONTENTS

DEDICATION

This book is dedicated to the great paleontologists who have contributed so much to our modern understanding of the evolution of fossil mammals, and taught me so much about them:

Dr. Michael O. Woodburne

Dr. Earl Manning

Dr. Richard Tedford

Dr. Malcolm C. McKenna

Dr. Robert J. Emry

Dr. Christine Janis

Dr. Spencer Lucas

Dr. Robert Schoch

Dr. Michael Novacek

Dr. J. David Archibald

Dr. Rich Cifelli

Dr. John Flynn

Dr. Bruce MacFadden

Dr. Richard Stucky

and

Dr. S. David Webb

PREFACE

This book came about when I first saw Greg Paul's *Princeton Field Guide to Dinosaurs* (2010) and felt that a similar book should be done for fossil mammals. It became a reality when Robert Kirk at Princeton University Press encouraged me to write such a book to complement their existing series of Princeton Field Guides.

However, in many ways this book cannot follow the format of Paul's book. He had only a few hundred species of dinosaurs to discuss; likewise, most field guides to living animals have only a few hundred species to list. By contrast, there are over 5,500 species of living mammals, and many thousands of species of fossil mammals. Most of these are known only from teeth and would not be suitable for the large-scale skeletal reconstructions that Paul did. McKenna and Bell (1997) required over 600 pages just to list all the genera of fossil and living mammals, giving each genus only one or two lines apiece, with no illustrations. In fact, there are more species of fossil rodents (again, largely known from teeth only) than there are of all named and described dinosaurs. In addition, the species-level taxonomy of many fossil mammals is still a mess. There are hundreds of invalid species that no paleontologist takes seriously but have not yet been revised. Clearly the scale and approach of this volume must be different, emphasizing the orders, families, and genera, and focusing on the genera that are known from partial or complete skeletal material, not just teeth.

On the other hand, this huge diversity of taxa and abundance of material of fossil mammals gives us the ability to look at some problems (e.g., detailed patterns of evolution through time, fine-scale biogeography, large population samples that allow statistical analysis and studies of variation and ontogeny) that could not be studied in the much scarcer fossils of dinosaurs. Thus, I will follow the broad format used in Paul's book and other Princeton Field Guides, but the detailed discussions will reflect the strengths and weaknesses of the mammalian fossil record.

Acknowledgments

I thank Robert Kirk of Princeton University Press for encouraging us in this project, his colleague Kathleen Cioffi for ushering the guide through production, and Amy K. Hughes for copyediting. I thank J. David Archibald, Christine Janis, and Spencer Lucas for their helpful reviews of the entire book, and Bill Sanders and Darin Croft for comments on specific chapters. I thank all the artists and people who generously sent me photographs and art; their work is acknowledged in the illustration credits section.

Finally, I thank my amazing wife, Dr. Teresa LeVelle, and my wonderful sons, Erik, Zachary, and Gabriel, for all their love and support during the project.

CHAPTER 1

THE AGE OF MAMMALS

Today, the earth is host to about 5,500 species of mammals. They range in size from the tiny bumblebee bat (a bit over an inch long) to the mighty blue whale, the largest animal ever to have lived on the planet, which can grow to 30 m (100 ft) long and weigh 190 metric tons (210 short tons). They occupy a wide variety of ecologies and habitats, and their forms range from burrowers to fast land runners to huge elephants to flying bats to a spectrum of marine creatures including whales, manatees, seals, and sea lions. Mammals eat a wide range of foods, from vegetation of every sort to a variety of prey, including other mammals, birds, fish, reptiles, and smaller vertebrates, as well as insects and even the plankton in the ocean.

Figure 1.1. The famous Ashfall Fossil Bed State Park near Orchard, Nebraska. Most fossil mammals are found as fragments of teeth and jaws, but there are rare complete examples from exceptional localities. Nicknamed the "Rhino Pompeii," Ashfall preserves the remains of hundreds of hippo-like rhinoceroses (*Teleoceras major*) that suffocated and died when they became trapped in and inhaled volcanic ash that covered the region 10 Ma. The complete skeleton of each animal, down to the tiniest throat bone and the remains of its last meal, is preserved. Some females had unborn babies in them or a calf nearby in nursing position. A small number of horses, musk deer, birds, and other mammals from the middle Miocene of Nebraska were also trapped and fossilized.

Mammals have been the largest and most dominant creatures on the planet ever since all the dinosaurs—except for their bird descendants—vanished 66 Ma (million years ago). The earth is home not only to wild mammals, found on every continent, in the skies, and in all parts of the ocean, but to domesticated animals (cattle, sheep, horses, goats, and pigs). Humans are now the dominant large species nearly everywhere. Mammals took over the niches for large land animals that the dinosaurs once occupied on land, as well as the large marine predator roles once occupied by marine reptiles.

To some people, prehistoric mammals may not seem as glamorous as the dinosaurs, but the study of mammal fossils offers many advantages over the study of dinosaur fossils. For one thing, fossil mammals are much more abundantly and completely preserved than dinosaurs, so we typically have hundreds of specimens of many kinds of mammals (Fig. 1.1), while most dinosaurs are known from at best a few fragments or partial skeletons. Thus, **paleontologists** (scientists who study fossils) can use mammals to examine complex ancient ecological communities, or look at how populations of ancient mammals behaved, or decipher their patterns of evolution through time. Even more important, mammals evolved rapidly and their fossils are typically abundant, so they are very useful for establishing the age of rocks, especially on land, of the past 66 m.y. (million years).

Finally, prehistoric mammals are just as amazing as any dinosaur, but in their own ways. Saber-toothed cats and giant mammoths are just as popular as dinosaurs in the public imagination, along with many other familiar Ice Age mammals such as gigantic ground sloths, enormous bison, and hulking mastodonts. The monstrous rhinoceroses known as indricotheres, the largest land mammals that ever lived, have been featured in many documentaries. How can you top the ultimate weirdness of the six-horned and fanged uintatheres, or the huge brontotheres, with the blunt paired horns on their noses, or the giraffes built like moose and the camels built like giraffes?

DATING ROCKS

The fossils of the common ancestors of all mammals are known as far back as 165 Ma, while the earliest known mammal fossils are from at least 225 m.y. in the past, and the earliest relatives of the mammal lineage go back roughly 315 m.y. How do we know this? How can we talk about the age of fossils?

The principles of establishing the age and sequence of layered rocks that yield fossils is known as **stratigraphy** ("study of layered rocks," from Greek). The first principle is known as **superposition**, first proposed by Danish doctor Nicholas Steno in 1669. In any layered sequence of rocks (such as layers of sedimentary rocks, or even lava flows), the oldest layers are at the bottom, and the layers get younger as you go up the sequence. This is simple common sense: you cannot put something on top of a stack unless the stack is already there. Think of a pile of papers on a messy desk. The ones you looked at last are at the top, while those you may not have seen for weeks are lower in the pile.

By 1795 pioneering geologists like William Smith in England showed that the fossil record shows a definite nonrepeating sequence of extinct animals through time; this is known as **faunal succession** or fossil succession. As Smith and later geologists realized, this sequence is key to establishing **relative age**, or age in relation to something else. In other words, we want to know whether a certain fossil is younger than one layer or fossil assemblage but older than another one.

The sequence of fossils and history of life through time was first worked out in fossil fields in England, then deciphered in fossil digs all over the world. This sequence can be seen in many places, including western North America, especially in the Rocky Mountains and the western High Plains. Layered rocks and their mammal fossils can be found in sequence (Fig. 1.2), starting with beds from the Late Cretaceous (end of the Age of Dinosaurs) in bowl-shaped sedimentary basins in the American West, and then through the first 25 m.y. of the Age of Mammals. In some places, we can find rocks that go from 70 Ma to 45 Ma in one sedimentary basin. The next part of the sequence (about 45–40 Ma) is best preserved in the Uinta Basin of northeastern Utah. The interval from 37 Ma to the last Ice Age is almost continuously represented by excellent exposures with beautiful fossils in western Nebraska, South Dakota (especially in the Big Badlands), and eastern Wyoming. We can objectively demonstrate that the history of mammalian fossils occurs in a certain order, because the entire **relative sequence** is well preserved just in the northern plains alone.

From the relative sequence of fossils all over the world, geologists over the last 200 years have pieced together the standard **geologic time scale** (Fig. 1.3). The time scale is broken down into a hierarchy of large units subdivided into smaller units. For example, the last 540 m.y. is known as the **Phanerozoic Eon**, and it is subdivided into three **eras**: the Paleozoic ("ancient life" in Greek, 540–250 Ma); the Mesozoic ("middle life," 250–66 Ma), also known as the Age of Dinosaurs; and the Cenozoic ("recent life," 66 Ma to present), also known as the Age of Mammals. The eras, in turn, are subdivided into smaller units known as **periods**. The Mesozoic Era is divided into the Triassic, Jurassic, and Cretaceous Periods. The traditional division of the Cenozoic Era was the Tertiary (65–2.6 Ma) and Quaternary (2.6 Ma–present) Periods, although more recently geologists have come to prefer a more balanced subdivision into Paleogene (66–23 Ma) and Neogene (23 Ma to present) Periods. Finally, the periods are divided into smaller units called **epochs**. The Paleogene Period includes the Paleocene (66–55 Ma),

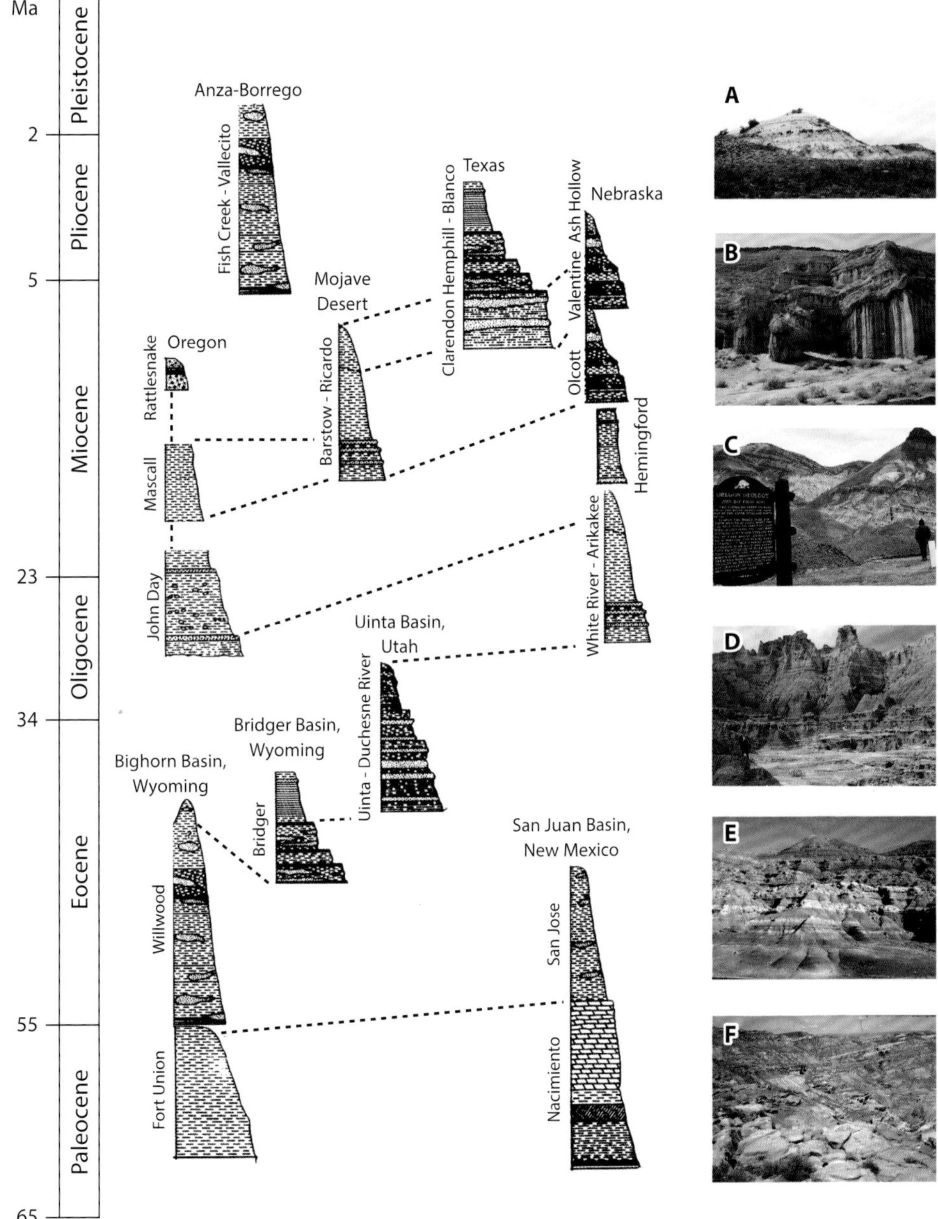

Figure 1.2. The overlapping sequence of fossiliferous sections found in the basins of the western United States. The rocks are correlated with one another and with distinctive fossil assemblages that changed rapidly through time. The Paleocene is represented by rocks in the Bighorn Basin (photo F) of Wyoming (Fort Union Group) and New Mexico (Nacimiento Formation). The early Eocene is represented by many sequences of rocks in the Rocky Mountain basins, including the Willwood Formation of the Bighorn Basin, the Wasatch Formation in the Wind River and Powder River basins of Wyoming, and the San Jose Formation in New Mexico. The early middle Eocene can be found at the top of the Willwood and Wasatch Formations, and also in the overlapping Bridger Formation of southwestern Wyoming. This section overlaps with the base of the late middle Eocene Uinta and Duchesne River formations of the Uinta Basin in Utah (photo E). The uppermost Utah rocks overlap with the upper Eocene rocks at the base of the White River Group (Big Badlands) in Nebraska, Wyoming, and the Dakotas (photo D). The early Oligocene is best demonstrated in the White River Group of the Big Badlands, and the late Oligocene–early Miocene in the Arikaree Group of Nebraska and South Dakota, and also the John Day Formation of central Oregon (photo C). The late early Miocene can be found in the Hemingford Group of Nebraska and the upper John Day Formation. The middle Miocene is well represented by the Barstow Formation in California, the Mascall Formation in Oregon, and the Olcott Formation in western Nebraska. The middle late Miocene can be found in the Ricardo Group of Redrock Canyon in California (photo B), the Valentine and Ash Hollow Formations of Nebraska, the Rattlesnake Formation of Oregon, and the Clarendon and Hemphill beds of the Texas Panhandle (photo A). The Pliocene can be documented from the Blanco beds of Texas and the long sequence of the Palm Springs Formation in the Fish Creek–Vallecito badlands of the Anza-Borrego Desert in California, which goes through the Pleistocene Ice Ages, as well.

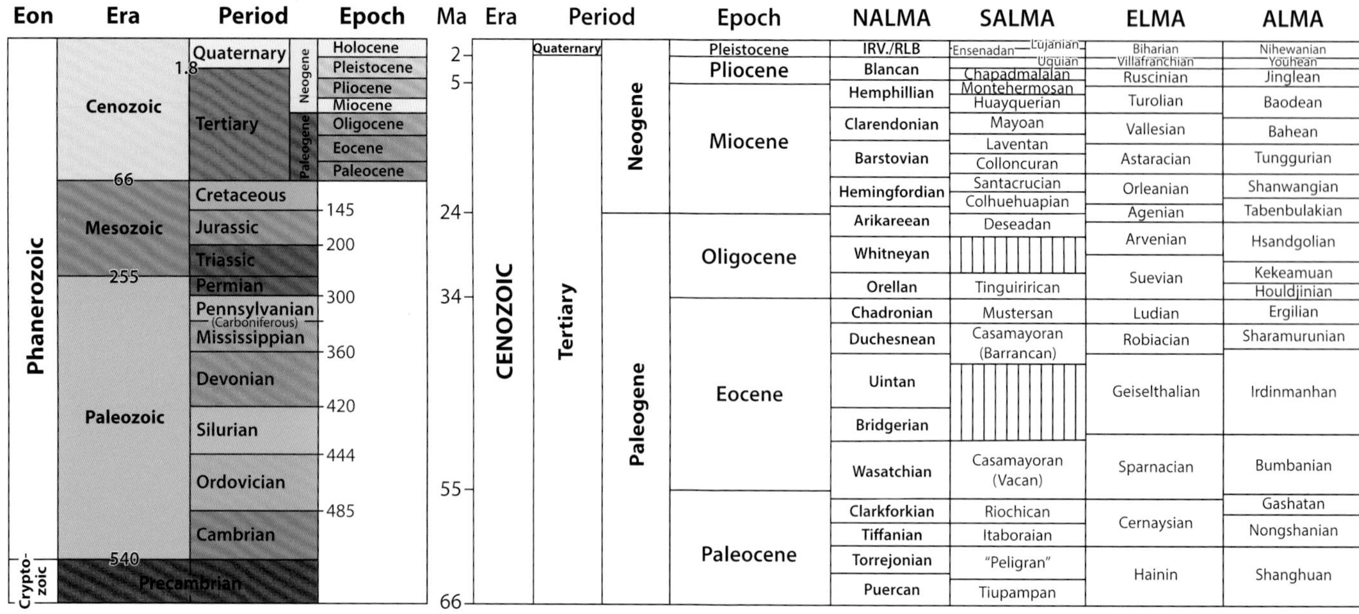

Eon
Era
Period
Epoch
Ma
Era
Period
Epoch
NALMA
SALMA
ELMA
ALMA
Phanerozoic
Cryptozoic
Cenozoic
Mesozoic
Paleozoic
Quaternary
Tertiary
Neogene
Paleogene
Holocene
Pleistocene
Pliocene
Miocene
Oligocene
Eocene
Paleocene
1.8
66
255
540
Cretaceous
Jurassic
Triassic
Permian
Pennsylvanian
(Carboniferous)
Mississippian
Devonian
Silurian
Ordovician
Cambrian
Precambrian
145
200
300
360
420
444
485
Age of the Earth = 4.5 billion years
2
5
24
34
55
66
CENOZOIC
Quaternary
Tertiary
Neogene
Paleogene
Pleistocene
Pliocene
Miocene
Oligocene
Eocene
Paleocene
IRV./RLB
Blancan
Hemphillian
Clarendonian
Barstovian
Hemingfordian
Arikareean
Whitneyan
Orellan
Chadronian
Duchesnean
Uintan
Bridgerian
Wasatchian
Clarkforkian
Tiffanian
Torrejonian
Puercan
Ensenadan
Lujanian
Uquian
Chapadmalalan
Montehermosan
Huayquerian
Mayoan
Laventan
Colloncuran
Santacrucian
Colhuehuapian
Deseadan
Tinguirirican
Mustersan
Casamayoran (Barrancan)
Casamayoran (Vacan)
Riochican
Itaboraian
"Peligran"
Tiupampan
Biharian
Villafranchian
Ruscinian
Turolian
Vallesian
Astaracian
Orleanian
Agenian
Arvenian
Suevian
Ludian
Robiacian
Geiselthalian
Sparnacian
Cernaysian
Hainin
Nihewanian
Youhean
Jinglean
Baodean
Bahean
Tunggurian
Shanwangian
Tabenbulakian
Hsandgolian
Kekeamuan
Houldjinian
Ergilian
Sharamurunian
Irdinmanhan
Bumbanian
Gashatan
Nongshanian
Shanghuan

Figure 1.3. Left: The geologic time scale. Right: The Cenozoic land mammal ages on different continents. NALMA = North American Land Mammal Ages; SALMA = South American Land Mammal Ages; ELMA = European Land Mammal Ages; ALMA = Asian Land Mammal Ages.

Eocene (55–34 Ma), and Oligocene (34–23 Ma) Epochs, while the Neogene Period contains the Miocene (23–5 Ma), Pliocene (5–2.6 Ma), Pleistocene or "Ice Ages" (2.6 Ma–10,000 years ago), and Holocene (the last 10,000 years) Epochs.

These subdivisions are global and universal, no matter what continent or ocean your fossils come from. The epochs, however, can be subdivided into smaller local units called **ages**, which are based on the local sequence of fossils. For example, the dramatic change in fossil mammal assemblages from one layer to the next allows paleontologists to define local land mammal ages. The local ages are different for each continent, so there are North American land mammal ages (NALMA), as well as systems of similar land mammal ages for South America (SALMA), Asia (ALMA), Europe (which uses a system of abbreviations, called "MP" for Mammalian Paleogene and "MN" for Mammalian Neogene), and Africa. Finally, in many places the land mammal ages are subdivided into even finer units, such as zones, based on the stratigraphic occurrence of key fossils. These are the finest-scale subdivisions for the Cenozoic, and often let us distinguish events less than 10,000 years in duration.

CLOCKS IN ROCKS

But how do we determine the ages of these rocks and fossils in terms of numbers, such as millions of years ago? This is the principle of **numerical dating** (once incorrectly called "absolute dating"). We cannot directly date most sandstones, shales, and limestones and other layered sedimentary rocks, so we must use a different principle: radioactive decay. The principle of radioactivity was first discovered in 1896, and by 1913 geologists found a way to interpret it to determine the numerical ages of rocks.

The dating process all starts with molten rock, or **magma**, which cools into crystals and forms what is known as an igneous rock. You have probably seen videos of magmas erupting to form igneous rocks, from the black basalts that form the lava flows in Hawaii to the explosive volcanic ashfalls that happened after the eruption of Mount Vesuvius above Pompeii and Mount St. Helens in Washington. Other magmas cool very slowly under the earth in huge magma chambers to form rocks such as granites. At first some radioactive elements, such as uranium-238, uranium-235, and potassium-40, float around in the magma, but eventually, as crystals form and cool, these elements are locked into the crystals, as if enclosed in a crystalline cage. Over long periods of time, these radioactive "parent" atoms begin to spontaneously decay to a different "daughter" atom. Uranium-238 decays to lead-206; uranium-235 decays to lead-207; potassium-40 into argon-40; and so on. The rate of this decay is precisely known, so all scientists have to do is measure the number of parent and daughter atoms in a crystal, and the ratio between them will indicate how long the radioactive elements have been "ticking" like a clock.

To make the principle of numerical dating work, however, we need these igneous rocks (such as ashfalls or lava flows) to occur in beds above, below, or between the fossil-bearing layers. Then the numerical date on the igneous layer will bracket the age of the fossiliferous layers above or below it. In the fossil-mammal-bearing rocks of the Rocky Mountains and High Plains (Fig. 1.2), for example, there are many deposits of volcanic ashes, which have helped us assign very precise numerical ages to most North American mammal fossils, down to the nearest 100,000 years for fossils that are tens of millions of years old.

There is one exception to the general rule that we need igneous rocks in order to date the sediments containing the fossils. This is when we use the carbon-14 dating method, the one most people have heard about. However, carbon-14 decays very rapidly (half of the original carbon-14 parent atoms decay every 5,370 years), so it is useful only for objects less than 60,000 to 80,00 years old. After that, the radiocarbon "clock" has run down and the material cannot be dated by this method. Thus, radiocarbon is useful only for samples younger than 80,000 years, dating from some time during the last Ice Age, or for events in the last 10,000 years since the last Ice Age. For that reason, radiocarbon dating is used only by geologists and paleontologists who work on the last Ice Age, and by archaeologists or anthropologists who work on human evolution over the past 80,000 years. There is one big advantage of carbon-14, though—it can be used to date the object directly. Any object with carbon in it (bone, wood, pottery, fabric, and many other materials) is suitable if it is young enough. By contrast, to use potassium-40/argon-40 dating for fossil dating, there must be a volcanic layer interbedded with the fossil-bearing sediments. But potassium-argon dating works for rocks as old as the earth (4.6 billion years old), and as young as just a million years, so it is the preferred method for most geologic problems and most of the dates in mammalian evolution.

WHAT'S IN A NAME?

The system of naming and classifying animals has been handed down since the days of Swedish naturalist Carl von Linné. He is better known as "Carolus Linnaeus," the Latinized form of his name, since all scholars wrote in Latin then. In 1758, Linnaeus published the tenth edition of his classification of animals, and this is the foundation of all classification schemes since then.

Every organism—animal, plant, fungus, bacterium, and other living form on the planet—has a formal, universally accepted **scientific name**, which is usually different from the common name that local people give it. Many animals have common names based on the language of the country they are found, so the *javelina* in Latin America is known as a "peccary" north of the U.S. border. Even in areas with the same language, there are often regional names. In much of the United States, "gopher" means a small burrowing rodent, while in others "gopher" describes a tortoise.

But scientific names avoid such problems, because they are global and agreed upon by all scientists, no matter what language they speak or what local names they use. In scientific terms, the rodent gopher goes by the scientific name *Geomys*, but the tortoise is *Gopherus*. The collared peccary (in the English-speaking world) or *javelina* (in Latin America) is always *Pecari tajacu*, no matter what language the scientist speaks. Most of the extinct mammals mentioned in this book do not have a common or popular name, so we have only their formal scientific names with which to refer to them.

Each organism on the planet has a scientific name made out of two parts, or a **binomen**. The first part is the *genus* (plural, *genera*) name, which is easy to recognize because it is always capitalized and always either underlined (when handwritten) or italicized (in print). The second part of the name indicates the *species* within that genus (*species* is both singular and plural). The name of the species is never capitalized (even if it comes from a proper noun), but it too is always italicized or underlined. Thus, our scientific name is genus *Homo* and species *sapiens*, making the binomen *Homo sapiens*. In zoology, the genus name can never be used for more than one kind of animal, but species names are reused all the time, appended to different genus names. Thus, the species name cannot stand by itself, but must always be accompanied by the genus name, so we can write *Homo sapiens*, or *H. sapiens*, but never just "sapiens." The scientist can base the genus or species name on anything, except that he or she cannot name a species after himself or herself. One can, however, name a species after someone else and have that person return the favor with a different new species.

A genus must contain at least one species, or it can have more. *Homo sapiens* is only one species in the genus *Homo*. There are also *Homo neanderthalensis*, *Homo erectus*, *Homo habilis*, the newly discovered *Homo naledi*, and several others. Each genus is contained within the next rank up, *family*; the family may contain one genus or a cluster of genera. In animals the scientific family name always ends with the suffix *-idae*. *Homo sapiens* belongs to the family Hominidae (which also contains other genera, such as *Sahelanthropus*, *Ardipithecus*, *Paranthropus*, *Australopithecus*, and so forth), while dogs are in the Canidae, cats in the Felidae, rhinoceroses in the Rhinocerotidae, and so on. Families are clustered into a bigger group, the *order*. The major divisions of the mammals are orders, such as the order Primates (lemurs, monkeys, apes, and humans), the order Carnivora (flesh-eating mammals like cats, dogs, bears, hyenas, weasels, raccoons, seals, and walruses), and the order Rodentia (rodents).

All the orders are clustered into a larger group, the *class*. The only class we will discuss in this book is the class Mammalia, but other animal classes group birds, reptiles, amphibians, and fishes of several kinds. Classes are clustered into an even bigger group, the *phylum* (plural, *phyla*). The phylum Chordata includes all animals with backbones (vertebrates like ourselves), but there are phyla for the snails, clams, squids, and octopus (phylum Mollusca); the insects, spiders, scorpions, millipedes, and their kin (phylum Arthropoda); the sea stars, sea urchins and their relatives (phylum Echinodermata); the corals, sea anemones, and sea jellies (phylum Cnidaria); and dozens of other types of invertebrates. Finally, the phyla are united into *kingdoms*. We are members of the kingdom Animalia, but there are also a plant kingdom, fungus kingdom, and several more for single-celled organisms.

An example of the Linnaean hierarchy for humans and for the Indian rhinoceros is given below:

KINGDOM	Animalia	Animalia	(animals)
PHYLUM	Chordata	Chordata	(vertebrates)
CLASS	Mammalia	Mammalia	(mammals)
ORDER	Primates	Perissodactyla	
FAMILY	Hominidae	Rhinocerotidae	
GENUS	*Homo*	*Rhinoceros*	
SPECIES	*sapiens*	*unicornis*	

There are a lot of rules governing how scientific names are created and what they mean, which we will not discuss at length. A few are worth mentioning, however, because they come up frequently in discussing fossils. The most important is the **rule of priority**: the first name given to an organism is the only valid name, unless there are problems with that name. For example, paleontologists no longer use the name *Brontosaurus*, because the same scientist had already named different specimens of the same animal *Apatosaurus* some years earlier. (Recently, a few paleontologists have tried to resurrect *Brontosaurus*, but this is still controversial). A paleontologist must keep track of names, and find all the older names applied to a group of fossils, to determine which has priority and is valid and which ones are **junior synonyms** that must be abandoned. This is mandatory, no matter how popular a name may be with the public, as *Brontosaurus* is. There are rare exceptions, but usually the only way to get rid of an obscure and inconvenient but older name is to get an international commission to rule on it.

The rule of priority is in place to prevent unnecessary and pointless arguments about which name is correct. The rule holds even if the name becomes inappropriate. For example, in 1878 the first name given to the Big Badlands rhinoceros was *Subhyracodon*, which falsely suggests that the creature was a hyracodont, a member of an extinct family distantly related to true rhinoceroses, family Rhinocerotidae. For years, paleontologists have preferred an 1880 name, *Caenopus*, for this true rhino fossil, but it does not matter. By rules of priority, the misleading name *Subhyracodon must* be used, and *Caenopus* cannot be used. Similarly, early scientists named a giant fossil *Basilosaurus* (which means "emperor lizard"), thinking it was a reptile, even though it later turned out to be an early whale, not a lizard or reptile of any kind.

The other rules establish what kinds of things must be supplied to make a scientific name valid. For the last century, a new scientific name must include a clear diagnosis of how to tell it apart from other similar species, a good description of the specimens, good illustrations, a list of specimens considered to be part of the species, a **type specimen** that is the basis for its designation as a species, the geographic range and time range of the species, and many other things. All of these must be published in a reputable scientific journal, not on a web page or unpublished dissertation or somewhere else. Otherwise, the new name of a genus or species is not valid, and other scientists will not recognize or use it.

These rules may seem boring and unnecessary, but they are essential to maintain order and stability in scientific naming. The scientific community set these rules up over a century ago; all other scientists follow these rules, and the scientific journals will not publish any work that violates them. Knowing these rules is like knowing the rules of the road before you take your driving test. The Department of Motor Vehicles, and all other drivers, must assume that you know the proper rules for driving, because they do not want you to cause a deadly accident if you suddenly break the rules. Thus, we have many cases where amateur fossil collectors often try to create new names, or even publish them in books and websites, but without following the rules properly. The rule book allows the professional scientists to quickly determine who is right and who is not, and whose work deserves attention and whose ought to be ignored.

HOW DO WE CLASSIFY ANIMALS?

We have established how we give names to animals, but how do we decide how to classify them? There are lots of ways that people classify things. We could sort them into categories like "good to eat" and "toxic and bad tasting," or "dangerous to humans" and "not dangerous" (as some cultures do). We could cluster them by color patterns or where they live or how they behave. The science of classification is called **taxonomy**, and any category or rank of organism or group of organisms (a genus, a species, a family) is a **taxon** (plural, *taxa*). *Homo*, *Homo sapiens*, and Hominidae are all taxa.

Before the time of Linnaeus, many natural historians realized that the best way to classify creatures was by unique anatomical specializations that distinguish them from other similar creatures. Some classifications clustered unrelated creatures like fish and dolphins together because they were aquatic or turtles and armadillos because they had a hard shell. By Linnaeus's time, scholars began to realize that the overwhelming number of anatomical specializations clump some animals together and not others.

For example, fishes and whales have superficial similarities, in that they are both swimmers with streamlined bodies and a tail fin. But if you look past these ecological overprints, you find that every other anatomical feature of fishes and whales is completely different. This similarity of external form but differences in every other part of their anatomy is a result of **evolutionary convergence**, and it has occurred often in the history of life. As we shall discuss in Chapter 3, the pouched mammals of Australia have converged in body form with many of the placental mammals of the rest of the world, since they evolved in isolation in Australia and did not encounter competition from placentals. Saber-toothed predators evolved at least four independent times in mammalian history (Fig. 1.4), including once in pouched marsupial mammals, once in an extinct group called creodonts, and twice in the order Carnivora (once in the "false cats," or nimravids, and once in the true cats, or felids).

To overcome the false trail of convergence, we need to find characteristics that are unique specializations for the group of animals we are classifying, not features left over from their remote past. These are known as **shared derived characters**, or **synapomorphies**. For example, the order Primates is distinguished by having grasping hands and feet, nails instead of claws, forward-pointing eyes with binocular vision, and good color vision. However, groups within the Primates are defined by their own specializations, so apes and humans (family Hominidae) share anatomical features such as the loss of a tail, complex nasal sinuses, five or six vertebrae in the hip, an elongate middle finger, and another dozen features in just the skeleton. We would not use the occurrence of grasping hands to define the apes and humans, because for them it is a **shared primitive character**, or **symplesiomorphy**; it is useful only to distinguish primates from other mammals, not groups within the primates. Nor would we use a very primitive feature, such as the occurrence of four limbs (Fig. 1.5). That is a shared primitive feature that apes and humans inherited from the first amphibians and not useful in defining the Hominidae. Nor would we mention the presence of a backbone, which we inherited from the earliest vertebrates.

This emphasis on basing classification on shared derived features, or shared evolutionary novelties, makes classification a reflection of the evolutionary branching history of life. This was apparent when Linnaeus's 1758 classification showed a branching pattern like a bushy "tree of life," but it was a century later that Charles Darwin emphasized that the pattern of classification was evidence for evolution. Since those days biologists

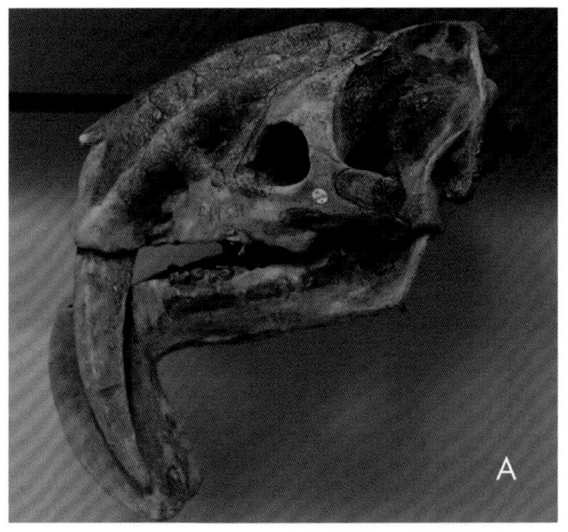

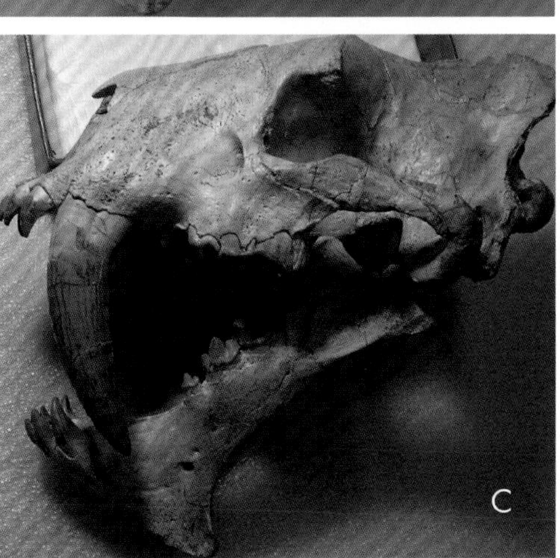

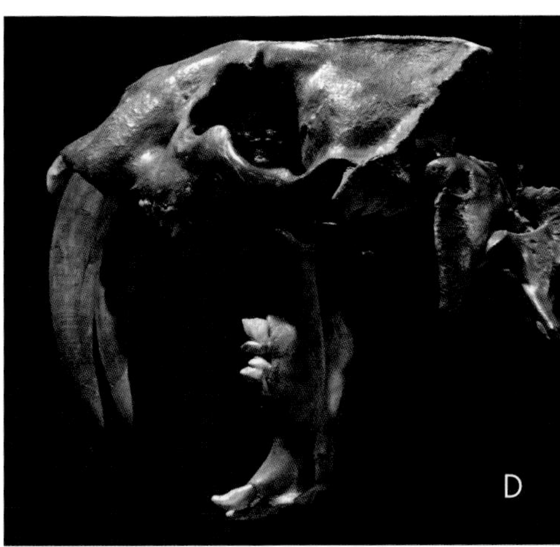

Figure 1.4. Convergent evolution on the saber-toothed skull shape has happened at least four or five times independently in fossil mammals. Even though the different groups look superficially similar, with their large stabbing canines, in the details of the skull it is clear that they are not closely related but independently evolved the saber-toothed shape from different types of ancestors. A. The saber-toothed marsupial *Thylacosmilus*. B. The saber-toothed oxyaenid creodont *Machaeroides*. C. The nimravid ("false cat") *Hoplophoneus*. D. The saber-toothed true cat *Smilodon*.

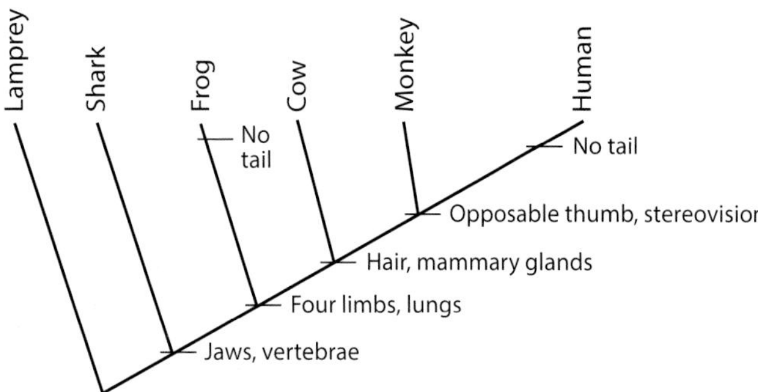

Figure 1.5. The branching diagram of the relationships of some animals, which is known as a cladogram. It shows that certain features are specialized, or derived, for some groups (such as four legs in tetrapods) but primitive in others. In this example, mammals are defined by hair and mammary glands. Primates are defined by their stereovision and grasping hands, not by more primitive features such as a backbone or four legs.

have argued over the best way to classify organisms. Traditionally, some groups of animals were clustered together based on a mixture of both shared evolutionary novelties plus primitive features. For example, the category "fish" is useful to anglers and grocers and restaurateurs and diners, but it has no meaning as a group of animals. Some "fish," such as lungfish and coelacanths, are actually part of the lineage that leads to land vertebrates. Others, such as lampreys and sharks (Fig. 1.5), are very primitive creatures not at all closely related to the bony fish, such as a tuna or a goldfish. In some circles, people talk about "jellyfish," "starfish," or "shellfish," broadly applying the name "fish" to *any* animal that lives in water, no matter what it is related to. Thus, over the past 40 years, biologists have been trying to avoid groupings of animals that are unnatural wastebaskets based on primitive similarities (like the aquatic body form of "fish").

This has long been a problem in the Mammalia. For example, many of the orders of mammals were once huge wastebaskets of creatures united only by shared primitive similarity. For a long time, the order "Insectivora" was such a wastebasket, including three kinds of mammals that were actually closely related (shrews, moles, hedgehogs) and a wide spectrum of other creatures that were insect eaters but not closely related, including tree shrews, elephant shrews, golden moles, tenrecs, and several others. As we shall discuss later, this wastebasket has long since been broken up, and only the original three groups are clustered any more. Another wastebasket was the "Condylarthra," long used for any extinct hoofed mammal that was not a member of a living group of hoofed mammals (see Chapter 13). That group was shown to be a useless wastebasket by Earl Manning, Martin Fischer, and myself in 1986. It was not only unnatural but, as my coauthors and I found when we sorted out the true relationships of its members, "Condylarthra" was covering up and obscuring what we did not know and hiding the problems that needed to be solved. Once "Condylarthra" was abandoned as a meaningless group, we made great strides in figuring out how all the hoofed mammals were interrelated. The same was true of another wastebasket group, the "Pantotheria," which was debunked in 1981.

Another problem we find with classification as we tease out the branching points of the tree is that now there are more splits that need names than we have names for. As an example, if we cluster mammals into class Mammalia and treat each order as a separate group, we find that there are lots of branching points between them. The living Mammalia first splits into three groups, the monotremes (platypus and echidnas), marsupials (pouched mammals, such as kangaroos and opossums), and placentals (mammals that give birth to well-developed young). Is each of those a subclass? Within the subclass Eutheria (placentals) we have numerous splits before we get down to the rank of order. We can use ranks such as "infraclass" and "superorder," but quickly we run out of ranks between superorder and order, and between subclass and infraclass. Consequently, the traditional ranks of classification are receiving less and less emphasis now, and there are lots of new ways of showing the branching pattern of evolution without creating formal ranks for each evolutionary branch point.

From the 1960s through 1980s there were a number of breakthroughs in thinking about classification and how to do it. Most taxonomists came to agree with the emphasis on shared evolutionary specializations and on avoiding wastebasket groups based on shared primitive similarity. Thus, classification should reflect the evolutionary branching sequence and nothing else. A group that includes all the descendants of a common ancestor is known as a **monophyletic** group (Fig. 1.6).

This change has been difficult for people accustomed to the traditional groups of animals. One of the main ideas of the new way of thinking is that if classification reflects evolutionary branching history and nothing else, then each monophyletic group should include all its descendants within it. Otherwise, it is an unnatural, arbitrary **paraphyletic** group (Fig. 1.6). Some biologists were scandalized when it became clear that birds evolved from a subgroup of dinosaurs resembling *Velociraptor*, and thus birds are descended from dinosaurs. To the modern taxonomist, birds are grouped within dinosaurs and are not a separate class, Aves, distinct from the class Reptilia. If we do not put birds within Reptilia, then "reptiles" becomes a wastebasket group of all nonmammalian land vertebrates that are not birds. Likewise, we use Amphibia to talk about salamanders and frogs, but reptiles are all descended from extinct groups sometimes placed within the Amphibia and thus a subgroup of them—so Amphibia becomes another unnatural wastebasket. Modern classification is gradually abandoning these ancient wastebasket categories that are well known but not natural by using a new set of names that are defined only by their shared evolutionary specializations. Thus, the Tetrapoda (four-legged vertebrates) includes all amphibians, reptiles, and their descendants. The Amniota (tetrapods in which the embryonic development takes place inside a membrane called an amnion, among other

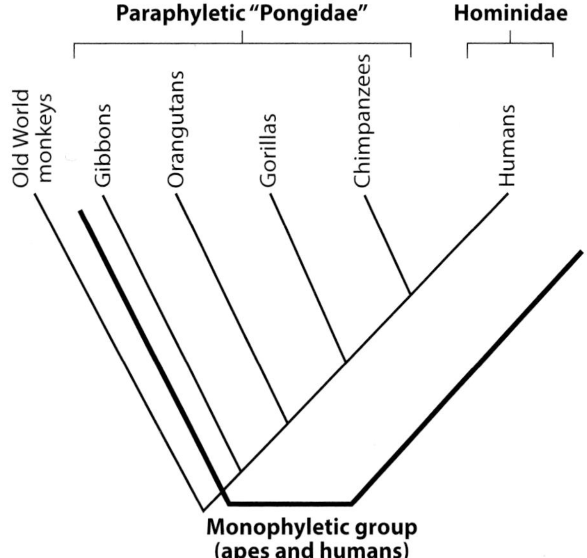

Figure 1.6. Traditional classifications emphasized the huge differences between groups, such as the difference between humans and other apes, by putting the rest of the apes in the paraphyletic family Pongidae, and separating humans in their own family of equal status, family Hominidae. Modern biological thinking, however, argues that "Pongidae" is a meaningless wastebasket group that arbitrarily excludes groups (e.g., humans) based on features of shared primitive similarity of most apes to each other. Today scientists try to recognize only monophyletic groups that include all descendants of a common ancestor without excluding any of them, no matter how different they look from their ancestors. Anthropologists now include all the great apes and humans in the family Hominidae, and the "Pongidae" is now obsolete.

features) includes all reptiles, birds, and mammals. Gradually, obsolete ideas such as "Reptilia" and "Amphibia" and "invertebrates," which are clumps of creatures defined by shared primitive specializations and not including all their descendants, are vanishing from biology or being redefined (even if the public still does not know it).

BONES VS. MOLECULES

But how are all the major orders of mammals interrelated? Since the beginning of scientific study, the primary line of evidence has been their anatomy, especially the bones and teeth, although the entire anatomy of every biological system (muscles, nerves, brains, and every other soft tissue) can be used. This evidence was first studied in detail in the 1800s by natural historians who noticed many anatomical features that we still use to classify mammals. Yet there were still lots of problems, including issues with groups of mammals that had little clear evidence as to who their closest relatives were.

In the 1980s and 1990s, however, the emphasis on shared evolutionary novelties and only natural monophyletic groups led to great strides in our understanding of mammalian relationships. The phylogeny of mammals based on anatomical features had more or less reached consensus.

At the same time, however, a new data set emerged from material that was not known to Linnaeus or Darwin or any biologists until the 1960s. These were data from molecular biology, especially the detailed sequence of biochemicals (amino acids in proteins, nucleotides in DNA) making up the genes and proteins found in any living mammal. Sure enough, these data also produced a branching sequence that closely matched the evolutionary pattern deciphered from the external anatomy, confirming that both kinds of data carried an original signal. In most cases, the branching sequences produced by anatomy closely matched the molecular branching sequence. We will talk about some of the exceptions in a later chapter.

BONES AND TEETH

We recognize mammals by a number of obvious features, including their possession of hair or fur, a warm-blooded active physiology, and females that have mammary glands to provide milk for their young. There are many other features of the soft anatomy of mammals, but paleontologists cannot use them, because such features are only rarely preserved in the fossil record. Thus, we need to find features that help us recognize and identify mammals by their fossilized hard tissues alone. For example, only mammals replace their baby teeth no more than once (reptiles replace their teeth many times and continuously), and mammals also have unique features in the ear bones and jaw seen in no other animals.

Mastering the details of skeletal anatomy is essential for mammalian paleontologists, who must be able to identify nearly any bone of any fossil mammal they find. For this book, I will use only the barest minimum of anatomical terms, and use the common English names for bones wherever possible. Many bones (such as the ribs and the vertebrae) are rarely fossilized and are not very diagnostic even when they are found, so mammalian paleontologists rarely study them (or collect them when they are found in isolation). The details of the limb bones are much more useful, not only because they are more often fossilized but also, and especially, because they can be used to infer things about how the animal walked, ran, climbed, or whatever it did.

Mammalian paleontologists are experts at seeing just a skeleton or a few skeletal pieces and trying to visualize the complete animal. However, most people see a skeleton as just a skeleton and do not notice the subtle anatomical clues. In a book like this, it is necessary to go beyond the objective reality of bones and teeth, and try to reconstruct what the creature looked like in life and what it did. Keep in mind, however, that all such reconstructions (no matter how skilled the paleoartist) are as much imagination as fact. With a few exceptions, we have no real idea what color most fossil mammals were, what their skin or fur looked like, how they behaved, how they sounded, what they thought, and many other details that are often presented as fact in documentary series, such as *Walking with Beasts*. When I have appeared in such documentaries, the directors and writers have always tried to push the limits of science, putting speculation in the script, or tried to get me to say things that I could not say with confidence as a scientist. There are guidelines for some general principles about how animals behave and what extinct species may have looked like (based on their modern analogues), but always keep in mind that Nature is much more surprising and unpredictable than most people realize.

Some people assert that they can reliably reconstruct the muscles and body contours of an extinct animal based on the skeleton, but there are limitations. By studying living animals, we find that many muscles leave no scars or ridges on their bones, and in other cases the ridges and scars on bones do not correspond to any significant muscles. By the skeleton and soft tissue alone, it would be impossible to guess that the Galapagos marine iguana (which is built just like any other iguana) spends all its time diving in the surf eating algae; it has no obvious adaptations for swimming or algae eating, and only the salt glands in its body give a clue to its marine existence. Or how about the tree kangaroo? Nothing in its anatomy tells us that it spends most of its life in trees and not on the ground. How about the bird known as the dipper or water ouzel? Although it looks like a perfectly ordinary bird, it actually does not fly in the air much but spends its time diving in rushing mountain streams and walking along the bottom to catch prey. Again, there are no obvious clues in the anatomy about its strange behavior. So when any paleontologist talks about things like color or muscles or behavior, always remember the caveat that most of it is speculative and beyond the limits of testable science.

By far, the most diagnostic parts of the mammalian skeleton are in the head. The head has nearly all the sense organs (for sight, hearing, smell, taste), the brain cavity, and all the features of the teeth and jaws, which are distinctive in most fossils. Many fossil mammals are known only from skulls (or parts of them), so details of skull anatomy are essential in identifying and understanding how fossil mammals lived and what they ate. Of the parts of the skull, by far the most durable are the teeth. Because they are covered in a coating of hard enamel (the most durable tissue in the body), they can survive being broken away from the skull and jaws, and eroded and tumbled around in the sediment in rivers or the ocean, and still survive in recognizable form.

For this reason, the vast majority of fossil mammals are known only from their teeth. Mammalian paleontology is heavily dependent on recognizing teeth, not only to identify animals and determine who their relatives were, but also because teeth preserve a record of what the animals ate, and many other clues to the environment can be found in the scratches on the teeth's surfaces and the chemistry of their mineral content. To keep things simple, however, I will minimize mentions of teeth and refrain from giving too many details of tooth anatomy. Paleontologists who do not work on fossil mammals often joke about how mammalian paleontologists act as if fossil teeth in one layer gave rise to fossil teeth in a higher layer, or that the fossil mammal is just a collection of teeth that evolve through time, but some of this is a necessity if teeth are all we have (Fig. 1.7). We are fortunate that of all the skeletal tissues that can survive, the durable teeth are the most information-rich, helping us both in identifying the animal and determining what it ate.

The basic anatomy of mammalian teeth is shown in Fig. 1.8. In front of the mouth there is a series of teeth that are usually small and pointed, the **incisors**, which are used for nipping or grabbing food (the four teeth in the front of your mouth). Behind the incisors is a pair of upper and lower fang-like teeth known as **canines**, which are the large teeth used for grabbing food and for fighting. These are very large and important in carnivorous mammals but completely lost in most herbivores, which do not need to grab their plant prey and kill it.

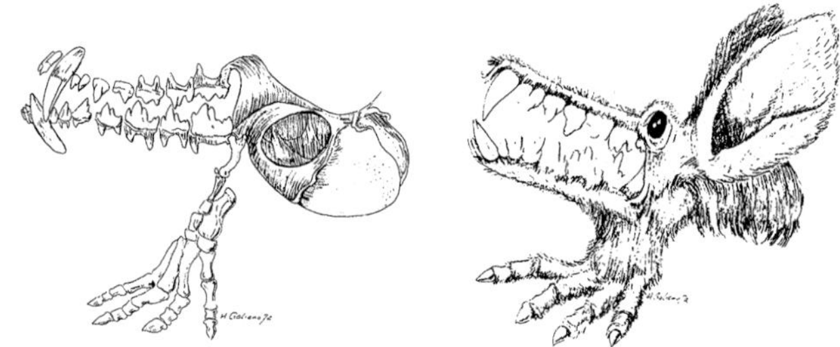

Figure 1.7. Most fossil mammals are known primarily from their durable teeth, while a few are known from the bones of the ear region or their foot bones. This satirical cartoon by paleontologist and artist Henry Galiano, commissioned for a paper by Malcolm McKenna, pokes fun at the idea that some paleontologists treat fossil mammals as if they were just teeth with ear bones and feet, and sometimes ignore the rest of the bones even when they are preserved.

Opossum – Marsupial

Full set of generalized incisors, canines, premolars, and molars for omnivorous diet

Gorilla – Primate

Fewer teeth but generalized for omnivorous diet

Wolf – Carnivora

Large stabbing canines and nipping incisors

Carnassial teeth for slicing meat

Dolphin – Cetacea

Numerous similar peg-like teeth for catching fish

Cow – Artiodactyla

Upper incisors lost

Lower incisors, diastema, grinding cheek teeth

Anteater – Xenarthra

Teeth lost; feeds on ants with tongue

Capybara – Rodent

Gnawing incisors

Diastema and cheek teeth

Horse – Perissodactyla

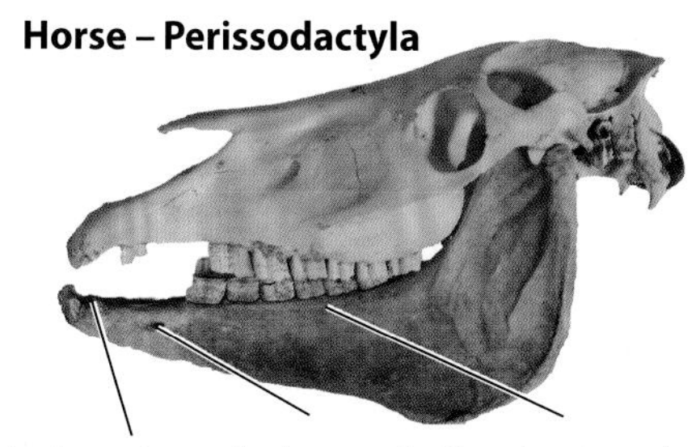

Nipping incisors, large diastema, grinding cheek teeth

Figure 1.8. The number and shape of mammalian teeth vary tremendously from group to group, depending on the group's diet and its ancestry. The opossum typifies the primitive mammal pattern, with nipping incisors in front, large stabbing canines, and a full set of premolars and molars in its cheek tooth row. Carnivorans, such as the wolf, have almost the same front teeth as the opossum, but the cheek teeth are narrow blades for slicing meat, and include a set of enlarged more powerful teeth called *carnassials*, used for breaking and crushing of prey. Primates, like the gorilla, have shortened jaws and a reduced number of teeth, and the crowns are rounded to accommodate omnivorous diets. Herbivores such as cows, horses, and rodents have lost their canines and most of their incisors and developed a gap (diastema) between the front teeth and the cheek teeth. Toothed whales, such as dolphins, have lost all specialized teeth and have only simple conical teeth for catching fish; they have multiplied their number of teeth as well. Anteaters have lost their teeth altogether and have a simple tubular snout with a long sticky tongue for lapping up ants and termites.

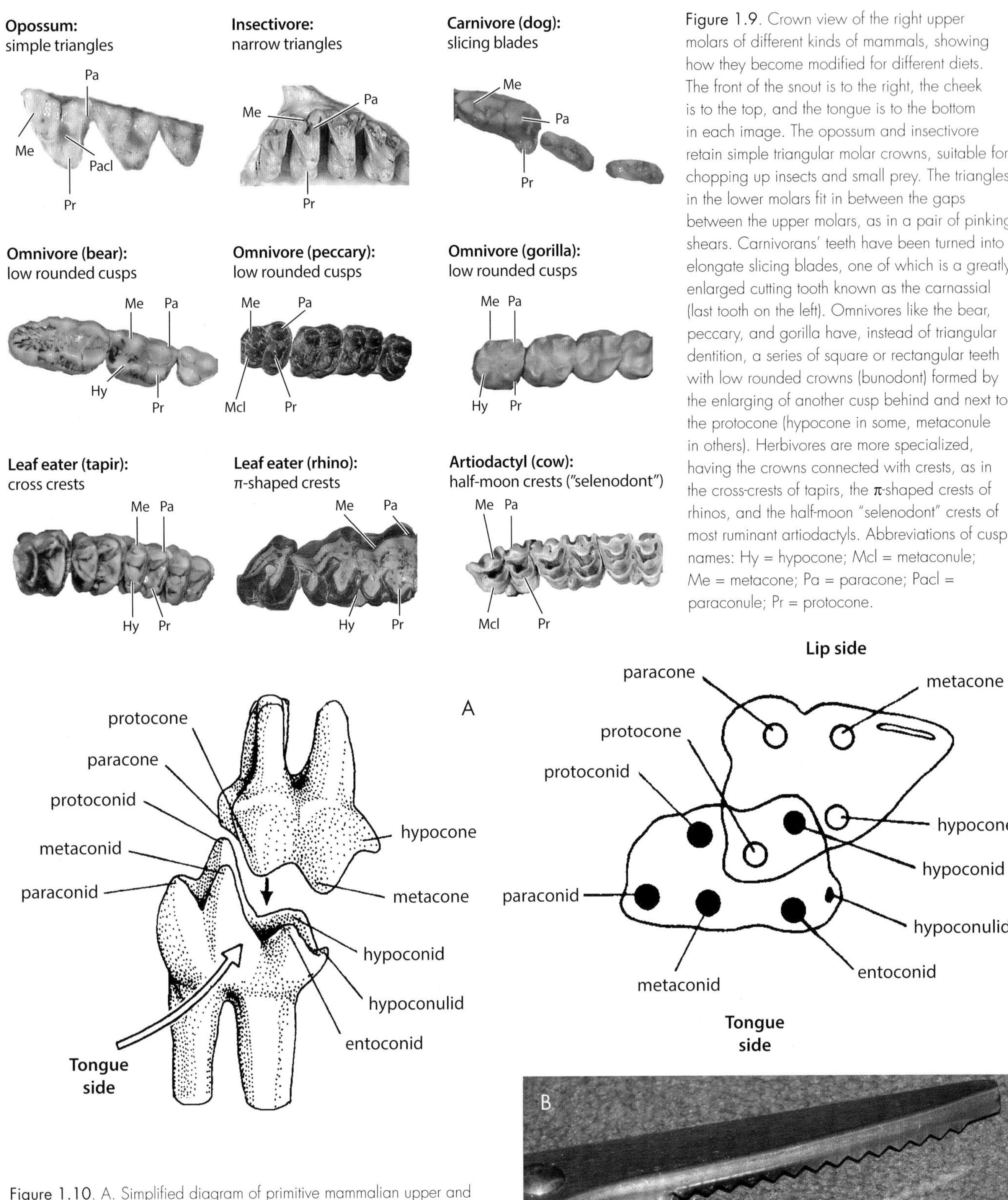

Figure 1.9. Crown view of the right upper molars of different kinds of mammals, showing how they become modified for different diets. The front of the snout is to the right, the cheek is to the top, and the tongue is to the bottom in each image. The opossum and insectivore retain simple triangular molar crowns, suitable for chopping up insects and small prey. The triangles in the lower molars fit in between the gaps between the upper molars, as in a pair of pinking shears. Carnivorans' teeth have been turned into elongate slicing blades, one of which is a greatly enlarged cutting tooth known as the carnassial (last tooth on the left). Omnivores like the bear, peccary, and gorilla have, instead of triangular dentition, a series of square or rectangular teeth with low rounded crowns (bunodont) formed by the enlarging of another cusp behind and next to the protocone (hypocone in some, metaconule in others). Herbivores are more specialized, having the crowns connected with crests, as in the cross-crests of tapirs, the π-shaped crests of rhinos, and the half-moon "selenodont" crests of most ruminant artiodactyls. Abbreviations of cusp names: Hy = hypocone; Mcl = metaconule; Me = metacone; Pa = paracone; Pacl = paraconule; Pr = protocone.

Figure 1.10. A. Simplified diagram of primitive mammalian upper and lower cheek teeth (such as those of an opossum or insectivore), showing how they occlude. The triangle of cusps on the upper molar (protocone-paracone-metacone) fits in the valley between the triangle of cusps on the lower molar (protoconid-paraconid-metaconid), like the teeth on a pair of pinking shears (B).

Behind the canines are the cheek teeth. The front cheek teeth start with baby teeth, then are replaced with adult teeth. These are known as **premolars** ("bicuspids" to a dentist), and they vary in shape, size, and number in many interesting ways. Finally, behind the premolars are the large crushing and grinding teeth known as **molars**, which never have baby teeth preceding them. In humans, the last of these molars erupt during the late teens and are called the "wisdom teeth," because people acquire them much later than childhood, as they do their wisdom. In some cases, people no longer develop their last molars, or they are so crowded in our shortened jaw they become impacted and must be surgically removed. The large durable molars are the most diagnostic of all, and most of the focus of mammalian paleodentistry is on the subtle changes in the shapes of crests and cusps on the molars, as well as features like the surface wear and the chemistry of the enamel, which tells us what they ate with their teeth (Fig. 1.9). The pattern and sequence of eruption of teeth is under very tight genetic control, a good reason teeth are useful in our quest to understand mammalian evolution.

As we shall see in the following chapters, we can tell a lot about which mammal we have, and what it eats, from the teeth alone. Insectivorous mammals have teeth that are formed into alternating triangles, which come together like pinking shears to chop up the external shell (cuticle) of their insect prey (Figs. 1.9, 1.10). Carnivorous mammals always have large canines for stabbing and grabbing their prey, and their cheek teeth are extremely modified into narrow blades that shear against one another when the mouth closes like the blades in a pair of scissors. Such precise shear is necessary to slice up flesh and break up bone. If their edges do not meet precisely, then they are useless for eating meat, just like a pair of scissors with a loose rivet in the hinge. Herbivorous mammals have reduced the canines (or lost them completely), and their cheek teeth are modified into broad grinding surfaces, although the details of their crests and cusps differ from group to group (allowing us to recognize them). Marine mammals such as dolphins and seals have their teeth reduced to simple conical pegs for grabbing their swimming prey. Omnivorous animals (such as primates, pigs, peccaries, bears, and a few others) have very simple rectangular crowns on their cheek teeth with low rounded cusps for processing all sorts of different foods). Many animals (such as baleen whales, platypuses, and the ant-eating mammals in several orders) have lost their adult teeth altogether, because they do not need them to chew on the tiny prey they ingest.

CHAPTER 2

THE ORIGIN AND EARLY EVOLUTION OF MAMMALS

SYNAPSIDS (PROTOMAMMALS OR STEM MAMMALS)

Mammals have an extensive fossil record going back to the beginning of the Age of Dinosaurs. But we can trace their ancestry even further back to fossils from the Carboniferous (Fig. 1.3), the "age of coal swamps." Although these early mammal relatives are often referred to as "mammal-like reptiles," the proper name for this group, which includes the ancestors of mammals and all their mammalian descendants, is Synapsida. If you prefer a casual term, some people call them "protomammals" or "stem mammals." The Synapsida split off from the reptiles in the Early Carboniferous, at 315 Ma. Both the earliest members of the synapsid lineage (*Protoclepsydrops* and *Archaeothyris*) and the earliest members of the reptile lineage (*Hylonomus* and *Westlothiana*) evolved side by side at this time, but *at no time were the ancestors of mammals ever reptiles*. People still use the obsolete term "mammal-like reptiles" out of habit, or from copying outdated sources, but no up-to-date paleontologist uses the term any more.

Most people are familiar with the "finback" fossils such as *Dimetrodon* and *Edaphosaurus*, which appear in many dinosaur books for kids and often in toy kits of dinosaurs as well (Fig. 2.1). Except these animals are *not* dinosaurs—they are among the earliest synapsids. These creatures are part of the first great evolutionary radiation of synapsids (Fig. 2.2) and are often called

Figure 2.1. Reconstructions of some protomammals or synapsids (formerly but incorrectly called "mammal-like reptiles"). On the right in the background is the fin-backed predatory "pelycosaur" *Dimetrodon*, and on the left is the fin-backed herbivorous "pelycosaur" *Edaphosaurus*. In front on the left is the huge predatory "therapsid" gorgonopsian *Gorgonops*, and behind it is the herbivorous "therapsid" dinocephalian *Moschops*. Behind the human is the dinocephalian *Estemnosuchus*, with the bizarre crests and tusks on its face. In the right front are the wolf-size predatory "cynodont" *Cynognathus*, and the cow-size herbivorous dicynodont "therapsid" *Kannemeyeria*, which has a toothless beak except for canine tusks.

"pelycosaurs," but that is a wastebasket group. They do not look much like modern mammals, but there are some key synapsid features that show they are our distant relatives, more closely related to us than any bird or dinosaur or reptile.

One of those features is a distinctive hole (called the temporal fenestra) in the side of the skull for attachment and expansion of the jaw muscles (Fig. 2.3). In synapsids there is a single hole, low on the side of the skull, below the bone behind the eye socket, whereas most reptiles have two temporal fenestra. "Pelycosaurs," such as *Dimetrodon*, also have large, stabbing canine teeth, rather than the unspecialized conical teeth seen in reptiles, and they have other subtle features of the skull and braincase and palate. These big "pelycosaurs" were the largest land animals of the Early Permian in places like western Texas. For example, the big predator *Dimetrodon* (Fig. 2.1) is one of the most common fossils in the red beds near Seymour, Texas. *Dimetrodon* reached 1.7–4.6 m (5.6–15.1 ft) in length, had a sail that reached about 1.7 m (5 ft) above the ground, and weighed up to 250 kg (550 lb). The sail, or fin, along the back of *Dimetrodon* and *Edaphosaurus* has long been a source of controversy. It is of the right size to be an efficient device for absorbing heat and warming these animals up when they turned their sides to the sun, and for dispersing heat and cooling their bodies when they faced the sun. However, no other "pelycosaurs" required such a device, so it was apparently not performing an essential function like this. More likely it served for recognizing other members of the same species and showing dominance among males, as the horns and antlers of antelopes and deer are used today.

By the Late Permian, the "pelycosaurs" had vanished and been replaced by a second wave of more advanced synapsids known as "therapsids" (Fig. 2.2), another wastebasket group for synapsids that does not include mammals. These incredible creatures (Fig. 2.1) dominated the landscape in the Upper Permian beds of South Africa, Russia, China, and elsewhere. They included not only large predators with stabbing teeth but also the first large land herbivores. Some of these herbivores, known as dicynodonts, had a toothless beak and big canine tusks. They were up to 3.5 m (11 ft) long and weighed up to 1000 kg (2200 lb). The second group of plant eaters was the dinocephalians. Their heads were covered by bumps and warts, and some had thick, bony battering rams on their heavily armored

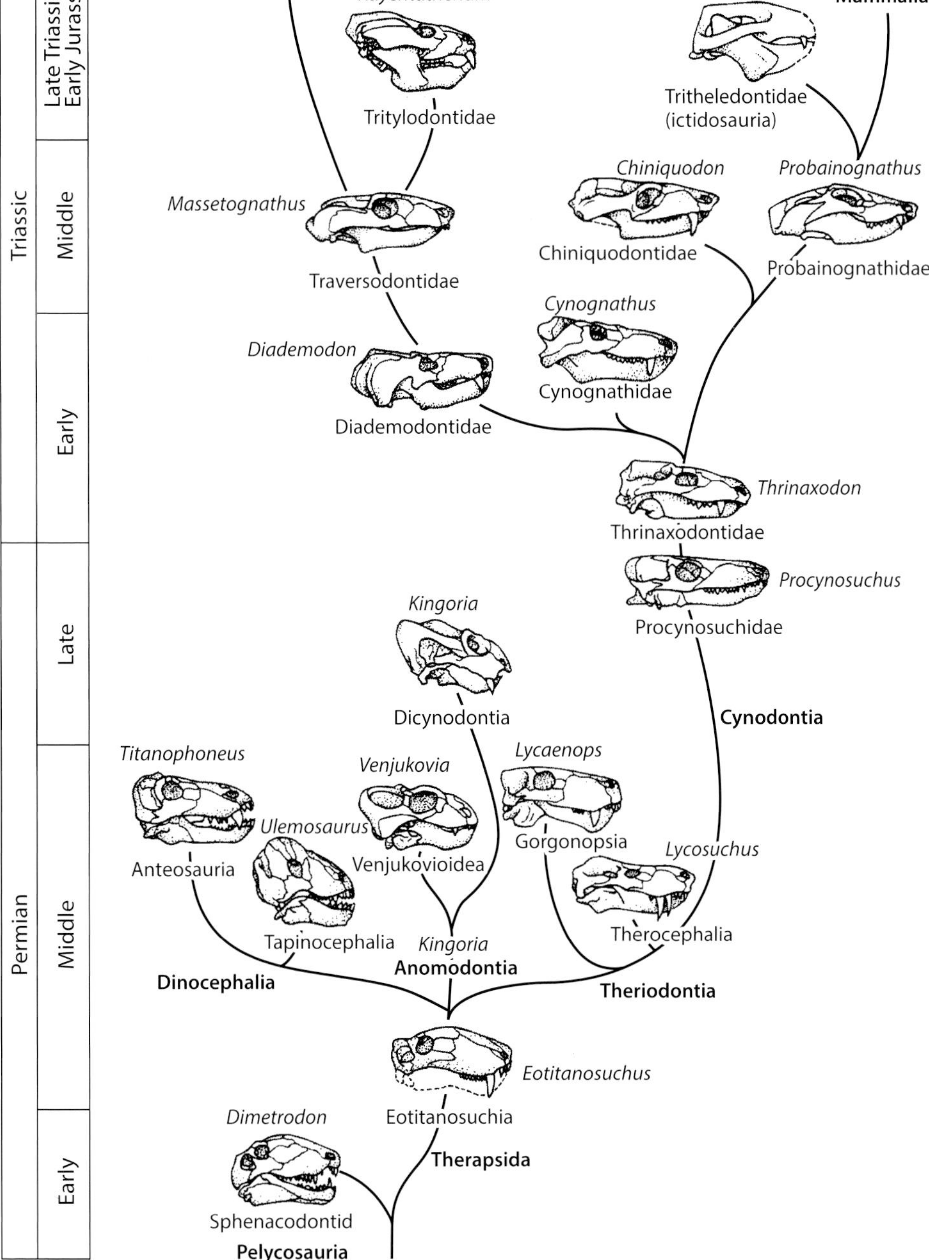

Figure 2.2. Evolution of the synapsids, from the primitive "pelycosaurs" of the Early Permian to the "therapsids" of the Late Permian, and finally the advanced "cynodonts" and mammals of the Early Triassic.

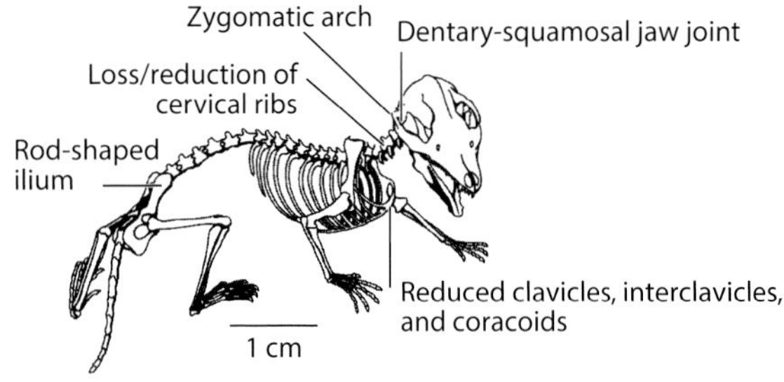

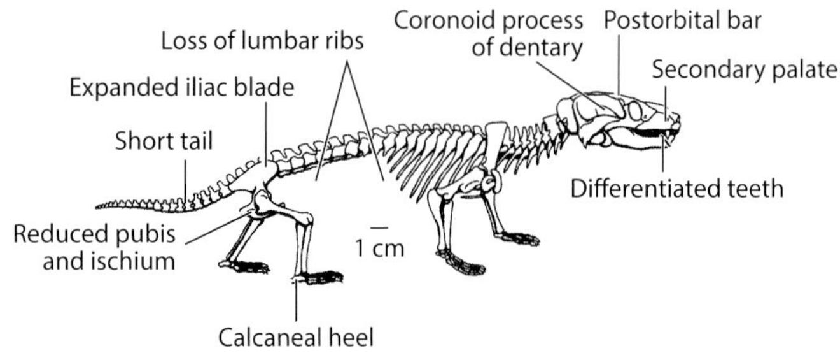

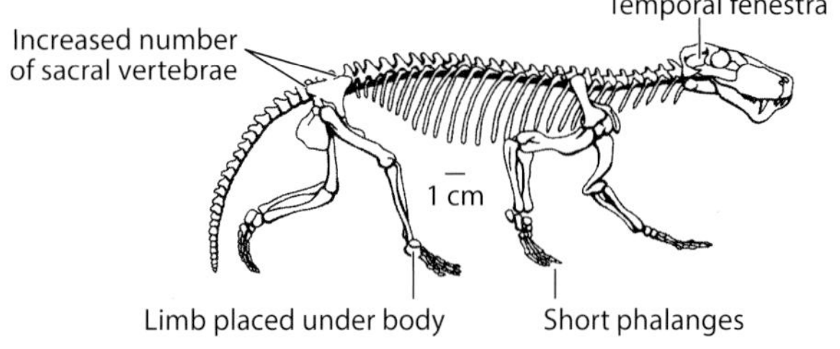

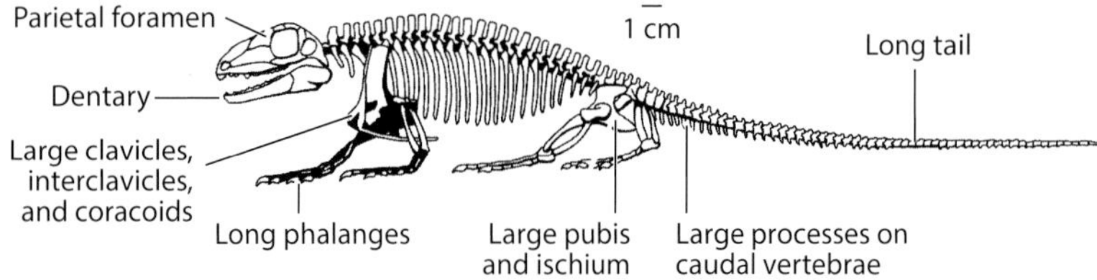

Figure 2.3. Skeletal changes in synapsids and mammals, from primitive "pelycosaur" *Haptodus* to the "therapsids" *Lycaenops* and *Thrinaxodon*, which are progressively more and more like mammals, to a true mammal, *Megazostrodon*. (Modified from Prothero 2007, p. 275, fig. 13.4.)

skulls. Many were up to 4.5 m (15 ft) in length and weighed up to 2000 kg (4400 lb).

These herbivores were hunted by a variety of predatory synapsids. The biggest of these, the gorgonopsians, had huge skulls with large canines, strong jaw muscles for chewing, and powerfully built bodies. Some were bigger than bears and had a skull 45 cm (18 in) long, saber teeth over 12 cm (5 in) long, and a long, sprawling crocodile-like body up to 3.5 m (11 ft) in length.

"Therapsids" showed progressively more and more features that look mammalian. The small synapsid opening on the skull turned into an expanded arch behind the eye from which powerful jaw muscles could bulge. This gave them powerful bite forces and even some chewing ability. Above their original reptilian palate grew a secondary palate, which enclosed the nasal passages. (If you run your tongue over the roof of your mouth, you can feel the line where the two halves of the palate grew together.) A secondary palate allowed advanced "therapsids" to chew a mouthful of food and breathe at the same time, which enabled a faster metabolism. Animals that must hold their breath until prey is swallowed, such as reptiles, have a slow metabolism. "Therapsids" developed many more mammal-like modifications, including a double ball joint at the base of the skull beneath the spinal column, a skeleton with a more upright and less sprawling posture, and many specializations of the teeth, jaws, and skull.

Most of the "therapsid" predators and herbivores died out during the great Permian extinction, dubbed the "Mother of All Mass Extinctions," when about 75% of land species and 95% of marine species vanished. However, a few lineages did survive and led to the third great radiation of synapsids in the Triassic (Fig. 2.2). These are known as "cynodonts," yet another wastebasket group. By the end of the Triassic, they were so much like the earliest mammals that the line between them is hard to draw (Fig. 2.3). "Cynodonts" had fully upright posture, with their legs completely beneath their bodies, rather than sprawling as in "pelycosaurs." They had teeth specialized into incisors, canines, and cheek teeth; a complete secondary palate that goes back all the way to the throat; a more mammalian shoulder and hip structure; a very large synapsid opening behind the eye; and dozens of other mammalian features. Some of them, like the weasel-shaped *Thrinaxodon*, had lost the lower back part of their rib cage. This suggests they had a muscular diaphragm beneath their lungs, so they used both the diaphragm and some rib-assisted breathing, the latter being the only kind of breathing found in reptiles. *Thrinaxodon* also had pits on its snout for whiskers, suggesting that "cynodonts" had at least some hair on their bodies. Most important, they had a jaw mechanism that was much more mammal-like, with just a few leftover reptilian jaw bones.

MAMMALS IN THE AGE OF DINOSAURS

The last of the Late Triassic "cynodonts," such as the tritylodonts and trithelodonts, are so much like the earliest undoubted mammals that paleontologists have long debated where to draw the line. But certain fossils, such as the Late Triassic *Morganucodon*, *Megazostrodon*, *Eozostrodon*, and *Adelobasileus*, are generally considered to be the earliest true mammals (Figs. 2.3, 2.4). Most are about the size of a shrew or a mouse, with a very shrew-like body as well. They had completely mammalian teeth and skull region, and their jaws and skeleton were more advanced than those of any primitive synapsid. Crucially, they show the signature of the mammals: a mammalian jaw joint made of the dentary and squamosal bones. In later Mesozoic mammals, the original reptilian jaw joint between the articular bone of the back of the jaw and the quadrate bone of the skull moved to the middle ear to become the "hammer" and "anvil" bones you hear with. We can see this gradual transition in a remarkable array of synapsid fossils, and it also happened when you were an embryo: two of your middle ear bones started out in your embryonic jaw.

From these early shrew-like creatures of the Late Triassic, mammals diversified into a number of groups in the Jurassic and Cretaceous (Fig. 2.5). Most remained small creatures (mostly shrew-size to rat-size, none larger than a dog), because they lived on a planet dominated by dinosaurs. They probably lived deep in the undergrowth and underground to hide from small predatory dinosaurs, and many probably were nocturnal to avoid predators as well. Mammals remained small and hid from dinosaurs for about 130 m.y. during most of the Age of Dinosaurs, twice the length of time as the 66 m.y. of the Cenozoic Age of Mammals (Figs. 1.3, 2.5). In other words, the first two-thirds of mammalian history occurred under the shadow of dinosaurs, a period during which mammals were not able to grow large or occupy many different niches. Only after the combined triple-whammy effect of gigantic volcanic eruptions in India, loss of massive shallow seas, and the later impact of an asteroid in Yucatan did the dinosaurs (except for their bird descendants) leave the stage. After 130 m.y. of hiding, finally the mammals had a chance to grow in size and occupy many different ecological niches vacated by the dinosaurs.

For centuries, very few mammal fossils were known from the Mesozoic, since the specimens from that era are usually just tiny shrew-size jaws and teeth. They were often overlooked during searches for big dinosaur bones, although by the 1880s mammal fossils had been found in the Jurassic dinosaur beds of England and Wyoming. When I published research in Jurassic mammals in the 1970s, there were still only a handful of teeth and jaws known, and only one or two genera were known from a skeleton. In 1979 only 116 fossil mammal genera were known from the Mesozoic, and these were mostly from the uppermost Cretaceous beds that produced *Tyrannosaurus* and *Triceratops*. Since then, however, the number of specimens (especially good complete skeletons) has exploded, and more than 310 genera were known by 2007, and that number continues to grow.

Figure 2.4. Reconstruction of the shrew-size *Morganucodon*, one of the earliest known true mammals.

MORGANUCODONTS

The most primitive and earliest known mammals are the morganucodonts, including *Morganucodon*, *Eozostrodon*, *Megazostrodon*, and about a dozen other genera (Figs. 2.3, 2.4, 2.5). Most specimens consist only of teeth and jaws, but a few species have nearly complete skeletons. They would have looked very much like modern shrews in both their size and shape, and their tiny teeth have triangular three-cusped crowns suitable for shearing up insects, as do the teeth of modern insectivorous mammals. Even though they had a mammalian jaw joint between the dentary and squamosal bones, they still had some of the ancestral jaw bones as vestiges in the inside back part of the jaw. They had large brains, another mammalian feature, compared to most of the later mammal-like "cynodonts." Unlike most primitive synapsids, which keep replacing all their teeth through their lives, the morganucodonts (and all other mammals) have only one cycle of tooth replacement, so only one round of baby teeth preceded their adult teeth.

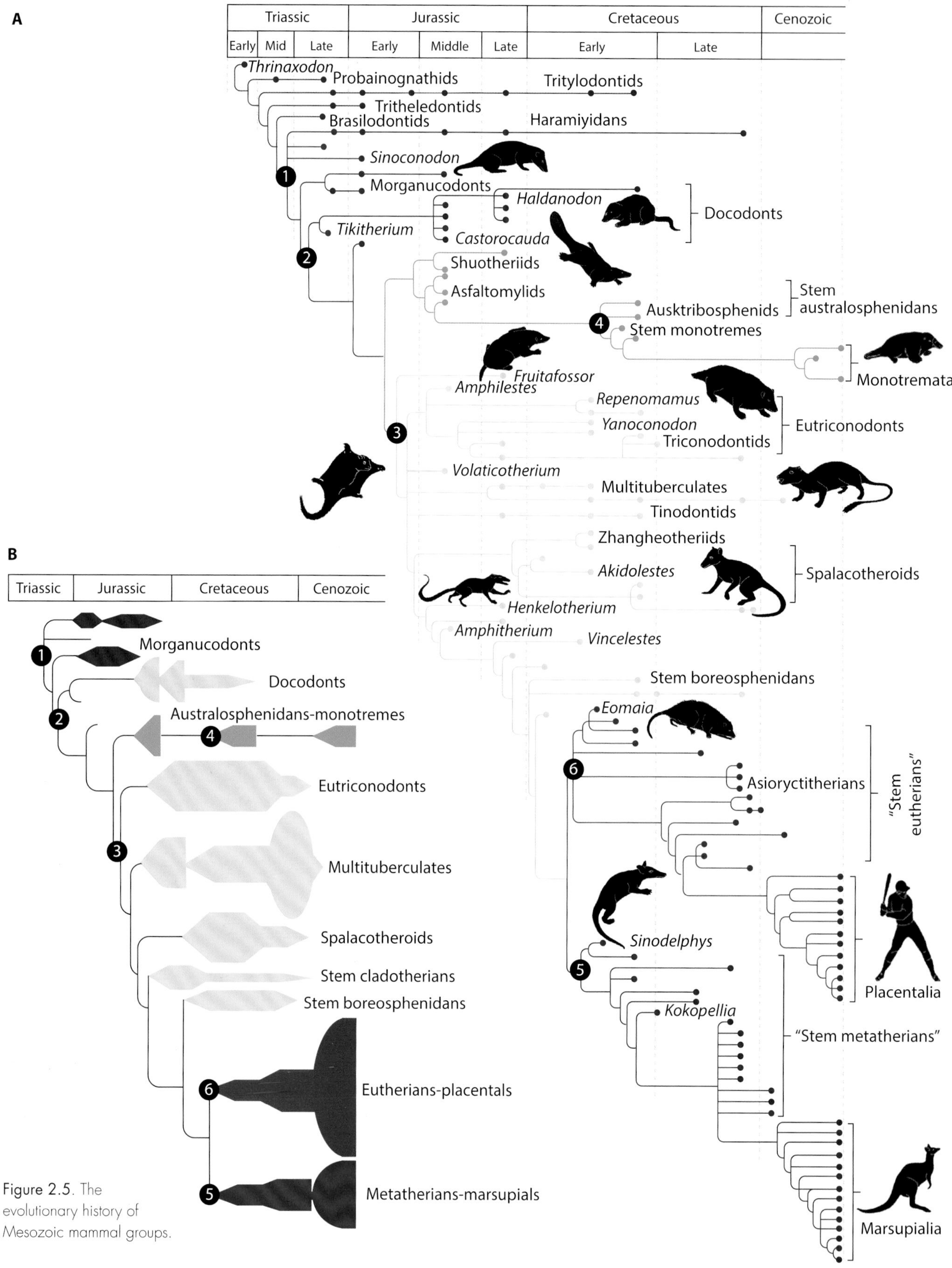

Figure 2.5. The evolutionary history of Mesozoic mammal groups.

DOCODONTS

For a century, jaws and teeth with strangely squared crowns (rather than the typical three-cusped pattern we saw in the morganucodonts and will see in the triconodonts) were common from the same Upper Jurassic beds that produced other Mesozoic mammals as well as huge sauropod dinosaurs. Known as docodonts (Fig. 2.5), these creatures had teeth suggesting they were slightly more omnivorous than most Mesozoic mammals, rather than fully insectivorous. Nothing was known about the skeleton of these animals until the 2000s, when more complete skeletons were found of fossils such as *Haldanodon* and *Castorocauda*.

Haldanodon, from the Upper Jurassic (150 Ma) Guimarota coal beds of Portugal, was about 38 cm (15 in) long and had a mole-like body and powerful forelimbs with short curved fingers and claws for digging. The snout bones had a roughened area for some sort of shield or nail-like hardening on the nose, consistent with the digging mode of life. Its jaws were very robust, with features showing that they had relatively powerful bite forces.

As notable as that discovery was, even more surprising was the recently discovered *Castorocauda* (Fig. 2.6A, B, C), known from a remarkably complete skeleton from the Middle Jurassic (164 Ma) Daohugou lake beds of Inner Mongolia. The specimen is so beautifully preserved that even the hair and soft tissues are visible. It is the oldest mammalian fossil to have hair preserved, although Triassic fossils such as *Thrinaxodon* (Fig. 2.3) suggest that hair first evolved much earlier in synapsid evolution. The body of *Castorocauda* was about 43 cm (17 in) long, and it weighed about 800 g (almost 2 lb), making it one of the largest mammals of its time. It was built a bit like an otter or platypus, with a broad tail adapted for swimming and covered with scales, like the tail of a beaver. This similarity to modern-day animals is how it got its name, *Castorocauda lutrasimilis* ("beaver tail like an otter"). Its vertebrae were flattened like those of an otter or beaver, and it even had webbed feet. Its forelimbs were very robust and strong, for digging or swimming, and look like those of the living platypus, which both digs and swims. Its docodont teeth have been highly modified for fish catching, so they resemble the teeth of seals and other fish eaters.

The discoveries of several Chinese Jurassic docodonts were announced in 2015, based on beautifully complete articulated skeletons with hair impressions from the Ganggou site in Hebei Province, China. One of them, *Agilodocodon* (Fig. 2.6C), was built much like a squirrel and spent its life in trees. Another one from the same beds is known as *Docofossor* (Fig. 2.6C); it had skeletal features suggesting that it burrowed like a mole.

Figure 2.6. The spectrum of adaptations of the docodonts. A–B. *Castorocauda*. (See also next page.)

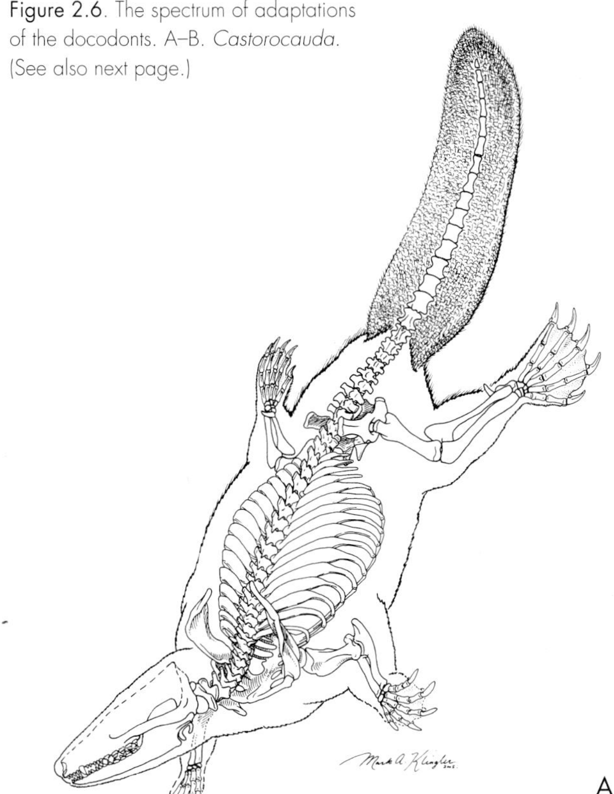

A.P. Neander 2014
C

Figure 2.6. The spectrum of adaptations of the docodonts. C. A scene from the Middle Jurassic of China, showing *Agilodocodon* (in the ginkgo tree), *Docofossor* (emerging from burrow), and *Castorocauda* (swimming). D. *Hadrocodium*.

These fossils indicate that docodonts were very different from most of the shrew-like insectivorous Mesozoic mammals: they had distinctly different teeth adapted for an omnivorous diet, and some skeletons show they were built more like moles, squirrels, or otters.

Another remarkable fossil just slightly more advanced than the docodonts was *Hadrocodium* (Fig. 2.6D). Known from a complete skull from the Early Jurassic found in China, it was only 3.2 cm (1.3 in) in body length, which makes it one of the smallest mammals ever known. Yet it has the enlarged advanced brain and middle ear features of the more advanced mammals, so it is not as primitive as its contemporaries, the morganucodonts and early docodonts. The fact that its jaw and ear bones were completely mammalian means the jaw no longer had to reach very far to the back of the skull, which was a factor limiting the size of the brain. As the mammalian jaw joint shifted forward, it gave the back of the skull room to expand the brain, and allowed mammals to have both large brains and teeth specialized for different diets.

MONOTREMES (PLATYPUS AND ECHIDNA) AND THEIR RELATIVES

When the platypus was first brought from Australia and shown to European scientists, they thought it was a hoax (Fig. 2.7A). A furry creature with a duck's bill and webbed feet? But the platypus is indeed a real animal, and even stranger than its appearance, it is one of only two groups of mammals that lay eggs. When the eggs hatch, the females (like other mammals) nurse their young with milk, but (unlike other mammals) their milk glands have no nipples. The mother extrudes milk from glands on her belly so the babies can lap it up. Platypuses also have no separation between their reproductive tract and their anus and urinary openings but have a common cloaca like that of reptiles. In their shoulder, they still have such reptilian bones as the procoracoids, coracoids, and interclavicle, seen in the earlier synapsids but reduced or lost in most mammals.

The only other modern-day mammals with this odd anatomy are the "spiny anteaters," or echidnas (Fig. 2.7B) of Australia and New Guinea. Even though they look somewhat like other ant-eating mammals, with a long sticky tongue and toothless snout, and are encased in sharp spines reminiscent of those of a hedgehog, echidnas are primitive egg-laying mammals, and their spines and ant-eating features are due to convergent evolution. Together, the platypus plus echidnas form a group known as the **monotremes**. The name means "single hole," in reference to the common cloaca used for both reproduction and excretion.

For a long time, the relationships of monotremes to other mammals and to fossil Mesozoic mammals were a mystery. This was especially true because adult monotremes have no teeth, so it was impossible to compare them to Mesozoic mammals

A

B

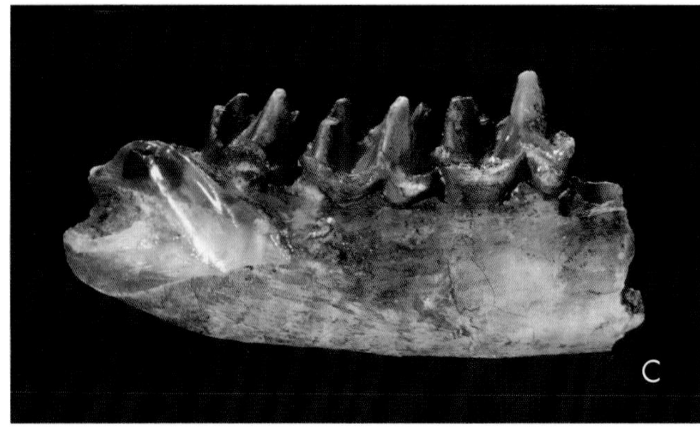
C

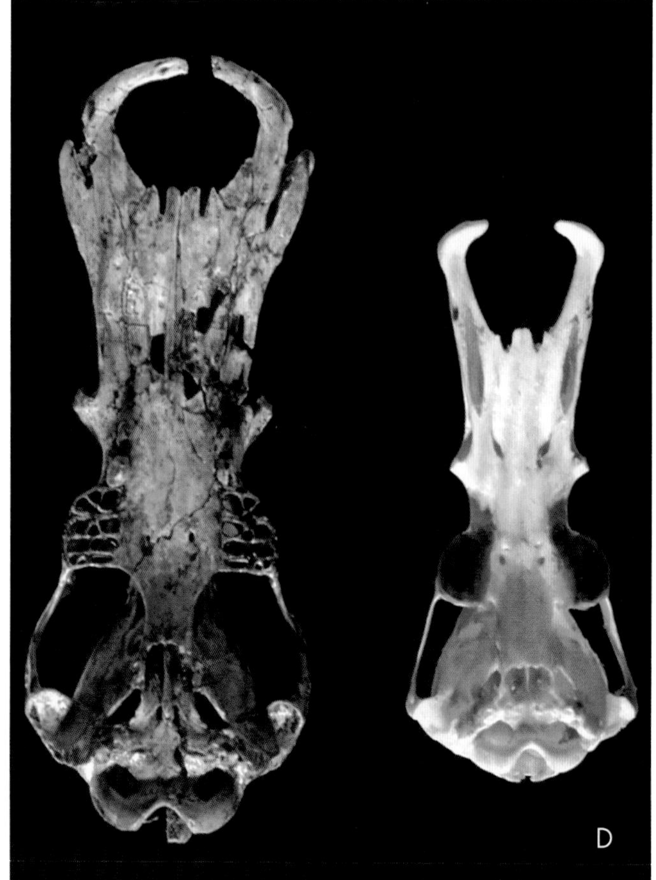
D

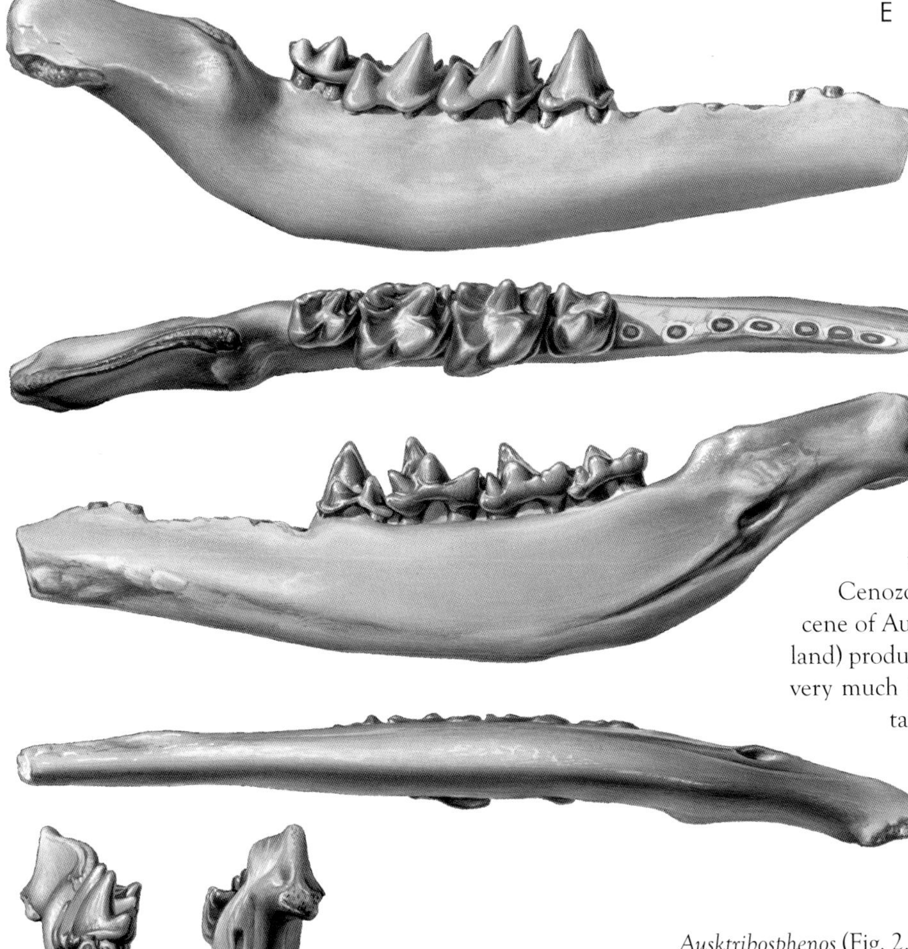

Figure 2.7. Monotremes A. Platypus. B. Echidna. C. *Steropodon,* a primitive monotreme from the Cretaceous Lightning Ridge opal mine in Australia. It was found completely replaced by opal and narrowly escaped being ground down to jewelry. D. A modern platypus skull (right) compared to the skull of the much larger Miocene Riversleigh platypus *Obdurodon*. E. The peculiar Mesozoic mammal *Ausktribosphenos*.

consisting only of tooth fossils. Only juvenile platypuses develop teeth, which they lose later in development. When I was a student, no fossil monotremes were known. However, in recent years more and more fossils of ancient monotremes, whose teeth resemble those of platypus pups, have come to light. The oldest known monotreme fossil is *Teinolophos* from the Early Cretaceous of Australia (about 121–112 Ma). The next youngest fossil was found in an Australian opal mine, and the entire bone of the fossil had been replaced with opal (Fig. 2.7C). Named *Steropodon* ("lightning tooth," after Lightning Ridge, the locality where it was found), it is also Early Cretaceous, but only about 105 m.y. in age. Another similar monotreme from Lightning Ridge is *Kollikodon*. Each of these latter two creatures had teeth that somewhat resemble those of a juvenile platypus and suggest an omnivorous diet. Both were relatively large for a Cretaceous mammal, reaching lengths of 1 m (3.3 ft) or more. Monotreme fossils remain rare through most of the Cenozoic, although early in the middle Miocene of Australia, the Riversleigh beds (in Queensland) produced the platypus *Obdurodon*, which looks very much like a modern platypus, although it retained teeth as an adult (Fig. 2.7D).

Monotremes and their relatives were thought to be restricted to Australia until 1997, when Tom Rich of the Museum of Victoria discovered, in Australia, a weird Early Cretaceous fossil given the name *Ausktribosphenos* (Fig. 2.7E). At first glance it seemed to resemble many other Mesozoic mammals, which have the crowns of their cheek teeth arranged in alternating interlocking triangles (tribosphenic) that chopped up insects like a pair of pinking shears (Fig. 1.10). But a closer look shows that the cusps of its cheek teeth face *backward* from those of every other Mesozoic group. A second similar fossil, *Bishops*, was found in the same locality. Then another puzzling surprise: the Middle and Upper Jurassic beds of Patagonia in South America have yielded similar fossils, *Ambondro* and *Asfaltomylos*.

Now it seems clear that the surviving platypus and echidnas are the tiny remnant of a great radiation of primitive egg-laying mammals, known as the Australosphenida (Fig. 2.5), that were once distributed across the Jurassic and Cretaceous of the southern supercontinent, Gondwana. Most of the fossil representatives died out at the end of the Cretaceous, but one of them survived in South America until the Paleocene.

MULTITUBERCULATES

The next branch point (Fig. 2.5) in mammalian evolution is a group known as the multituberculates (Fig. 2.8). Most of the groups mentioned so far (except for living monotremes) had a low diversity through only part of the Mesozoic and then vanished. The multituberculates, by contrast, lasted for upwards of 165 m.y., the longest duration of any mammal group, living or extinct (Fig. 2.5). If the Triassic fossils known as haramyids are considered very primitive multituberculates (which is controversial), the lineage first showed up in the Late Triassic (200 Ma). Undisputed multituberculates appeared in the Early Jurassic and lasted through the entire early Cenozoic, vanishing at the end of the Eocene (35 Ma). In addition to their long duration, multituberculates survived the mass extinction event at the end of the Cretaceous (which decimated the marsupials and many other groups), and then became incredibly abundant, diverse, and successful in the jungles of the Paleocene and early Eocene of North America and Europe. They are among the most common fossils recovered from the beds of these epochs, and they were so diverse that there are at least 200 species known.

So what were multituberculates? In shape and size, they resembled squirrels (Fig. 2.8A, B, C). Some of them were ground dwellers and may have looked like marmots. Some had long prehensile tails for grasping branches (as an opossum has), feet and hands well suited to climbing trees and even to climbing down the trunk headfirst. They had a rodent-like skull with long chisel-like front incisors and a toothless gap (**diastema**) between the

Figure 2.8. Multituberculates. A–B. The very primitive multituberculate known as *Rugosodon* from the Early Cretaceous of China. C. The Paleocene squirrel-like *Ptilodus*, which had ankle joints that allowed it to descend a tree trunk headfirst, as well as a grasping tail. D. The largest of the multituberculates, the Paleocene beaver-size *Taeniolabis*.

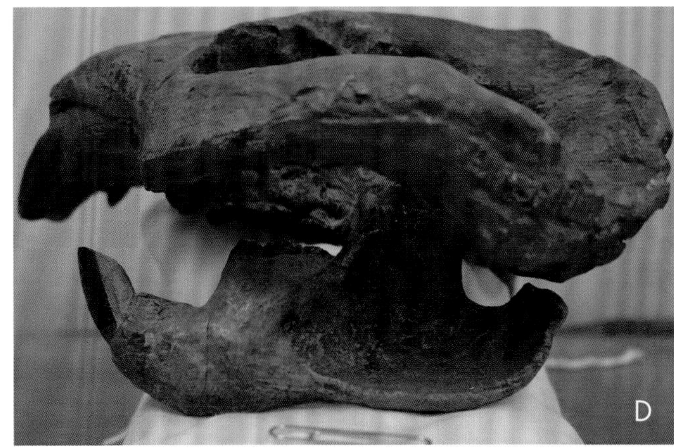

incisors and grinding cheek teeth. These cheek teeth, however, look like those of no other group of mammals: they are relatively long and narrow, with multiple rows of cusps, or tubercles (hence the group's name). Even more remarkable is that the last premolar on the lower jaw on many multituberculates was developed into a distinctive large, chisel-like slicing blade, apparently used for cracking open seeds and nuts or slicing insects. This single blade-like tooth is enough to identify many species of multituberculates in the Paleocene.

Multituberculates first originated with the Jurassic Chinese form *Rugosodon*, the earliest true multituberculate (Fig. 2.8A, B). Through the Jurassic and Cretaceous, multituberculates flourished on the northern continents, especially North America and Eurasia. However, there was another group of mammals, the poorly known Gondwanatheria from the Cretaceous of South America, that may be relatives of the northern multituberculates. The Lower Cretaceous beds of Australia have yielded a single tooth, *Corriebataar*, which may establish the multituberculates group in that continent as well.

After the Cretaceous extinctions wiped out the non-bird dinosaurs and decimated the marsupials, the multituberculates took over the role of the small seed- and nut-eating tree-climbing mammal in the jungles of the Paleocene of North America and Europe. They were so abundant that their fossils can be found in any beds of this age, with their distinctive blade-like lower premolars occurring in large numbers. Some of the multituberculates, such as *Taeniolabis*, were big robust ground dwellers as large as a modern beaver (Fig. 2.8D).

By the end of the early Eocene, however, multituberculates declined in numbers and diversity. They were rare in the middle Eocene and vanished completely at the end of the Eocene. The reasons for this are not completely known, although some paleontologists have suggested that they were outcompeted in their niche by the rapidly diversifying rodents and early primates, which apparently fared better as the small seed-and nut-eating herbivores. Thus, after a long, diverse run lasting over 165 m.y., the longest-lasting group of mammals that ever lived finally vanished.

TRICONODONTS

Another group of Late Jurassic and Cretaceous mammals originally known from only teeth and a few jaws was the triconodonts. Described as early as 1861, these animals were distinctive in having cheek teeth with the crowns formed into a triangle with three conical cusps (hence their name, Triconodonta, "three-cone tooth"). Since this is the primitive mammalian tooth condition seen in even the earliest mammals, the archaic morganucodonts were once included in the group as well (Fig. 2.5).

In the late 1990s a number of triconodont specimens with complete skeletons were found in China. They include the tiny (13 cm/5 in long) *Yanoconodon*, from the Yan Mountains, from about 122 Ma. It is remarkably primitive for a Cretaceous mammal, with middle ear bones that are not fully mammalian and ribs in its lower back (a feature lost in nearly all mammals). Even better known is *Jeholodens*, from the Middle Cretaceous Jehol beds (about 125 Ma) of China, based on a complete specimen with hair and soft parts preserved. It was a small, furry, shrew-like creature with a long skinny tail (Fig. 2.9A), much like most Mesozoic mammals, but it was more advanced than *Yanoconodon* in having completely mammalian middle ear bones and no ribs in its lower back.

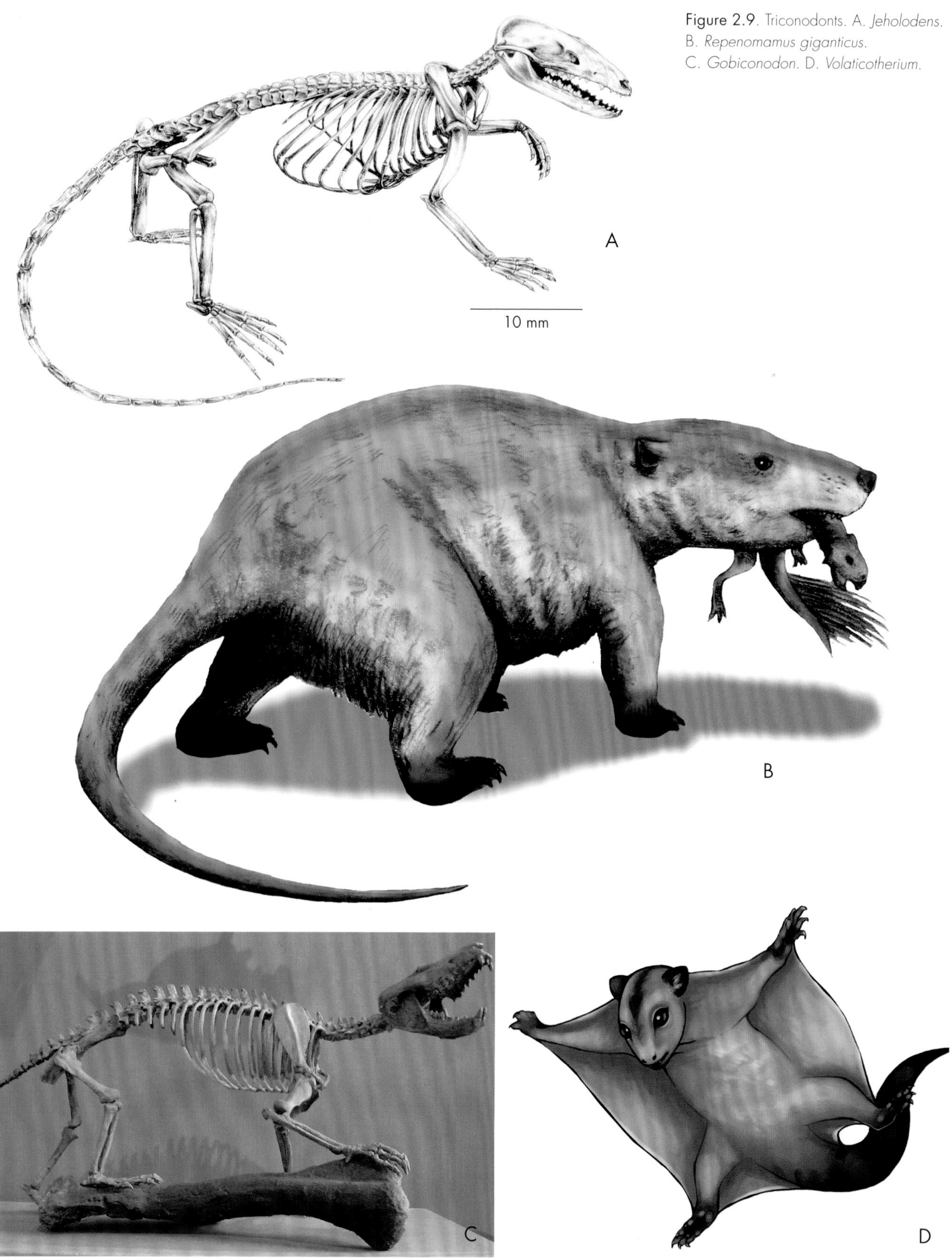

Figure 2.9. Triconodonts. A. *Jeholodens*. B. *Repenomamus giganticus*. C. *Gobiconodon*. D. *Volaticotherium*.

A big departure from these more primitive shrew-like creatures was the discovery of *Volaticotherium* from the Jurassic of Mongolia (Fig. 2.9D). It was a nearly complete skeleton that showed very long, delicate limbs and body. There is even a dark carbon film of a gliding membrane, like that of a flying squirrel or the Australian marsupial known as the sugar glider. Just as the docodonts, the triconodonts seemed to come in variety of shapes and fill a range of ecological niches (Fig. 2.10), all evolving convergently on body forms that therian mammals (marsupials and placentals) would rediscover many millions of years later.

In 2005 came the discovery of *Repenomamus giganticus* (Fig. 2.9B) from the Lower Cretaceous beds of Liaoning, China, which is about 123–125 m.y. in age. It is one of several genera of triconodonts known as gobiconodonts, largely known from good specimens from the Cretaceous of Asia. Gobiconodonts tended to be large by Mesozoic standards; *Gobiconodon* itself (Fig. 2.9C), from the Early Cretaceous of the Gobi Desert in Mongolia, and also from Russia, China, and the Cloverly Formation of Wyoming, reached 5.4 kg (12 lb) and 51 cm (20 in), about the size of a large opossum. Much more robust and shorter-legged than most Mesozoic mammals, it was probably not a good climber but purely a ground predator.

Repenomamus giganticus is known from complete specimens with soft parts and fur impressions, and as the species name implies, it was the largest mammal known from the Mesozoic. Big specimens were about the size of a large dog, reaching 1 m (3 ft) in length and weighing about 12–14 kg (26–31 lb). *Repenomamus* had a massive, hulking body, a long tail and snout, and robust sprawling limbs, with large hands and feet that were plantigrade (meaning they walked with their palms/soles and heels in contact with the ground). Most surprisingly, they were larger than some of the small dinosaurs (such as *Graciliraptor*) from the same beds. Some specimens were found with baby *Psittacosaurus* (a distant relative of *Triceratops*) in their stomachs! So although most Mesozoic mammals were tiny and hid from the dinosaurs during the Jurassic and Cretaceous, *Repenomamus giganticus* managed to turn the tables on the dominant reptiles (Fig. 2.10).

The relationships of triconodonts are controversial. Most research shows that they were more advanced than morganucodonts, docodonts, or monotremes, but not as advanced as multituberculates or therian mammals (Fig. 2.5). However, other scientists place them just outside the branch point for the monotremes, or outside the monotreme-therian mammalian group altogether.

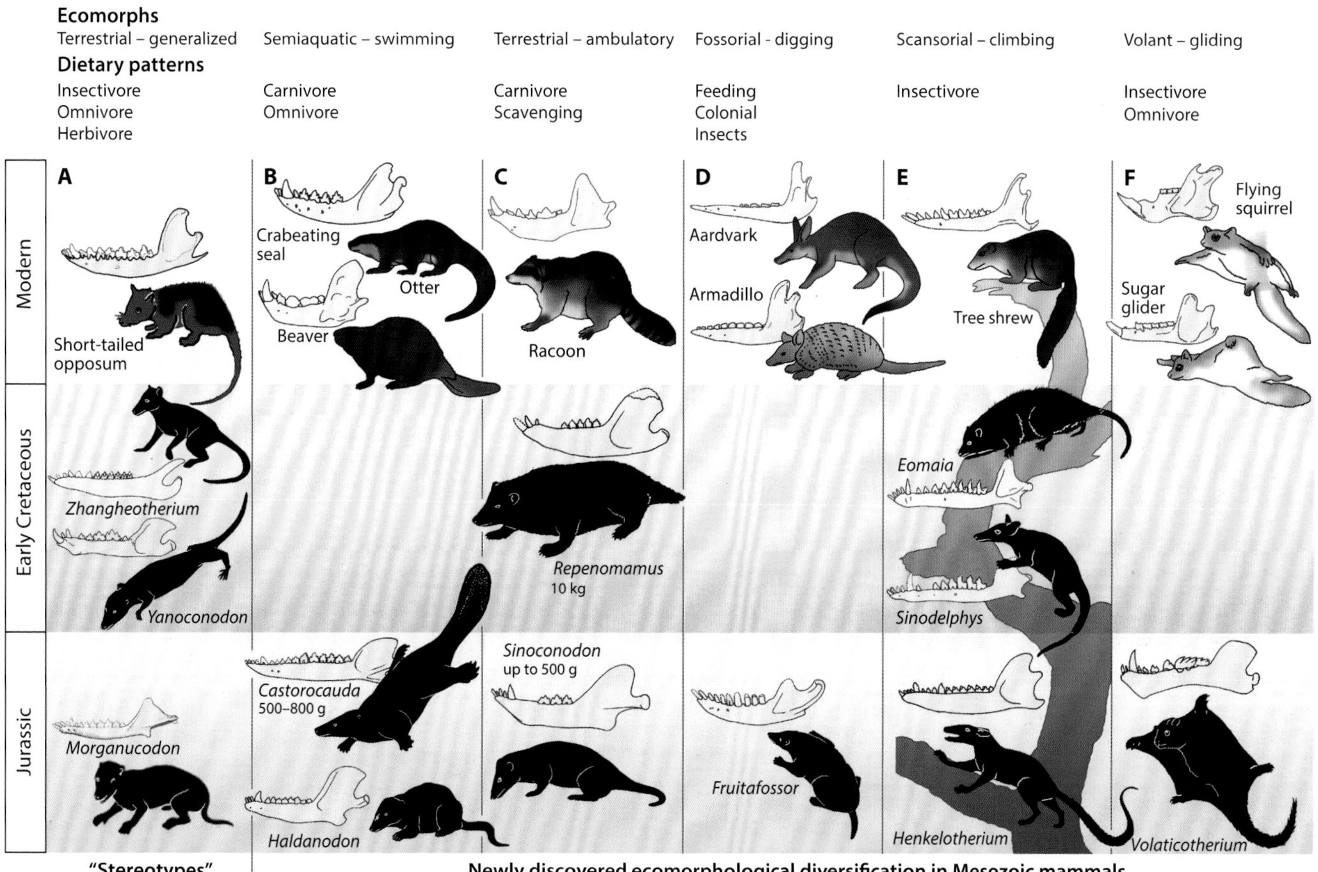

Figure 2.10. Ecological niches of Mesozoic mammals.

THERIA

Monotremes, multituberculates, docodonts, triconodonts, and morganucodonts are largely extinct side lineages on the mammalian tree; only one group, monotremes, still survives. The main surviving lineages of living mammals fall into two groups: the Metatheria (marsupials) and the Eutheria (placentals), both of which are known from excellent complete specimens with fur and soft parts found in Chinese beds about 125 Ma in age. One of these fossils, *Sinodelphys*, is the oldest known relative of the marsupial lineage (see Chapter 3), and it is found in the same beds as *Eomaia*, one of the oldest known relatives of the placentals (see Chapter 4). Together, the marsupials and placentals form a group known as the Theria, which is distinguished by the loss of the last of the reptilian bones in the shoulder (including separate coracoids in adults and interclavicles), found in many Mesozoic mammals, and by a distinctive type of ankle structure. In more advanced therians, the cheek teeth are formed into a "reversed triangle" pattern of three cusps (the tribosphenic tooth). These teeth work something like pinking shears, with their triangular cusps sliding in the V-shaped valleys between one another (Fig. 1.10). Such teeth are common in insectivorous mammals, as they are very effective for chopping up insect cuticle. In addition, modern therian mammals give birth to live young, have mammary glands with distinct nipples, and possess many other features not found in monotremes or most other Mesozoic groups.

Kuehneotheres and Symmetrodonts

The earliest known possible therian fossils include fragmentary jaws from creatures of the Late Triassic of Eurasia known as the Kuehneotheria (Fig. 2.11A). Kuehneotheres look vaguely like triconodonts in having three conical cusps on their triangular cheek tooth crowns, but unlike triconodonts, their cheek teeth occlude in a precise fashion, like pinking shears (although they are not yet fully tribosphenic). Traditionally kuehneotheres have been considered the most primitive of all the therian mammals, although more recently they have been excluded from the therians and considered to just be primitive mammaliaforms.

In the Jurassic, two large groups of therians that were more primitive than either placentals or marsupials appeared. One group, the Symmetrodonta (called the "spalacotheroids" in Fig. 2.5), had three-cusped cheek teeth (Fig. 2.11A) that vaguely resembled those of triconodonts but clearly had the "reversed triangle" pinking-shears style of occlusion. Although more than 30 genera are known from the Jurassic and Early Cretaceous of Eurasia and North America, most are based on just fragmentary jaws and teeth. Only *Zhangeotherium* is known from a complete skeleton with fur and soft tissues (Fig. 2.11B), found in the Lower Cretaceous beds of Liaoning, China. It was a small shrew-size creature, with robust, sprawling limbs like those of a reptile or early mammal, and bones in its ankle similar to the arrangement seen in monotremes. Specimens of *Zhangeotherium* have been found in the stomach region of the feathered non-bird dinosaur *Sinosauropteryx*.

Dryolestoids

The second group, the Dryolestoidea, was among the most common mammals in the Upper Jurassic beds of North America and Eurasia. When I did research on them in the 1970s, only a handful of jaws were known, many of which had been found and briefly described in the 1880s or the 1920s, and nothing more had been done with them since then. They were lumped into a wastebasket group called "Pantotheria," which included nearly all therian mammals from the Mesozoic that were not placentals or marsupials (represented by *Henkelotherium* in Fig. 2.5). As the "Pantotheria" has been banished from our thinking (although the term "Eupantotheria" still persists for some dryolestoids), we have come to realize how important the dryolestoids are for understanding the evolution of therian mammals.

Dryolestoids were very distinctive (Fig. 2.11A, bottom two jaws). Known from several dozen genera (mostly Late Jurassic of Europe and North America), their upper teeth had a triangular crown pattern that was very compressed front-to-back, so as many as eight or nine molars (most mammals have only three or four), and as many as 11 or 12 cheek teeth (most mammals have no more than eight) could be squeezed into the back of the jaw. Their lower cheek teeth also had very compressed, narrow crowns, with a hook-like shelf in the back of each. The only dryolestoid known from a skeleton is *Henkelotherium*, from the Upper Jurassic Guimarota beds of Portugal. It was a shrew-size creature with a long tail, and limbs that suggest it was largely arboreal.

In the Cretaceous of South America there was another radiation of mammals, the Meridiolestida, which have been assigned to the dryolestoids. About a dozen genera are known, some of which (such as *Mesungulatum*) had highly specialized teeth that superficially resembled those of hoofed mammals, and reached the size of a small dog. Another specimen, *Cronopio*, vaguely resembled the saber-toothed rodent Scrat from the *Ice Age* movies. One fossil, *Peligrotherium*, is from the Paleocene Salamanca Formation of Argentina. It is very incomplete, and in most ways it resembles an archaic hoofed mammal. However, if it is a dryolestoid, then it is the only member of a lineage other than placentals, marsupials, multituberculates, and monotremes to survive past the end of the Mesozoic.

Then there is the weird fossil known as *Necrolestes* (whose name means "grave robber"). This strange creature had a long,

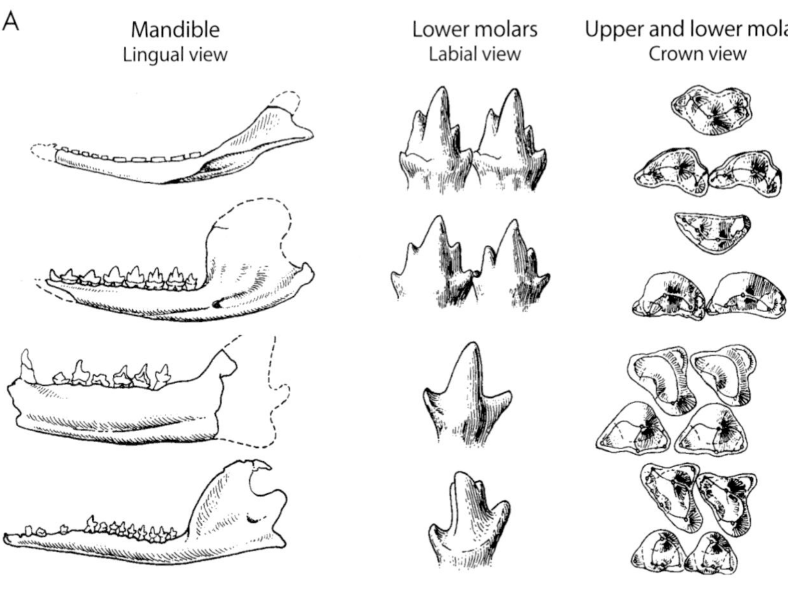

Figure 2.11. Primitive therian mammals. A. Lower jaws and teeth of the most primitive fossil therians. From top to bottom: *Kuehneotherium*, a symmetrodont, and two dryolestoids. B. The crushed fossil of *Zhangeotherium*, one of the few symmetrodonts known from a complete skeleton (including fur and stomach contents).

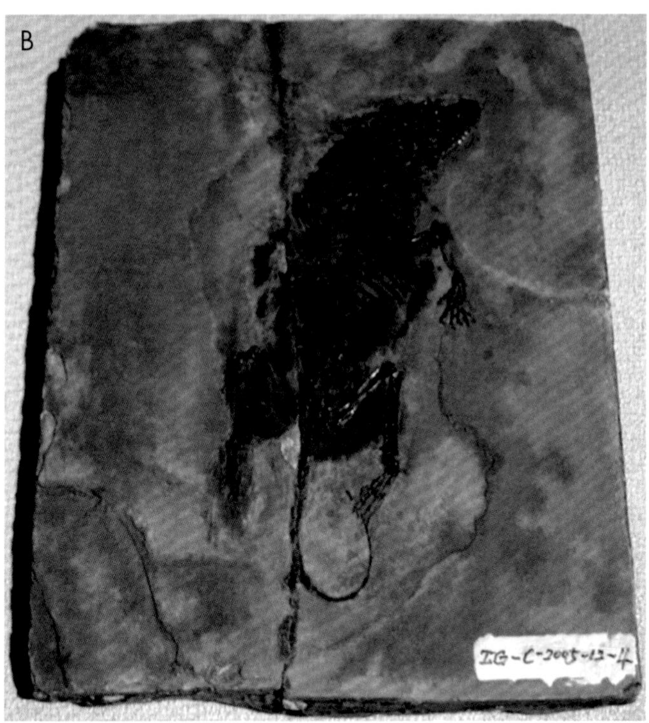

upturned snout and very primitive teeth (Fig. 2.12). Most reconstructions make it look much like a mole with a long, fleshy, upward-pointing proboscis. Described in 1894 by the legendary Argentinean paleontologist Florentino Ameghino from the early Miocene beds of Patagonia, *Necrolestes* was a complete mystery for over a century. Then in 2012 several analyses suggested that it was a very late surviving meridiolestid, most similar to *Cronopio* from the Jurassic. If so, it is a remarkably late survivor of a group that flourished mostly in the Jurassic and was thought to have vanished in the Late Cretaceous. It survived over 40 m.y. after its closest relatives had vanished.

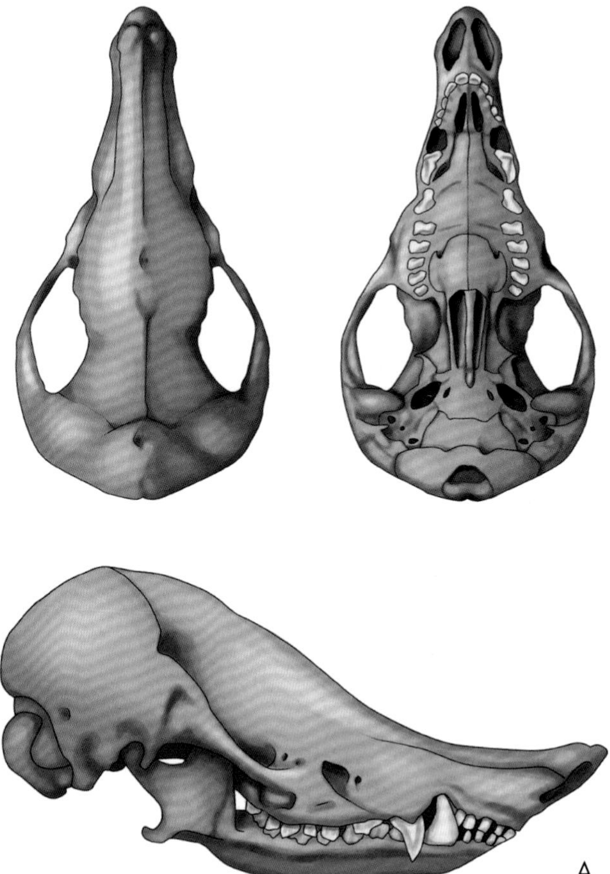

Figure 2.12. *Necrolestes*. A. The only known skull of this fossil mammal, from the Miocene Santa Cruz beds of Argentina. (See also next page.)

Figure 2.12. *Necrolestes*. B. Restoration of *Necrolestes*

Tribosphenida (or Boreosphenida)

By the Early Cretaceous, all the archaic lineages (docodonts, triconodonts, monotremes, multituberculates, dryolestoids, symmetrodonts, and several others) had become established, and many were living side by side in Eurasia and North America (Fig. 2.5). In the Lower Cretaceous beds of Texas and England, however, more advanced fossils were discovered that had not only the reversed-triangle pattern of occlusion in their cheek teeth (which worked like pinking shears) but also a new cusp on the inside point of their upper molar crowns known as the protocone (Figs. 1.9, 1.10). The presence of this cusp, and the effects on its rearrangement of the triangular crowns of the teeth (making a true tribosphenic dentition), defines a much more advanced group, the Tribosphenida (also called the Boreosphenida). This group includes not only the living marsupials and placentals, but also a few primitive transitional fossils such as *Vincelestes* from Argentina, *Butlerigale* from Texas, *Amphitherium*, *Aegialodon*, and *Peramus* from England, and a few others. These creatures are not yet either marsupials or placentals but very close relatives of them, showing how the marsupials and placentals arose from dryolestoid-like ancestors.

By the Late Cretaceous, the mammals had changed quite remarkably. Most mammals from the latest Cretaceous beds that produce *Tyrannosaurus* or *Triceratops* are clearly opossum-like marsupials or placentals that seem to be related to hoofed mammals (*Protungulatum*), to carnivores (*Cimolestes*), or to numerous insectivorous forms. Few are known from more than teeth or jaws, however, so it is difficult to place these fossils within any living orders of mammals. Multituberculates are also extremely abundant, and all three groups made it through the end-Cretaceous extinctions (although marsupials were decimated). Meanwhile, the fossil record tells us that the ancestors of monotremes were still hanging on in Australia, as were late-surviving descendants of the dryolestoids in South America. But most Mesozoic mammal groups (docodonts, triconodonts, symmetrodonts, most dryolestoids, and many others) did not survive to the latest Cretaceous. Instead, the world was dominated by marsupial and placental mammals, which will be the subject of the remaining chapters.

CHAPTER 3

MARSUPIALS

Pouched Mammals

MARSUPIAL VS. PLACENTAL

The marsupials, or pouched mammals (also known as metatherians), include opossum, kangaroos, koalas, wombats, and other familiar creatures, as well as a variety of extinct species (Fig. 3.1). Many (but not all) female marsupials have a fur-lined pouch on the belly in which they carry their young after giving birth. The tiny (bee-size), blind marsupial baby (a joey) is born prematurely and has almost no developed organs except its front limbs and mouth. It has to crawl up its mother's belly fur from the uterus to the pouch, where it clamps on to a nipple and finishes its embryonic development. This is because the fetus is just beginning to develop its own immune system, and the mother's immune system would reject it if it were not born prematurely. Marsupials lack the hormones (found in placental mammals) that would block rejection of the embryo.

The marsupial reproductive system has advantages and disadvantages. The young are born helpless, and it is a dangerous trip for the tiny fetus to reach the pouch. Marsupial young also have a slower growth rate, because transferring nutrients to a baby via milk is much less efficient that doing so through a placental membrane. On the other hand, the marsupial reproductive strategy allows the mother more flexibility in survival. If times are hard, the mother can abort her baby without losing her own life. In good times, marsupials reproduce by having multiple generations at once. In the case of kangaroos, the mother can

Figure 3.1. A spectrum of extinct marsupials. In the left foreground is the wolf-like sparassodont *Proborhyaena*. In the extreme right foreground is the saber-toothed sparassodont *Thylacosmilus*. In the center foreground is the marsupial "lion" *Thylacoleo*. The giant short-faced kangaroo is *Procoptodon*. The huge creature behind the human is *Diprotodon*. The sloth-like creature with the proboscis in the left background is *Palorchestes*.

have one fetus in the uterus, a joey in the pouch, and a third joey hanging around nearby before it is ready to go off on its own.

By contrast, about 95% of the 5,500 living species of mammals are placentals (also known as eutherians). In addition to the amniotic membrane found in all other amniotes, the fetus is surrounded by the placental membrane, which protects it in the uterus. In addition, placental mammals develop a large area of tissue made of the fused allantoic and chorionic membranes outside the amnion, called the **placenta** (expelled after delivery, when it is sometimes referred to as the "afterbirth"). The placenta serves as an organ for nourishment, gas exchange, and waste disposal for the developing fetus. Through hormonal control, the placenta also prevents the mother's immune system from rejecting it as it grows and matures. Although monotremes and marsupials have the allantoic and chorionic membranes, only in the placental mammals are these membranes fused together and attached to the uterine wall.

Unlike the poorly developed, prematurely born marsupial babies, the young of placental mammals are born almost fully developed. In some mammals (zebras and antelopes, for example), the newborn has to be able to stand and run within minutes of being born. Otherwise, it is food for predators. These animals have **precocial** development, meaning the baby is born capable of many important tasks. Other placentals (such as humans) give birth to young that are still very immature and require much more parental care as they develop (**altricial**). But in both cases, there is a trade-off for placental animals: the mother has to invest more of her resources into the young, and if she is being chased by predator or is starving, she cannot easily abort her young and live to breed another day, as a kangaroo can.

There is another disadvantage to marsupial reproduction. Due to the short development time, the fetuses are born with a smaller neocortex area of the brain compared to many placentals, so marsupial brains can never be as developed as those of some placentals. There are other limitations as well. There cannot be any marsupial whales, because the embryo is attached to a nipple in a pouch, which would prevent the female from swimming with babies attached to her. In addition, marsupial babies require front claws to survive the crawl to the pouch, so they could never develop hooves, as some placentals did.

Marsupials only dominate in places such as Australia, where there was no competition from advanced placentals. In some cases when marsupials and placentals competed for the same resources (such as when North American placentals invaded South America in the Pliocene), marsupials seem to have lost the competition—although most of the large extinct marsupials of South America were already gone before most of the placental competition arrived. This lack of dominance does not mean that marsupials are necessarily inferior to placentals, but there are some circumstances in which they are favored, and some in which they are not. In fact, there are many other things that argue against "marsupial inferiority" as an explanation. South America's smaller marsupials are very prolific and diverse, despite a wave of invasion of advanced placentals. In Australia, a number of placental invaders (such as rodents at 5 Ma and dingoes more recently) arrived without displacing the native marsupials—although the huge influx of sheep, rabbits, rats, and other placentals that humans have introduced have hurt most native marsupials.

In fossils, we cannot determine an animal's reproductive system based on just bones and teeth alone, so we look for other clues. All primitive mammalian skeletons except those of placentals have an extra bone (the epipubic, or "marsupial bone") in the front of the hips that supports the abdominal muscles. In marsupials, the bone in the back of the lower jaw is curved inward in a distinctive pattern that does not appear in other mammals. Most marsupials have three cheek teeth with simple crowns typical of premolars, and four teeth with complex crowns that look like molars. Placentals, by contrast, have four premolars and three molars. In primitive marsupials, the outer area of the crown of the upper molars (stylar area) is greatly enlarged, and the inner part of the molars is small, but in primitive placentals this stylar area is small and the molar crowns (Figs. 1.8, 1.9, 1.10A) are enlarged toward the inside of the mouth (around the cusps known as protocones). These and other more subtle differences help a paleontologist distinguish a fossil marsupial from a fossil placental in most cases.

MARSUPIAL EVOLUTION

The oldest known fossil relative of the marsupials is *Sinodelphys*, a beautifully preserved complete skeleton from the Lower Cretaceous beds of Liaoning, China, dated to about 125 Ma (Fig. 3.2). Even though the specimen was smashed flat when it was fossilized, it preserves the fur and soft tissues, and you can still see the characteristic hip bones and teeth and jaw features of marsupials. It was rat-size, about 15 cm (6 in) long, and weighed about 30 g (1.05 oz). Its hind limbs suggest it was a squirrel-like tree climber. For the rest of the Cretaceous, marsupials are known mostly from tooth and jaw fossils, mainly in Asia and North America. At the end of the Cretaceous, they were very common mammals (outnumbering placentals) in the *Tyrannosaurus-Triceratops* beds of Montana and Wyoming. However, marsupials were nearly wiped out in North America during the Cretaceous extinctions. Primitive opossums managed to hang on in North America and Eurasia through the early Cenozoic, then vanished. During the Ice Ages, modern opossums reinvaded North America after a long absence.

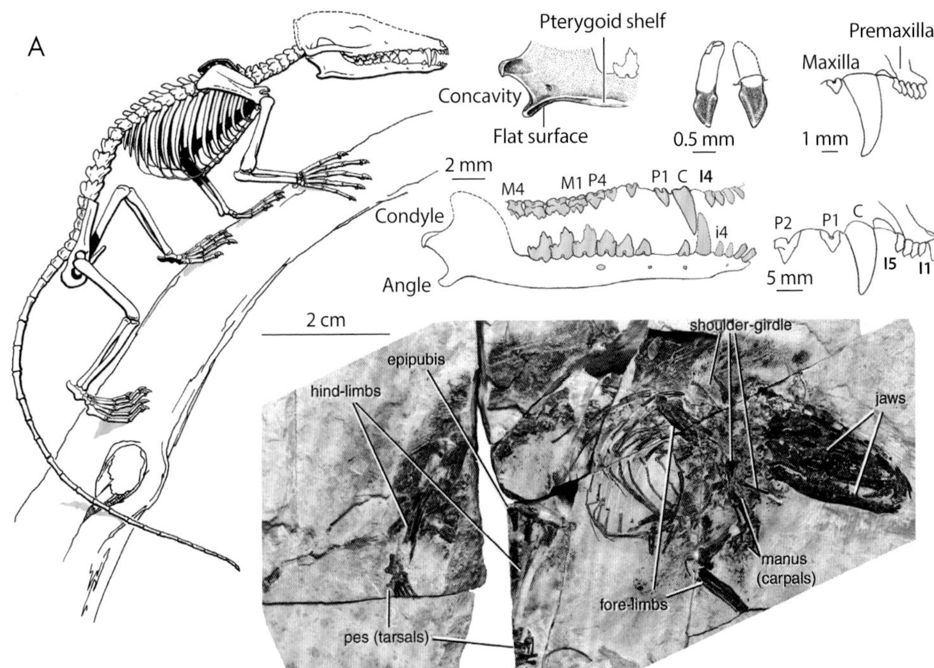

Figure 3.2. The earliest known marsupial fossil, *Sinodelphys*, from Lower Cretaceous (125 Ma) beds of China. A. Complete skeleton. B. Restoration of *Sinodelphys*.

AMERIDELPHIA

The major division of the living marsupials splits them into two groups: the Ameridelphia, comprising the New World opossums and their extinct relatives, and the Australidelphia, encompassing the great Australian radiation of marsupials, plus their close relatives, the microbiotheres of South America (Fig. 3.3).

Sparassodonts

Marsupials became established in South America as it separated from the northern continents during the Cretaceous. While they nearly vanished from the northern continents during the Paleocene, and placentals became dominant there, in South America marsupials evolved a great diversity of different descendants. South America was isolated from the north before any placental predators evolved, so marsupials performed that role instead. Eventually these primitive marsupials evolved into a group of predators known as sparassodonts (or borhyaenids), which evolved into body forms that resembled placental hyenas, wolves, bears, and other smaller predators (Figs. 3.1, 3.4). One of them was a saber-toothed marsupial, *Thylacosmilus*, which was amazingly convergent with the saber-toothed placental cats of the north (Figs. 1.4A, 3.1).

Without placental carnivorous mammals in South America, the only competition these marsupial predators faced was from giant snakes as long as a school bus (*Titanoboa*), huge crocodilians, and large, 3 m (10 ft) tall predatory birds, the phorusrhacids. Eventually, all the large marsupial predators vanished

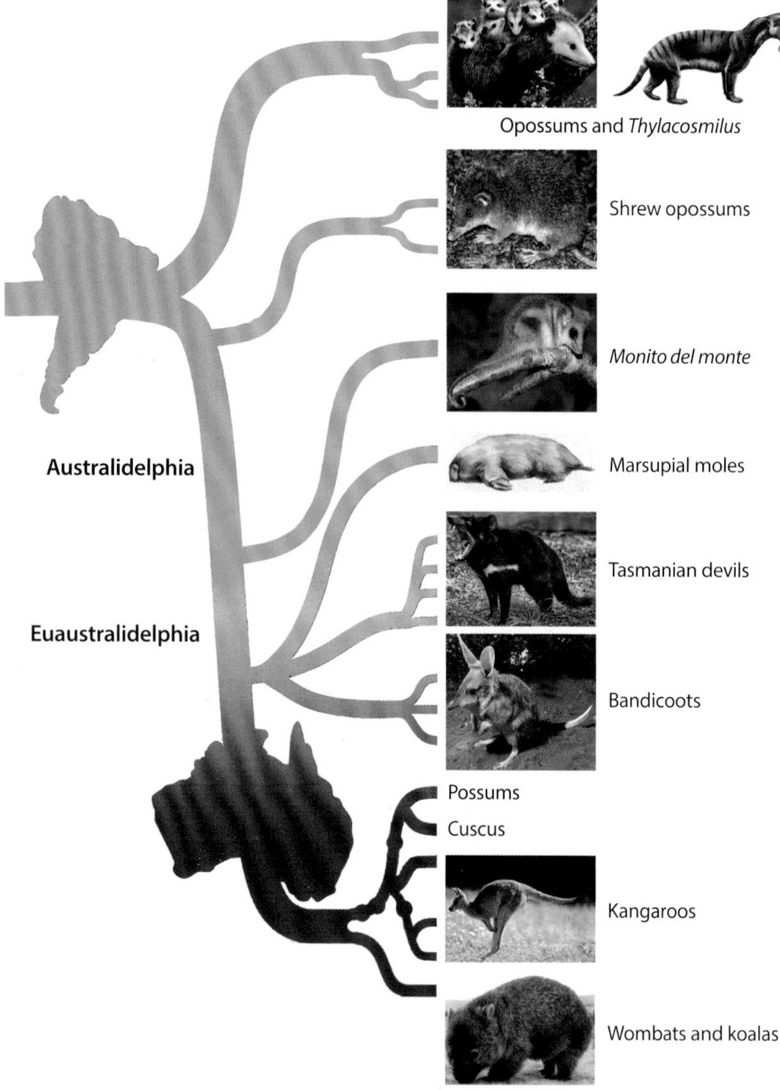

LEFT: Figure 3.3. The evolutionary history of the marsupials, showing their origin from the South American groups and then their spread to Australia.

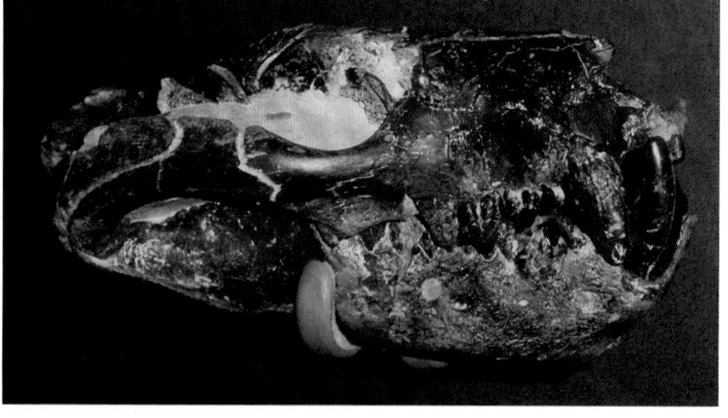

BELOW LEFT: Figure 3.4. A skull of the South American marsupial *Borhyaena*, a hyena-like or wolf-like predator.

from South America, and now cats (jaguars and ocelots), dogs (maned wolves and other uniquely South American dogs), bears (the spectacled bear), raccoon relatives (coatis and kinkajous), and other placental carnivorans occupy the main role as predators.

Didelphimorphs

Although the predatory sparassodonts (or borhyaenids) and saber-toothed marsupials are gone, the opossums and their relatives (the Didelphimorpha) are still very diverse, with about 93 species alive today. These include bushy-tailed opossums, mouse opossums, water opossums, four-eyed opossums, and many more (Fig. 3.3). Most are typically about the size and shape of the North American opossum, although the mouse opossum and some others are much smaller. Opossum relatives are found all over South America, especially in the forests of the Andes Mountains and in the Amazonian rain forest. They have lots of interesting peculiarities. For example, the "four-eyed" opossum has two white spots just above its eyes that look like an extra set of eyes to confuse predators. The thick-tailed opossums look and also live much like otters as well, swimming and diving for their food in the water.

The yapok, or water opossum, has short dense water-repellent fur, long whiskers on the face for feeling its way in muddy water, webbing on its hind feet, and a strong tail for swimming. Yapoks feed on aquatic animals: fish, frogs, crustaceans, and other fresh water creatures. Because yapoks swim so much, the female yapok can close her pouch tightly so her young will not drown while she is swimming and hunting. The male has a similar pouch that protects his genitals while swimming.

Caenolestidae (Shrew Opossums)

Another group, the shrew opossums (Caenolestidae, or Paucituberculata), is a relict of the great radiation of large marsupials in South America. Caenolestids are rat-size creatures confined to the high elevations of the northern Andes (Fig. 3.3). They are part of a surviving group numbering only three genera and seven species that has a fossil record going all the way back to the Paleocene. As recently as the Miocene, there were seven genera of caenolestids that we know of. Some paleontologists think these animals filled the shrew/tiny ground insectivore niche in

South America, and that their decline might be due to the relatively late invasion of shrews from the Northern Hemisphere.

As their name implies, most shrew opossums are about the size and shape of a rat or a shrew, having slender limbs, a long pointed snout, and a long rat-like tail. Like shrews, they are active predators, voraciously feeding on worms, insects, and small vertebrates found in the dense undergrowth and leaf litter. They are largely nocturnal predators, using their sense of touch, with the long snout and sensitive whiskers, plus hearing, and smell. Unlike many nocturnal animals, they have small eyes and relatively poor eyesight.

AUSTRALIADELPHIA

The second great branch of the living marsupials is the Australidelphia, found mostly in Australia, as its name suggests. These mammals are united not only by their geography, but also by a number of anatomical similarities, and molecular similarities as well. In addition to the Australian groups, the Australiadelphia also includes the South American microbiotheres.

Microbiotheria (Microbiotheres)

One of the surviving lineages of South American marsupials, the Microbiotheria, is closely related to the entire evolutionary radiation of Australian marsupials, which took place in the absence of any native land placentals (Fig. 3.3). Microbiotheres are mouse-size, but look a bit like a mixture of opossum and monkey. Like some nocturnal primates (such as tarsiers), microbiotheres have big eyes for night vision and grasping hands for catching insects and leaping from branch to branch. Like some arboreal monkeys and opossums, they have a long, prehensile tail with a naked underside that helps them to grasp branches.

Only one species of this "living fossil" from the early Cenozoic still survives in the forests of the Andes in Chile and Argentina, where it is restricted to high mountains. This peculiar and rare South American marsupial is locally known as *monito del monte* (little mountain monkey). Microbiotheres have lived in South America since at least the earliest Paleocene, and are also found in the Eocene rocks of Seymour Island in the Antarctic Peninsula. Even more interesting, the discovery of microbiothere fossils in the Eocene of Australia suggests that about 50 Ma, they once lived across all of South America, Antarctica, and Australia.

Australian Marsupials

In mammalian terms, Australia has always been a land dominated by marsupials. Except for a single Eocene tooth of a possible placental, known as *Tingamarra*, there is no evidence that placentals ever got established there. Without placental competition, Australian marsupials evolved to fill many of the ecological niches that placentals occupied elsewhere (Fig. 3.5). There are marsupial equivalents of placental moles (notoryctids), wolverines (Tasmanian devils), cats (quolls), dogs (the extinct Tasmanian "wolf" or thylacine), flying squirrels (phalangers), mice (dasycercids), anteaters (numbats), woodchucks (wombats), rabbits (bilbies), kinkajous (cuscuses), and large herbivores (kangaroos and wallabies). There were also a few specialists (such as koalas) that have no close counterpart among placentals except possibly the tree sloths.

These are just the living Australian marsupials. During the later Cenozoic, the marsupials evolved into some spectacular forms, such as those we see in the famous Miocene fossil beds in Riversleigh, Queensland. Riversleigh includes some amazing fossils, including the huge flightless land bird *Dromornis*, which weighed up to 500 kg (1100 lb), the heaviest bird ever known. The platypus *Obdurodon* (Fig. 2.7D) was up to 60 cm (24 in) in length, much larger than the living platypus. Among marsupials, there were the snail-eating *Malleodectes*, a variety of primitive relatives of the wolf-like thylacine, a number of bandicoots, and several large koalas, plus many different kinds of wombats, possums, and primitive kangaroos.

Dasyuromorphia (Dasyuromorphs)

Just as most of the carnivorous mammals of South America were sparassodonts, most of the carnivorous marsupials of Australia are dasyuromorphs. However, they are mostly tiny, not wolf-size like sparassodonts. There are three families with about 74 species in the Dasyuromorphia. Seventy of these species are in the family Dasyuridae, which includes a wide spectrum of small-bodied predators that are similar to wolverines, cats, weasels, shrews, mice, and comparable placental predators from other continents. The most familiar of these is the Tasmanian devil. Contrary to the portrayal of its Looney Toons counterpart, the Tasmanian devil is not a huge monster that moves in whirlwinds and eats almost anything. Instead, it has the build of a wolverine (Fig. 3.6A) but is slightly smaller, preying mostly on smaller mammals, birds, and reptiles. Tasmanian devils also eat carrion, including roadkill, and raid trash cans and steal food from domestic pets (or eat the pets themselves). With the extinction of the Tasmanian wolf, the Tasmanian devil is currently the largest predatory marsupial alive. It is stocky and muscular in build yet capable of running steadily over long distances, and climbing trees and swimming rivers as well. Its reputation comes from

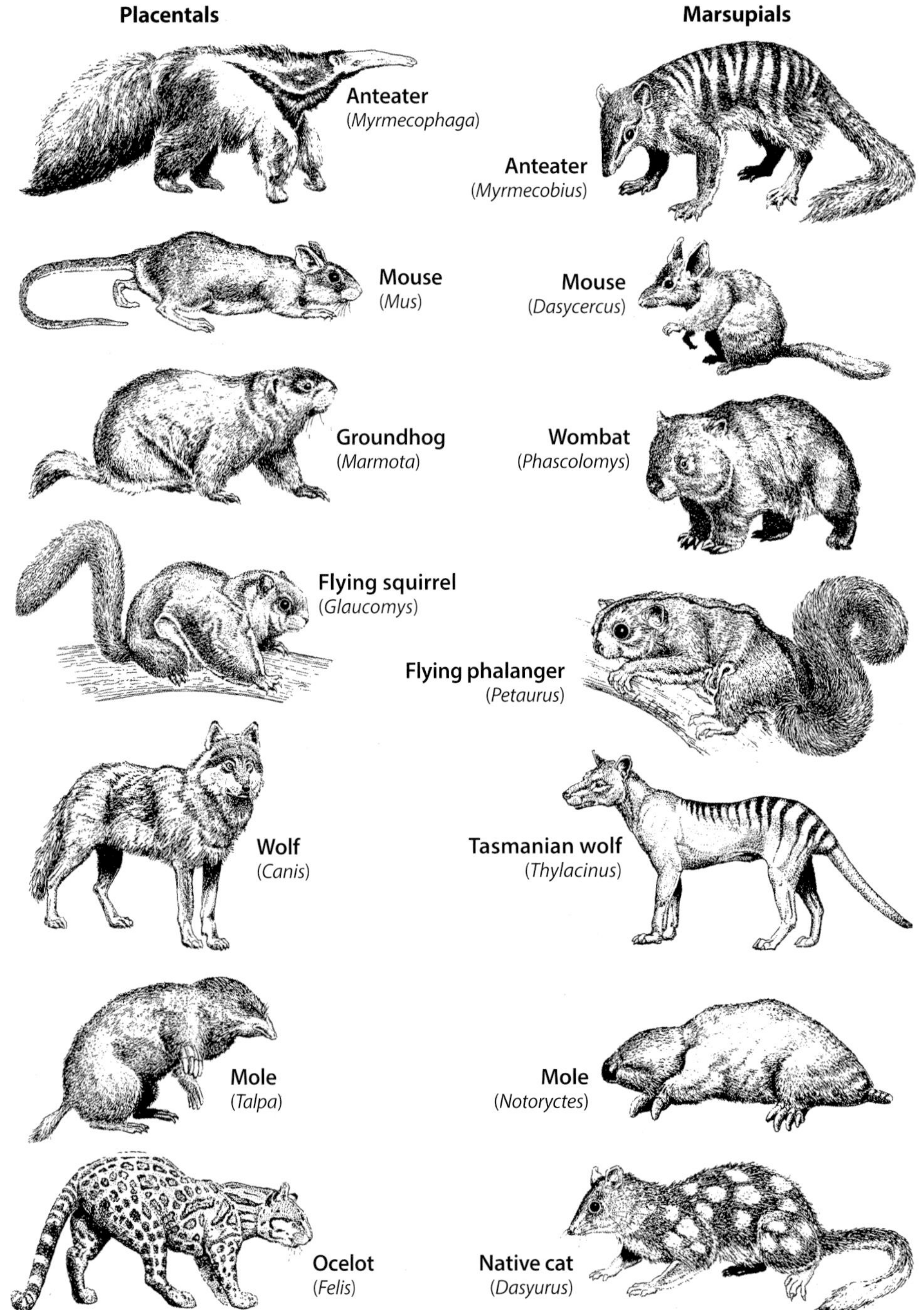

Figure 3.5. The convergent evolution of many different marsupials with their placental counterparts.

the ferocity it shows when fighting (especially when cornered) or feeding. It has the strongest bite for its size of any mammal alive. Devils are also infamous for the foul pungent odor they can spray and for their blood-curdling screech.

At one time devils were distributed widely over Australia as well; a Miocene species recovered from an Australian cave was about 15% larger and 50% heavier than the modern species. Apparently the mainland devils were driven to extinction by the introduction of dingoes. In Tasmania, they are dying off rapidly because of devil facial tumor disease, and are now endangered in their last known refuge. Fewer than 10,000 are thought to remain in Tasmania today.

In comparison to the Tasmanian devil, the rest of the Dasyuridae are much smaller in body size, ranging from the weasel-size quoll to more than 80 species of shrew-like and mouse-like forms. Most of the smaller species are insectivorous, although

Figure 3.6. Dasyuromorphs. A. Photograph of the last-known living Tasmanian "wolf" individuals in the Hobart Zoo. B The living Tasmanian Devil.

the larger ones also catch mammals, lizards, and birds, and eat fruit and flowers as well. Dasyurids are the most primitive group of living Australian marsupials, lacking many of the specializations seen in the other groups, such as kangaroos. Some dasyurids lack a full pouch for their young, and the females instead have just a shallow fold of skin around their nipples.

Although no large predatory marsupials remain, at one time there was a whole family of marsupials that mimicked dogs and wolves, the thylacines. Their earliest fossils go back to the late Oligocene, with *Badjcinus*, a cat-size predator. Another half-dozen genera of thylacines are known from the Miocene, including distant relatives of *Thylacinus*, the modern Tasmanian "wolf" or Tasmanian "tiger"—so called because of the stripes on its back (Fig. 3.6). Built like large short-haired dogs, with a long straight tail and striped back, these creatures were the apex predators of Australia and Tasmania after the large Ice Age mammals vanished about 30,000 years ago. They were still roaming in large numbers in Australia until competition from true dogs (dingoes brought by the Aborigines) reduced them to small populations about 2,000 years ago. In Tasmania, the destruction of the habitat by humans (especially the European settlers) further diminished thylacines' numbers, as did direct hunting of them for sport and to prevent them from eating the domestic chickens and sheep of the Europeans. Finally, they were decimated by the spread of marsupi-carnivore disease before vanishing from Australia by 1900. A few were still roaming around Tasmania, but the last known individual died in the Hobart Zoo in 1936 (Fig. 3.6B). This is one of the most famous and tragic of the recent extinctions of an important group of large mammals, one barely known to science before it vanished.

Yet another family of Dasyuromorphia is the Myrmecobiidae, the numbats or "marsupial anteaters" (Fig. 3.5). Numbats look vaguely like tiny versions of placental anteaters in having a furry body (with stripes on their backs), a long furry tail, and a long snout. Like other ant-eating mammals, they have a long, sticky tongue in the snout and specialized ridges on the palate for scraping off the termites as they lap them into their mouths. They feed exclusively on termites, and an individual consumes about 20,000 per day. However, numbats do not have the large powerful front claws seen in anteaters, aardvarks, pangolins, and other termite eaters, so they cannot rip a termite mound open, but must wait until the termites are active and lap them up from the entrances to their mound. Numbats are easy prey to a number of creatures (especially animals introduced by humans), and today they are endangered, with fewer than 1,000 living in the wild.

Peramelomorpha (Bandicoots and Bilbies)

The next most advanced group of Australian marsupials is the Peramelomorpha, or the bandicoots and bilbies, which includes 21 living species and many genera of fossils from the Miocene,

Non-syndactylous feet

Hallux (opposable) Four independent digits

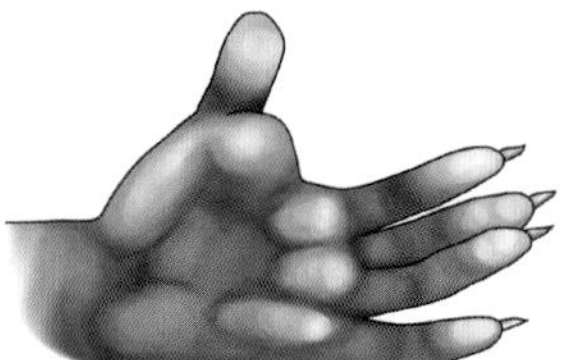

Didelphis virginiana
Virginia opossum

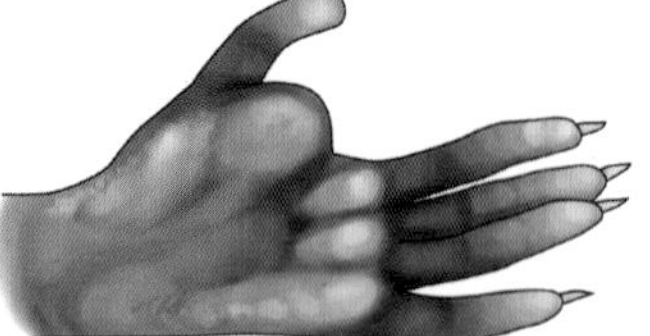

Philander opossum
Gray four-eyed opossum

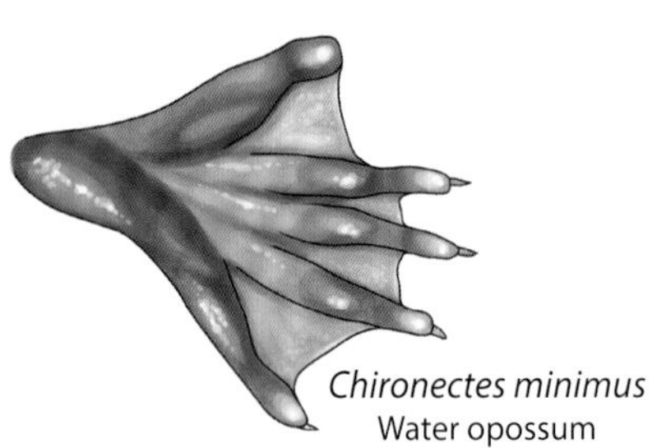

Chironectes minimus
Water opossum

Syndactylous feet

Hallux (opposable) Fused digits 2 & 3

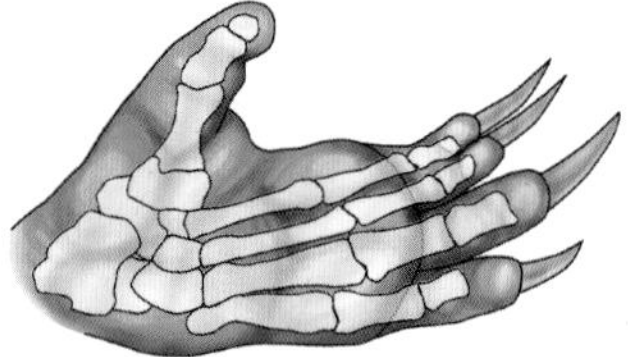

Phalanger
Cuscus

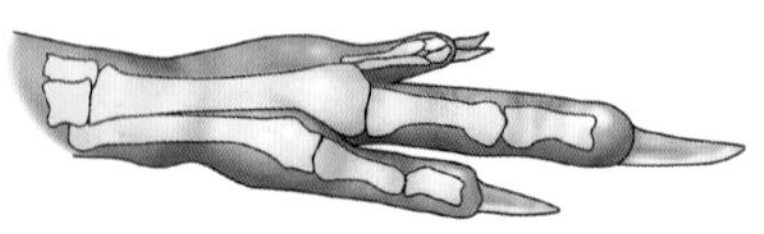

Potorous
Long-footed potoroo

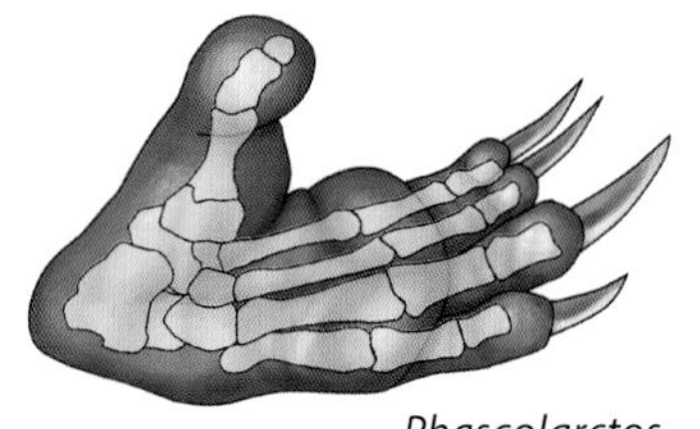

Phascolarctos
Koala

Figure 3.7. On the left are the primitive marsupial feet of opossums and dasyurids, which have most of the toes of the standard proportions. On the right are the syndactylous feet of many marsupials, which have the tiny second and third toes fused together. This feature is found in all the Diprotodontia and also in the Peramelomorpha.

Pliocene and Pleistocene. Many people have heard of the video-game character Crash Bandicoot, but there is no similarity to it and the real thing (any more than the Looney Toons character Taz resembles its namesake). In a form of scientific jest, the oldest-known fossil of the group (from the Miocene Riversleigh beds of Queensland) was actually given the formal genus and species name *Crash bandicoot* after the video-game character.

Bandicoots are mostly small mouse-size to rabbit-size creatures with large ears, a plump arch-backed body, relatively long, thin legs suitable for hopping and running, and other similarities to rabbits. Unlike rabbits, they have a long, pointed snout and a long rat-like tail (Fig. 3.3). Also, bandicoots are not primarily herbivores, as rabbits are, but eat mostly insects and other small invertebrates, plus seeds, fruits, and fungi. More than 25 species in five genera are known, found all over Australia and even New Guinea, although many of these species are now threatened or endangered, due to habitat destruction and predation from animals introduced by humans. In addition, the Peramelomorpha include the bilbies or "rabbit-bandicoots" (family Thylacomyidae), which have even longer ears (standing upright like those of a jackrabbit) and tails and softer fur than the bandicoots.

The relationship of Peramelomorpha to the rest of the Australian marsupials has long been controversial. In most of their anatomy, they are relatively primitive and do not differ much from the most primitive group, the Dasyuridae. However, they have two of their toes fused together (**syndactyly**), which seems to be a highly specialized feature, since it is only found in one other group of marsupials, the Diprotodontia. The syndactylous foot has the second and third toes reduced and tightly fused together well up to the ankle and down to the base of the claws (Fig. 3.7). Syndactylous feet are also modified in losing their first toe (the "big toe") but enlarging their fourth and fifth toes.

The molecular evidence has not been clear-cut yet, although some studies suggest that peramelomorphs are indeed related to higher marsupials (the Diprotodontia). If so, their highly specialized syndactyl feet are a synapomorphy (shared derived trait) that evolved only once in the common ancestor of Peramelomorpha and Diprotodontia, not an example of convergent evolution.

Diprotodontia (Kangaroos, Wallabies, Wombats, Koalas, Phalangers, and Their Relatives)

By far the largest and most advanced group of Australian marsupials is the Diprotodontia, containing more than 117 living species and many fossil species as well. These include the majority of the familiar groups: kangaroos, wallabies, potoroos, rat-kangaroos, tree kangaroos, wombats, koalas, possums, the gliding phalangers and sugar gliders, cuscuses, and many others (Fig. 3.3). The name Diprotodontia means "two front teeth," and refers to the large pair of protruding lower incisors, which occlude with three pairs of upper incisors. They have other specialized dental features not shared with most other marsupials, including reduced canines or the loss of canines altogether (except for the saber-toothed wallaby from Riversleigh known as *Balbaroo fangaroo*). Many have highly modified cheek teeth for eating a variety of tough types of vegetation. All the living Diprotodontia are herbivorous, although there are extinct forms, such as the marsupial "lion" (Fig. 3.1), that were carnivorous.

The other distinctive feature of the Diprotodontia is the syndactylous foot, shared with the Peramelomorpha and apparently a unique adaptation that unites both groups (Fig. 3.7).

In addition to the wide variety of living species, the Diprotodontia have an extensive fossil record going back to the late Oligocene (about 28 Ma). There are dozens of genera of different kinds of kangaroos, wallabies, wombats, phalangers, and koalas in the famous Miocene deposits of Riversleigh and in numerous other Miocene and Pliocene localities, since Diprotodontia have always been the most dominant and numerous group in Australia.

During the Ice Ages, some Australian animals reached immense size. In addition to *Megalania*, a Komodo dragon relative that reached 7 m (23 ft) in length and weighed about 2000 kg (4400 lb), and huge land birds (dromornithids), the Diprotodontia were spectacularly large as well. The family Diprotodontidae included 2000 kg (4500 lb) wombat relatives the size of hippopotamuses (Figs. 3.1, 3.8). There were gigantic kangaroos twice the size of living forms. One of these, the short-faced giant kangaroo *Procoptodon* (Fig. 3.1) was 3 m (10 ft) tall, and weighed about 230 kg (500 lb). The huge omnivorous or carnivorous kangaroos of the genus *Propleopus* reached 3 m (10 ft) in length and weighed about 70 kg (155 lb). There were large koalas as well as strange, clawed marsupials built like ground sloths and bearing a tapir-like proboscis known as palorchestids (Fig. 3.1).

Even more impressive was the marsupial "lion," *Thylacoleo* (Figs. 3.1, 3.9). It was the size of a modern leopard, weighing about 165 kg (365 lb) and wielding retractable claws (convergent with the condition found in cats) and a thumb that could help it climb and grasp prey. It had large fangs, like most predators (although they were incisors, not canine teeth), but instead of multiple sharp cheek teeth it had huge shearing blades in its mouth that enabled it to slash its prey to pieces.

However these huge Ice Age marsupials lived and flourished, they all vanished toward the end of the last Ice Age. This was

Figure 3.8. A skeleton of the rhino-size giant Ice Age wombat-like marsupial *Diprotodon*.

Figure 3.9. Skeleton of the Ice Age marsupial "lion" *Thylacoleo*.

probably due to the drying of the climate and also to the pressures of the Aboriginal peoples that arrived about 60,000–46,000 years ago from Southeast Asia, who hunted them and burned their landscape (see Chapter 18).

Today, many of the surviving marsupials are endangered, as their habitats continue to vanish, and they face pressures from animals that humans have released into the wild: sheep, cattle, horses, camels, and especially cats, mongoose, rats, and rabbits. These animals either prey directly on the marsupials, disturb their habitat, or else crowd them out of their preferred habitats and eat all their food. Thus, the great experiment in marsupial evolution that took 66 m.y. to develop is ending in a few hundred years, as humans and their introduced species wipe them out.

CHAPTER 4

PLACENTAL MAMMALS (EUTHERIA)

As we discussed in Chapter 3, the fundamental split between most living mammals is the pouched mammals (marsupials), which give birth to immature young that grow up in a pouch, and the placental mammals, which give birth to fully developed young. These differences in reproduction are hard to detect in fossils, however, so we must resort to features visible in the skeleton. There are quite a few that are useful. Living placentals have lost the epipubic bone (marsupial bone) in the hip joint. They have only three molars but four premolars in their cheek teeth. In their upper molars, the outer portion (the stylar area) is greatly reduced compared to that in marsupials, while the inner area of the crown is emphasized. These features, and several others, allow most fossils to be confidently identified as marsupial or placental.

The oldest known fossil that has been identified as a placental is *Juramaia* from the Late Jurassic of China, about 160 Ma (Fig. 4.1). It is a beautifully preserved specimen (although much of the back of it is missing), but it clearly shows the characteristic features of placental teeth and other anatomical features. Another probable placental fossil is *Eomaia*, from the Early Cretaceous of China, about 125 Ma (Fig. 4.2). It is a well-preserved complete specimen (though smashed), with impressions even of the fur and other soft tissues. Although the creature was only 10 cm (4 in) long and weighed only 20–25 grams (0.7–0.9 oz), about the size of a shrew, it is remarkably advanced for its time.

By the Late Cretaceous, placentals began to evolve rapidly (Fig. 4.3), and the evidence from this time includes numerous complete uncrushed skeletons of primitive placentals from Mongolia and China known as kennalestids and asioryctids. In the latest Cretaceous, we begin to

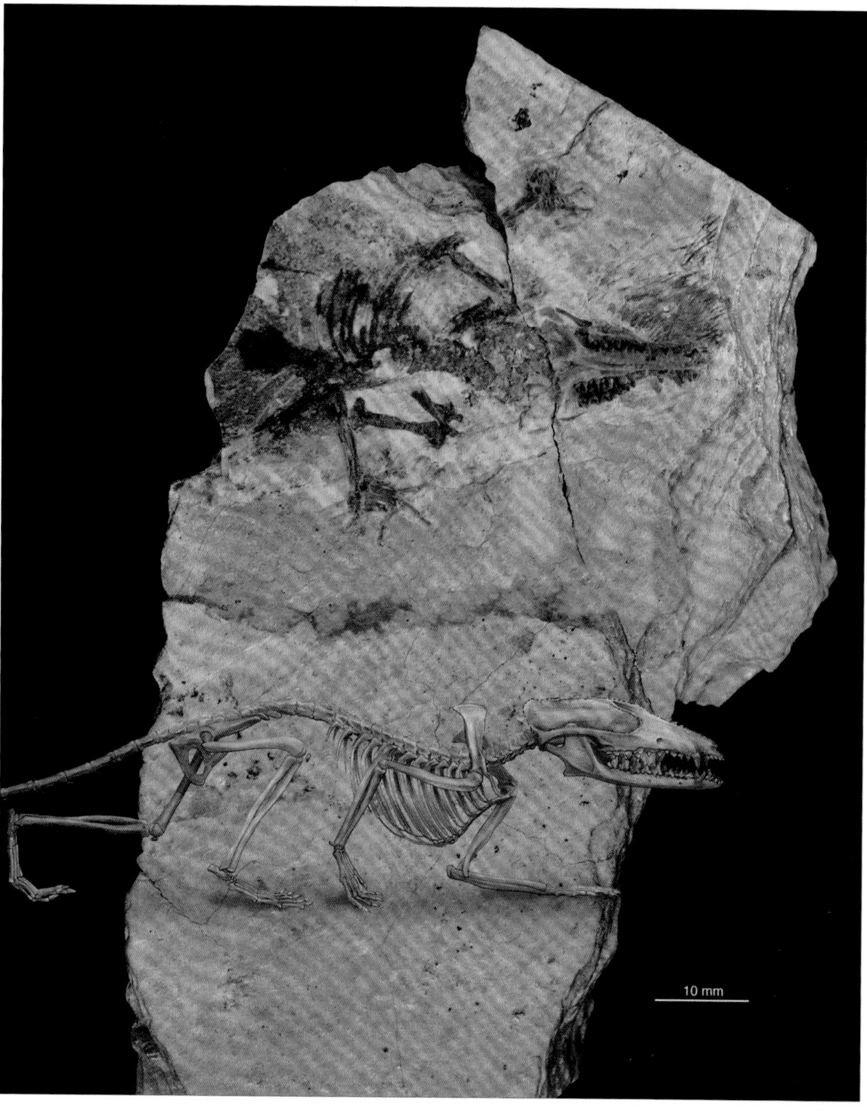

Figure 4.1. The Late Jurassic relative of placental mammals known as *Juramaia*.

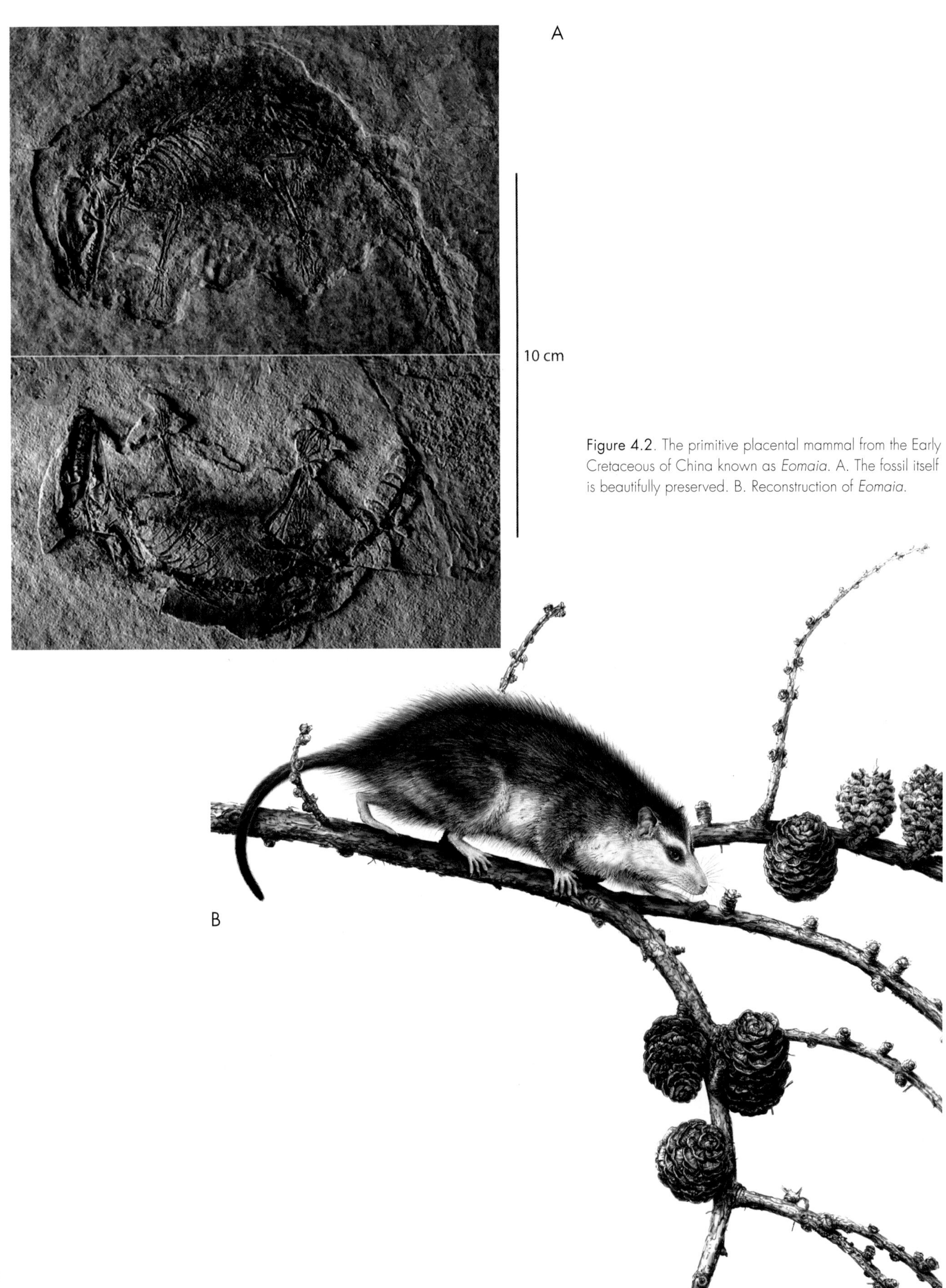

Figure 4.2. The primitive placental mammal from the Early Cretaceous of China known as *Eomaia*. A. The fossil itself is beautifully preserved. B. Reconstruction of *Eomaia*.

see the first evidence of some of the major groups, including hoofed mammals (*Protungulatum* from Montana) and carnivorous mammals (*Cimolestes*). Once the dinosaurs (excluding birds) vanished at the end of the Cretaceous, the placentals rapidly evolved to fill many of the niches in a landscape vacated by the large-bodied dinosaurs.

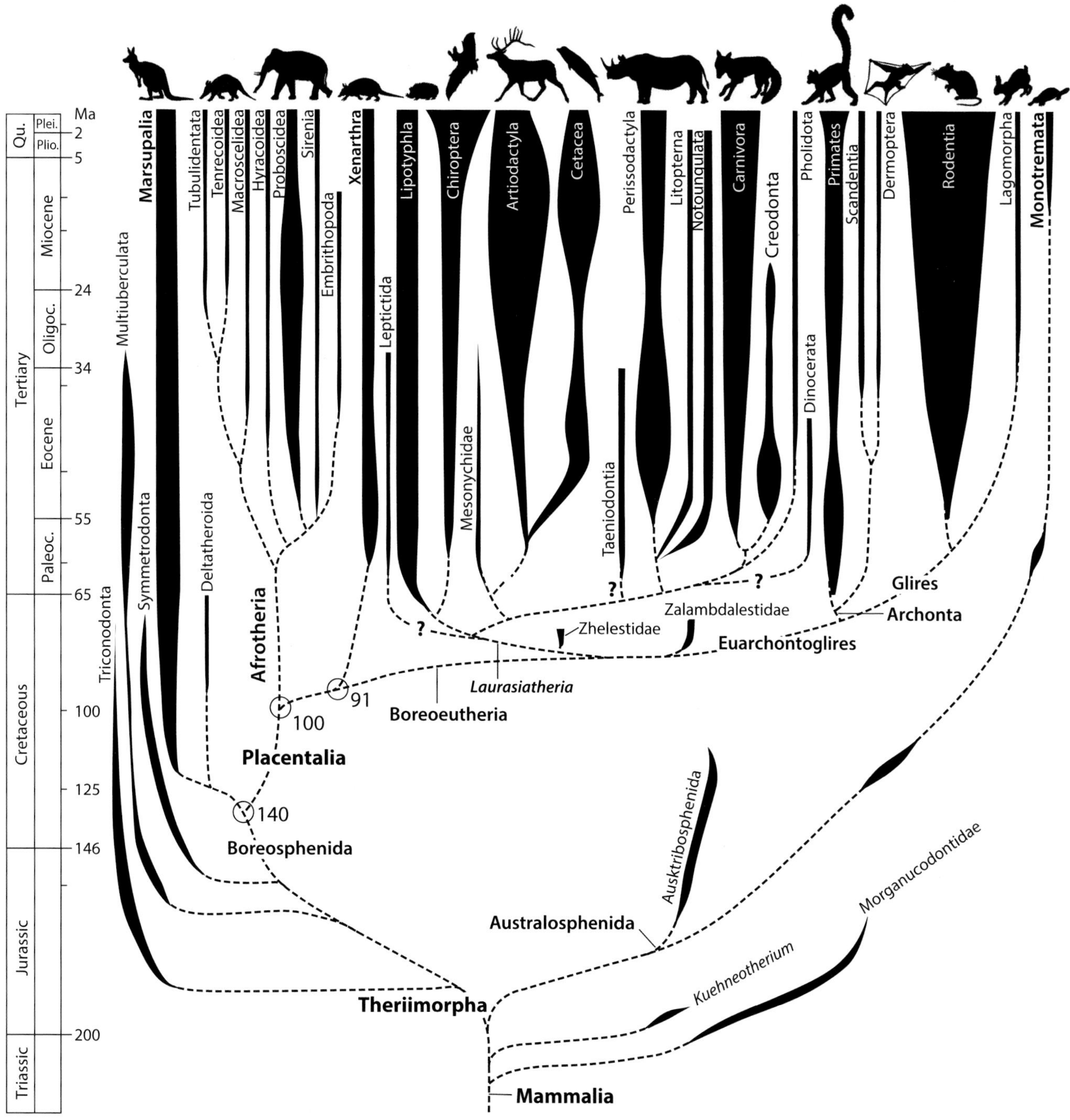

Figure 4.3. The evolutionary tree of mammals, focusing on the radiation of the placental mammals.

THE INTERRELATIONSHIPS OF PLACENTALS

For more than a century one of the great unsolved mysteries of zoology was the interrelationships among the orders of placentals. A famous 1910 study by the legendary anatomist and paleontologist William King Gregory made huge strides in the field, yet the mammal classification published by his colleague George Gaylord Simpson in 1945 still had lots of problems and wastebasket groups. But a rethinking of the methods of animal classification, starting in the 1960s and 1970s, spurred a renewed effort to tease out every useful feature of the anatomy of mammals and decipher their interrelationships by using only synapomorphies (shared evolutionary novelties), not primitive characteristics, and by recognizing only natural, monophyletic groups. By the 1980s and 1990s Malcolm McKenna, Michael Novacek, and other scientists at the American Museum of Natural History had published a number of family trees (**phylogenies**) and great progress had been made.

In the 1980s and 1990s molecular biology began to challenge the traditional ways of sorting out the interrelationships of placental mammals. Many of the groupings proposed by paleontologists from the 1970s through 1990s were confirmed. However, in some cases the molecular sequence data gave different answers than the anatomy did. In certain examples—such as the suggestion that guinea pigs were not rodents—there were clearly problems with molecular data, most of which were later deciphered and corrected. But the molecular phylogenies also suggested clusters of mammals that had never been supported by the anatomy. As more and more studies were published, the molecular evidence for such clusters became overwhelming, so in most cases mammalian paleontologists have come to accept these new groups. Keep in mind, however, that nearly all of these groupings are supported *only* by molecular data, and so far there is no other evidence (anatomical or otherwise) for them. In a few cases, anatomical features disagree with the molecular data, and this type of conflict has still not been fully resolved.

Nevertheless, the molecular evidence seems to get stronger and stronger over the years, so most mammalian paleontologists have begrudgingly come to accept it and to use the groups defined on molecular features only (Fig. 4.3). They are:

Xenarthra: The sloths, armadillos, and anteaters were long recognized as a natural group, since they have many peculiarities in their skeletons, and very primitive features of the anatomy and metabolism as well. Since 1975, paleontologists have realized that they diverged from the common placental root very early. Some molecular phylogenies place them slightly closer to the majority of the placentals that is shown in Fig. 4.3, while others suggest that they are one of the first groups to branch off.

Afrotheria: A group of mammals originally confined to Africa and nearby parts of the Middle East. The core members (elephants and sea cows, plus the extinct arsinoitheres) have long been connected as a group called the Tethytheria, first proposed by Malcolm McKenna in 1975. The hyraxes also tend to cluster with the tethytheres. More recently, a number of insectivorous groups previously lumped into the "Insectivora" wastebasket have proven to be afrotheres. These include the elephant shrews, tenrecs, golden moles, and aardvarks.

Boreoeutheria: All the remaining orders of placentals (i.e., all but the xenarthrans and the afrotheres) cluster together in the Boreoeutheria. This group is separated into two main branches:

Euarchontoglires: This cluster contains the euarchontans (primates, colugos, and tree shrews), and the glires (rodents and lagomorphs).

Laurasiatheria: Almost all the remaining orders of placental mammals belong to the Laurasiatheria, including the hoofed mammals (perissodactyls and artiodactyls, including whales and dolphins) and the carnivorous mammals (carnivorans and creodonts), plus the pangolins, the true insectivores (shrews, moles, and hedgehogs), and the bats (chiropterans).

CHAPTER 5

XENARTHRA

Sloths, Anteaters, and Armadillos

"EDENTATE" VS. XENARTHRAN

For more than a century, zoologists have recognized that the Xenarthra (Fig. 5.1) were unusual and primitive placental mammals and not closely related to any other group. They are peculiar in having many strange anatomical features seen in no other placentals but retained from earlier mammals, and they lack other features that all placental mammals have. They also have unusually slow and poorly regulated metabolisms and small brains, and female xenarthrans have a uterus that is divided by a septum and has no cervix. For many years, these animals were placed in the order "Edentata," which means "toothless" in Latin, although only the anteaters are completely toothless; sloths and armadillos have simple peg-like teeth made mostly of dentin with little or no enamel. Regrettably, the order "Edentata" once included pangolins and aardvarks as well, mammals that are not closely related to xenarthrans. Thus we no longer use "edentates," retaining only the group's formal name, Xenarthra ("strange joints" in Greek). This name refers to all the additional bony ridges and joints these animals have between the vertebrae in their back and to the unique way their hip bones are fused to their spine.

Figure 5.1. Examples of some of the extinct giant xenarthrans. The gigantic sloth in the center background is *Megatherium*; the sloths in the foreground are *Megalonyx* (right) and *Mylodon* (left). The armored creature with the spiky tail is *Doedicurus*, and the giant armadillo relatives in the back are *Glyptodon* (right) and the pampathere *Holmesina* (left).

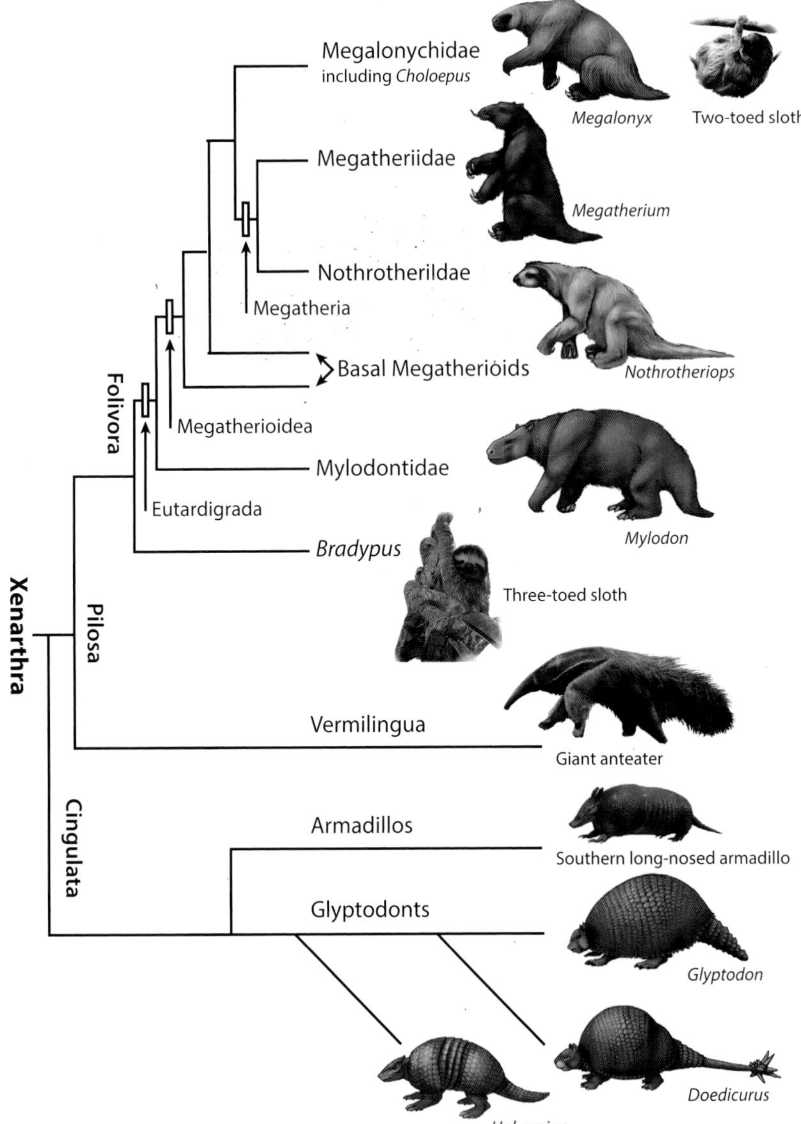

Figure 5.2. Family tree of the xenarthrans. The sloths plus anteaters form a group known as the Pilosa. The sloths (Tardigrada) have been mostly large ground-dwelling creatures. The living three-toed tree sloth (*Bradypus*) branched off early and is not closely related to the living two-toed tree sloth (*Choloepus*), which is descended from the huge megalonychid ground sloths and secondarily became small and arboreal. The armadillo branch (Cingulata) splits into most of the living armadillos and the glyptodonts. Within glyptodonts, the two main branches are the pampatheres (*Holmesina*) and the advanced glyptodonts (*Glyptodon* and *Doedicurus*).

In 1975 Malcolm McKenna of the American Museum of Natural History first suggested that xenarthrans are the oldest living branch of the placentals; in the 1980s and 1990s Michael Novacek of the American Museum of Natural History and others confirmed the distinctiveness of xenarthrans by clustering all the non-xenarthran placentals into a group called the "Epitheria." The "epitherians" were the mammals with a stirrup-shaped stapes bone in the middle ear and a few other unique features of the skull, which only xenarthrans lack among all living placentals. Under this scheme, Xenarthra is the most primitive branch of placental mammals, even more primitive than the insectivorous mammals that make up most of our Late Cretaceous record of placentals. Since then, the concept of "Epitheria" has not been confirmed, but most molecular phylogenies still place the xenarthrans as either the most primitive branch among placentals (the "Epitheria" concept) or one of the early branches and sometimes paired with the Afrotheria.

For decades, many classification systems clustered the xenarthrans with other ant-eating animals, such as pangolins and aardvarks. These similarities, however, are a classic case of convergent evolution, of an ant-eating or termite-eating lifestyle. Recent molecular analyses suggest that aardvarks are afrotheres, while pangolins tend to cluster with Carnivora. They are discussed in other chapters.

Xenarthrans are unusual in another way: they originated in South America and underwent nearly all their evolution on that island continent, where they became very diverse. Their earliest fossils date to the Paleocene (59 Ma) in South America, so they were there almost from the beginning of the Cenozoic. Only late in their history did they escape northward to Central and North America. This is yet another reason to believe they evolved their insect-eating lifestyles independently of pangolins (mostly found in Africa and southern Asia) and aardvarks (Africa and Middle East).

The three groups of living xenarthrans (Fig. 5.2) have anatomical characteristics that cluster the two hairy, non-armored groups (sloths and anteaters) in the order Pilosa ("hairy" in Greek), and put armadillos and their relatives in a separate order, Cingulata. Molecular data confirm these two groups.

ORDER CINGULATA (ARMADILLOS)

Only one living family of armadillos survives, the Dasypodidae, which has about 10 genera and 20 species. They are familiar to us for the armor covering their bodies, head, and tail. It is made of tiles of bone called **osteoderms**, which grow from the dermis layer in the skin, and it is flexible, allowing the shell to bend and curl up. It protects them from predators when they roll up and especially when they burrow down leaving only the armor exposed. Living dasypodids range in size from the pink fairy armadillo, only 13 cm (5 in) long, to the giant armadillo, up to 150 cm (59 in) long and weighing up to 54 kg (119 lb). Almost all of the species are known from South America, but a few have managed to move north and inhabit Central America and parts of the southern United States.

Most armadillos have powerful forelimbs with long claws on their hands, making them very strong diggers. They use this ability to dig up ant nests and termite mounds and to find underground grubs and roots, on which they feed, as well as to dig burrows for safety. They also can run surprisingly fast for such small (mostly), short-legged creatures, especially through thorny scrub where predators without armor cannot follow. When they are startled, armadillos jump up vertically rather than run, which is why they are so often roadkill in Texas and throughout the American South.

In addition to the living Dasypodidae there are two families of extinct cingulates: the pampatheres and the glyptodonts (Figs. 5.1, 5.2, 5.3). Both were built somewhat like armadillos but generally much larger, with big domed shells over their bodies. They could not roll up in a ball but had to crouch down below it for protection from predators. Pampatheres had osteoderms with distinctive round and hexagonal shapes, and three broad bands of hinged armor plates across the back that gave them a bit of flexibility (Fig. 5.3A). After originating in South America, they got larger through their evolution. They can be traced from the relatively small *Vassalia* (Miocene–Pliocene of South America) to the medium-size *Kraglievichia* (Pliocene–early Pleistocene, South America and Florida) to the bigger *Pampatherium* (middle to late Pleistocene, South America). Finally, there was the huge *Holmesina* (Pleistocene, North America), which was the size of a small car, reaching 2 m (6.5 ft) in length, and weighing about 227 kg (500 lb), almost four times as big as the living giant armadillo (Figs. 5.1, 5.3A). The size and slow habits of the larger pampatheres made it impossible to subsist on an insectivorous diet, so they apparently switched to grazing in the grasslands of the Americas with their primitive peg-like teeth (which became ever-growing to compensate for the heavy wear imposed by their diet).

Figure 5.3. Giant armadillos, glyptodonts, and pampatheres. A. The pampathere *Holmesina*, in front of the huge limbs of the giant ground sloth *Eremotherium*. B. Skeleton of the glyptodont *Panochthus*. C. A life-size restoration of *Glyptotherium*.

The other big group of extinct cingulates was the glyptodonts (Figs. 5.1, 5.2, 5.3B, C). Fossils of these animals were first discovered in 1833 by Charles Darwin on his *Beagle* voyage to Argentina, and the group was formally described by British zoologist Richard Owen in 1839. Like pampatheres, glyptodonts also had a big tortoise-like dome of armor over their backs, but its osteoderms were fused into a single shield that was not flexible. They also had rings of bone covering their tails, and some (such as *Doedicurus*) even had mace-like spikes on the tip for striking opponents (Fig. 5.1).

Glyptodonts first appeared in the early Miocene of South America (*Parapropaleohoplophorus* from Chile), and most of the dozens of genera are known only from south of Panama. A few glyptodonts (like the pampatheres and armadillos) headed north over the Panama land bridge (Fig. 5.4) in the early Pliocene but got only as far north as Texas and Florida in the United States. They apparently preferred moist subtropical habitats and may have been good in the water. Chemical analysis of their teeth shows they were grazers of typical savanna grasses. *Glyptodon* was the size of a Volkswagen Beetle and even larger than the pampathere *Holmesina* (Fig. 5.1). Some specimens were immense, reaching 3.3 m (10.8 ft) long and 1.5 m (5 ft) tall, and the living organism would have weighed 2000 kg (4400 lb).

All the glyptodonts and pampatheres died out from the Americas at the end of the last Ice Age, about 10,000 years ago, during the great megafaunal extinction. The cause is discussed in Chapter 18.

Figure 5.4. In what is known as the Great American Biotic Interchange, mammals migrated between the Americas beginning about 10 Ma in the late Miocene, and especially at 3 Ma in the mid-Pliocene when the Panama land bridge became complete. Some South American natives (such as ground sloths, glyptodonts, pampatheres, native hoofed mammals, capybaras, porcupines, and giant predatory phorusrhacid birds) managed to move north, but the vast majority of the migrations were North American natives heading south (mastodonts, horses, llamas, tapirs, peccaries, bears, cats, raccoons and coatis, dogs, deer, and many others).

ORDER PILOSA (ANTEATERS AND SLOTHS)

The Pilosa includes the two non-armored groups of xenarthrans. They can be broken down into two groups, the Vermilingua ("worm tongue" in Latin) or anteaters (four species in three genera alive today), and the Folivora ("leaf eaters" in Latin) or sloths (six living species in two genera). The anteaters range in size from the pygmy anteater, about 35 cm (14 in) long, to the giant anteater, which is about 1.8 m (6 ft) long and strong enough to kill a jaguar with its sharp claws. All anteaters have a long toothless snout and jaws that form a tube-like mouth (Fig. 1.8). Their long sticky tongue, which is longer than the length of the animal's head when extended, can be protruded deep into ant and termite burrows to trap their food. Anteaters also have long curved fore claws useful for ripping open ant and termite nests. Because of the claws' length and curve, giant anteaters must walk on their knuckles with the claws curved backward. Their dense, long fur protects them against the bites of insect soldiers.

The fossil record of anteaters is very limited. Although sloths and armadillos appeared in the Eocene or earlier, the oldest known anteater fossil is *Protamandua* from the early Miocene of Argentina. The articulated skeleton of *Eurotamandua* from the middle Eocene lake beds of Messel, Germany, is now thought to be a pangolin (Chapter 11).

The sloths (Folivora) are the third great branch of the xenarthrans (Fig. 5.2). Today there are three species of the three-toed sloth genus, *Bradypus*, and two of the two-toed sloth genus, *Choloepus*. These small animals are legendary for hanging upside down from trees their entire lives, moving very slowly, and feeding on leaves and digesting them slowly during long naps. Although sloths can fight with their long fore claws, their main defense is stillness and concealment. They even harbor symbiotic algae in their fur that make them even more invisible in the tree canopy. But they must crawl down to the ground to defecate, always in the same latrine, about once a week, and at this time they are vulnerable to predation by jaguars. Sloths are also decent swimmers when they need to cross bodies of water for food or shelter.

Sloths have large complex stomachs that make them almost as effective in bacterial digestion of plants as the ruminants are. They eat nothing but leaves, which are low on nutrition (though high in water content), so they must eat slowly and allow a long time for digestion (especially with their slow metabolism). Their long tongues reach as far as 30 cm (12 in) from their mouths, so they can gather food farther than their reach. Sloths compensate for the low nutrition of the foliage they eat by having a very slow metabolism and low body temperature, so they do not need much energy or nutrition.

But these tiny living species hanging from trees are a pathetic remnant of a huge evolutionary radiation of ground sloths dating back to the early Eocene. The Folivora once comprised dozens of species of big ground sloths in 86 genera among four extinct families. The two living species diverged about 40 Ma and are not that closely related (Fig. 5.2), which indicates that many of their similarities are due to convergent evolution. Two-toed sloths are actually related to the giant ground sloths and probably represent a secondary dwarfing of these ancient giants.

Ground sloths were truly remarkable creatures (Fig. 5.5). Their skulls looked like something out of a science-fiction movie, with huge bony protuberances from their cheek bones, a few simple peg-like teeth made of soft dentin, and a toothless "drool spout" on the front of the lower jaw through which the long flexible tongue slid back and forth, ripping down leaves. They had strangely shaped shoulder blades, powerful forelimbs, and long fore claws that curved inward when they knuckle-walked. The front claws were probably used to haul down limbs so their long tongues could strip the branches of leaves. Studies of the claws showed they were not for digging but may have been some help in defense. They had long hind limbs with such long claws they could not walk in a normal way—due to these hind claws, ground sloths walked on the side of their strangely deformed-looking ankle, with the hind claws curved inward. Their strong tail helped them balance in a tripod stance when they reared up on their hind legs to reach high branches. Although they had no armor plating like that of the Cingulata, they had thousands of pea-size nodules of bone (osteoderms) embedded in their thick hides. These bony pellets may have made their hides difficult for a predator to pierce when its canines struck them.

Four families of ground sloths are represented in the fossil record. The Mylodontidae family of small to medium-size ground sloths (Fig. 5.1) has dozens of species in about 28 genera scattered among several subfamilies. Most were restricted to South America, but *Thinobadistes* managed to island-hop across Central America and reach North America about 8 Ma, long before the Panama land bridge formed about 3.5 Ma (Fig. 5.4). *Paramylodon* (Fig. 5.5C), the biggest sloth found at La Brea tar pits, was about 3.8 m (10 ft) tall on its hind legs and weighed about 1000 kg (2200 lb).

The Megalonychidae (Fig. 5.1) includes 28 genera of medium-size sloths in several subfamilies and is the ancestral group of the living two-toed sloth. Most megalonychids were South American, but *Pliometanastes* was the first sloth to island-hop through Central America, about 9 Ma, even earlier than the mylodontid *Thinobadistes*. Future president Thomas Jefferson himself described *Megalonyx* ("giant claw" in Greek), in 1796, from claw fossils brought to him from a Virginia cave by Col. John Stuart. Jefferson was convinced they belonged to a giant lion that still roamed in the West and even asked Lewis and Clark to look for it during their expedition in 1803–05. Only in the 1820s, after the description of other fossils, including *Megatherium* (discussed

Figure 5.5. Giant ground sloths. A. An old illustration of the immense skeleton of *Megatherium* in the Natural History Museum of London, originally collected in Argentina by Charles Darwin and described by Richard Owen. B. Life-size model of the gigantic ground sloth *Eremotherium*. C. Skeleton of *Paramylodon harlani*, a mylodontid, the largest ground sloth at La Brea tar pits. D. Skeleton of the smaller nothrotheriid ground sloth *Nothrotheriops*.

later), did people realize the claws were from not a giant lion but a giant ground sloth.

The Nothrotheriidae was a group of small-bodied sloths with about 10 genera in two subfamilies. These species date back to about 11.5 Ma and were not able to spread from their South American base to North America until about 2.6 Ma. Several mummified specimens of *Nothrotheriops* with skin and hair have been found in dry caves in Mexico, New Mexico, and Arizona (Fig. 5.5D). Many of these caves also have yielded dried sloth dung that has been preserved for over 10,000 years.

The most remarkable of the known nothrotheriids was *Thalassocnus*, a late Miocene genus from Peru and Chile that returned to the sea (Fig. 5.6). The oldest species from the late Miocene were apparently semiaquatic, and four successive species show the gradual transition in the skeleton from terrestrial to fully marine modes of life. The bones of each successive species show an increasing density, necessary for ballast in a marine mammal, and their limbs gradually changed shape as the animals became more committed swimmers. They apparently used their long claws to cling to rocks in the surf (as do modern marine iguanas). The teeth of earlier specimens show wear and chemistry that results from eating vegetation mixed with sand from beaches and the sea bottom, but later forms show no such wear and instead are adapted for eating sea grasses and algae.

The most spectacular of the ground sloths are those in the family Megatheriidae (Figs. 5.1, 5.5A, B), which contains 20 genera in three subfamilies. *Megatherium* ("huge beast" in Greek) was first found in South America in 1788 and then described by the pioneering French anatomist and paleontologist Georges Cuvier in 1796. *Megatherium* and another giant, *Eremotherium*, reached truly elephantine proportions, weighing up to 4000 kg (8800 lb) and stretching 6 m (20 ft) from head to tail (Fig. 5.5A, B). They were about the size of modern elephants, and only certain mammoths and the gigantic rhino *Paraceratherium* were larger. Nearly all the megatheres remained in South America, except for *Eremotherium*, which managed to migrate up Central America (Fig. 5.4) and reach Florida and Georgia about 2 Ma.

All four families of ground sloths vanished at the end of the last Ice Age, about 10,000 years ago, although some specimens suggest even younger dates. They probably died out due to the combination of rapid climate change and human overhunting, as did most of the Ice Age megafauna.

Figure 5.6. The swimming ground sloth *Thalassocnus*, from the Miocene beds of Peru.

CHAPTER 6

AFROTHERIA

Elephants, Hyraxes, Sea Cows, Aardvarks, and Their Relatives

TETHYTHERES AND AFROTHERES

For decades, the relationships among the major living groups of African mammals—especially elephants, hyraxes, aardvarks, and a few minor groups, such as tenrecs, golden moles, and elephant shrews—were controversial (Fig. 6.1). However, a number of biologists had noticed anatomical similarities between the order Proboscidea, which includes the elephants and their fossil relatives, and the order Sirenia, the sea cows (manatees and dugongs), which swim primarily in tropical and subtropical waters around the world. In 1975 Malcolm McKenna proposed the taxon Tethytheria for proboscideans and sirenians, plus two

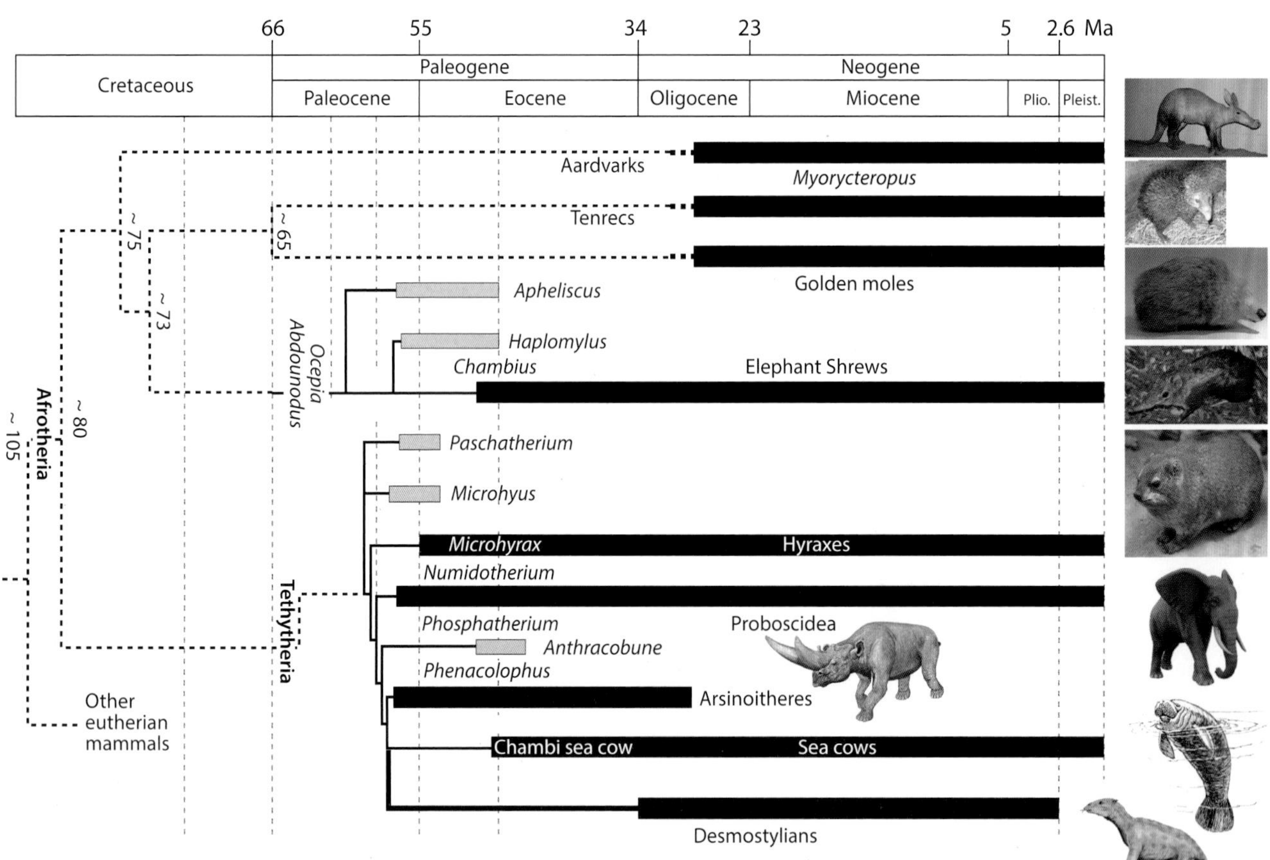

Figure 6.1. Phylogeny of living and extinct afrotheres.

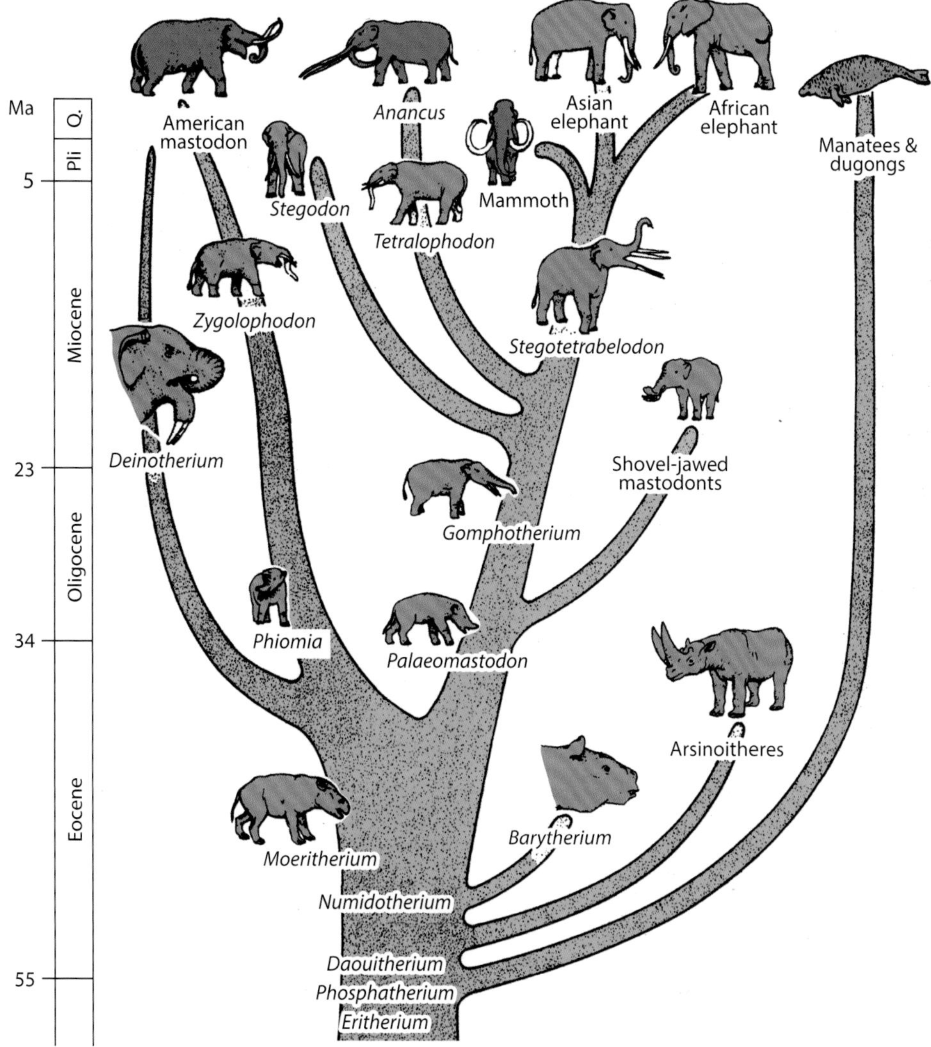

Figure 6.2. The evolution of proboscideans and their extinct relatives.

other extinct groups we will discuss in this chapter, the desmostylians and arsinoitheres. They were given this name because all four groups apparently evolved in and around the ancient Tethys Seaway, which used to stretch from Gibraltar across what is now the Mediterranean to Indonesia. This ancient seaway was a dominant feature during the Mesozoic, but by the Eocene the Tethys had been closed by the collision of India with Asia, and the collision of Africa and the Arabian Peninsula with Asia, which formed the eastern Mediterranean.

The group known as Tethytheria is based on extensive evidence from both anatomy and fossils. Both elephants and sirenians share a number of unusual specializations in the bones of their middle ear and in the arrangement of the bones of the skull around the eye sockets, and even show some similarities in configuration of the crests on their cheek teeth. Once it was proposed, Tethytheria became widely accepted by both paleontologists and mammalogists.

Tethytheria is part of the Afrotheria, a group that encompasses a wider spectrum of African mammals and is based entirely on evidence of molecular biology and similarity of DNA and RNA sequences. Although Afrotheria was formally proposed in 1998, biochemical features had already suggested that elephants, sirenians, hyraxes, and a few other groups were closely related. Since 1998, almost every molecular study published further supports the idea of the Afrotheria, and it has become a standard feature of mammalian classification and phylogeny.

Nevertheless, many paleontologists and mammalogists are still reluctant to accept Afrotheria. Despite diligent research, so far we have found no undisputed anatomical features shared by all the known Afrotheria (beyond the well-supported evidence of the relationships of the Tethytheria). Several features have been suggested, such as the number of vertebrae in the upper and lower back, the condition of the testes in males, and peculiar patterns of eruption of the teeth, but most of these also occur in other mammalian groups; furthermore, all of these features do not occur in the most primitive known afrotheres, so they are not strong evidence for the Afrotheria.

However, African Paleocene and Eocene rocks have recently produced a number of fossils that seem to be very primitive tethytheres or even afrotheres. Such fossils include *Ocepeia* from the middle Paleocene of Morocco, a very primitive afrothere that cannot be assigned to any later order; *Eritherium*, *Phosphatherium*, and *Daouitherium*, very primitive proboscideans from the late Paleocene and early Eocene of Morocco; *Chambius*, an early Eocene fossil from Tunisia that seems to be a primitive elephant shrew; another specimen from the early Eocene of Africa, yet unnamed, that is the earliest known sirenian; *Microhyrax*, *Seggurius*, and *Dimatherium*, the earliest hyraxes, from the early Eocene of Algeria; and numerous other fossils. The earliest specimens of most afrotheres that have a fossil record are entirely African (except for the arsinoitheres; discussed later).

This fossil evidence supports the idea that the entire lineage arose in isolation when Africa was an island continent in the Late Cretaceous. Through much of the Paleocene and Eocene, Africa remained isolated, with a fauna of mostly afrotheres until some immigrant groups (such as the primitive anthropoid primates, creodonts, carnivorans, and anthracotheres) managed to reach it from Eurasia. Most of the afrotheres remained restricted to Africa until the Oligocene and Miocene, when hyraxes, gomphothere mastodonts, and a few other groups managed to spread to Eurasia. Many afrotheres, such as the tenrecs, elephant shrews, and golden moles, remained restricted to Africa. The marine sirenians swam away from Africa via the Tethys and then to mainly tropical waters around the world soon after their origin in the late Paleocene.

ORDER PROBOSCIDEA (ELEPHANTS, MAMMOTHS, MASTODONTS, AND THEIR RELATIVES)

Elephants are the largest living land mammals on the planet today, represented by two species in Africa and one in southern Asia. But their order, the Proboscidea, has a long and excellent fossil record going back to the middle Paleocene of Africa (Figs. 6.2, 6.3). They remained an exclusively African group through most of their history until about 19 Ma, when the Arabian Peninsula docked with Africa and the eastern Mediterranean closed, allowing them to escape to Eurasia and eventually to North and South America.

The oldest known proboscidean relative was a fox-size creature from the late Paleocene (60 Ma) of Morocco, discovered in 2009 and christened *Eritherium*. Its teeth have the four rounded cusps typical of primitive tethytheres, with just a hint of the cross-crest pattern seen in later proboscideans. The front teeth and lower jaw also show other features of the proboscideans, and the eyes are far forward on the skull, typical of the proboscideans. From the earliest Eocene (56 Ma) was *Phosphatherium* (Fig. 6.3), a creature with many proboscidean features, including molars that had better-developed cross-crests than those of *Eritherium*. Scratches on its teeth suggest that *Phosphatherium* ate a wide variety of plants, though mostly leaves.

The next step is the early Eocene fossil *Daouitherium* from Morocco, from 53 Ma, and then *Numidotherium* from Algeria, from about 51 Ma (Fig. 6.3). The skull of *Numidotherium* (though incomplete) already shows the signs of the tall, vertical forehead so characteristic of elephants; it also shows a high and narrow nasal opening, so it did not yet have a short trunk or proboscis. The upper incisors are beginning to elongate, but *Numidotherium* has not lost all the

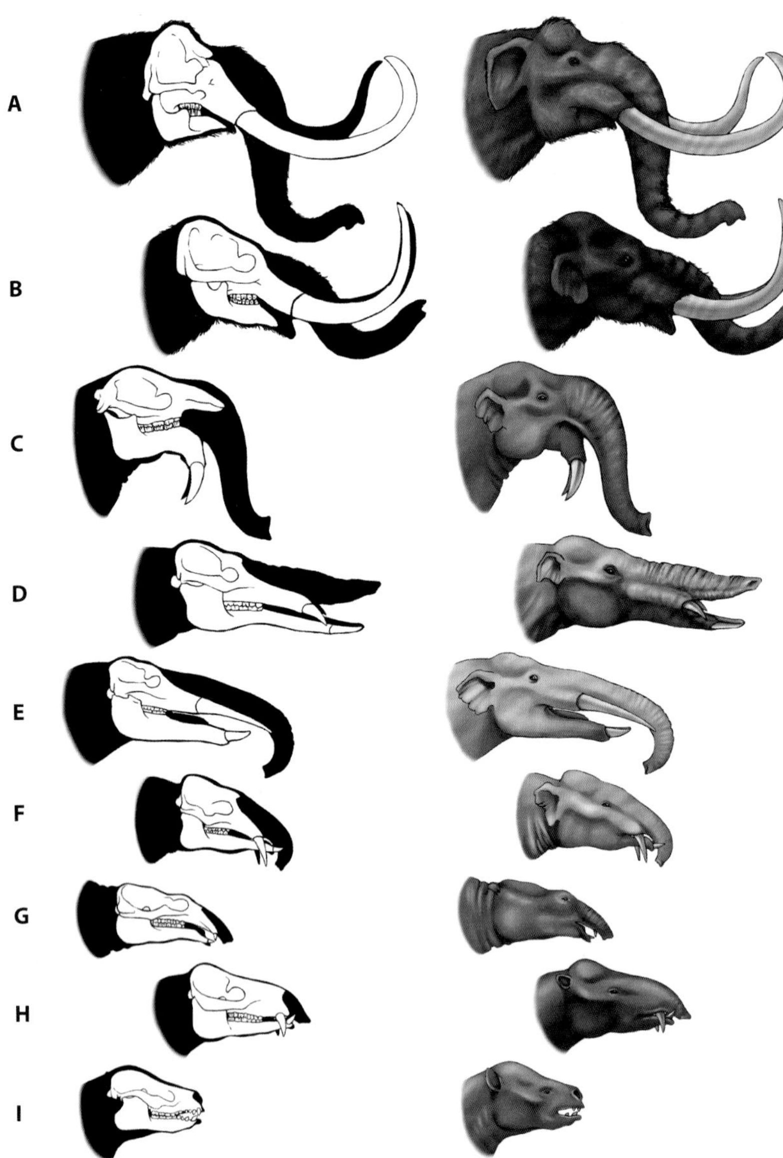

Figure 6.3. The evolution of proboscidean skulls from the Eocene (bottom) to the Pleistocene (top): I. *Phosphatherium*, H. *Numidotherium*, G. *Moeritherium*, F. *Palaeomastodon*, E. *Phiomia*, D. *Gomphotherium*, C. *Deinotherium*, B. *Mammut* (American mastodon), and A. *Mammuthus*, the mammoth.

rest of its front incisors, as the later proboscideans did. The lower front teeth are beginning to form a scoop shape typical of early mastodonts. *Numidotherium* was only tapir-size, standing about 1–1.5 m (3–5 ft) tall at the shoulder, but it had the robust limbs typical of all proboscideans.

At the dawn of the 20th century, a British expedition led by paleontologists C. W. Andrews and H. J. L. Beadnell collected a treasure trove of bones from upper Eocene and lower Oligocene rocks of the Fayûm Depression in Egypt, in the desert just west of the Great Pyramid of Giza. They found numerous teeth of a hippo-size beast called *Barytherium* ("heavy beast" in Greek), which turned out to be a slightly more advanced proboscidean with fully developed cross-crests on its molars, a high forehead, and a short trunk. But the most spectacular discovery was several complete skeletons of a late Eocene fossil named *Moeritherium* ("beast of Lake Moeris"), for a dry lake near Fayûm. *Moeritherium* had a barrel-shaped body and short stout legs. It was the shape of a pygmy hippopotamus or a tapir, and similar to those animals in height, about 0.7 m (2.3 ft) tall, but it was almost 3 m (10 ft) long. Its upper and lower canines were short tusks, and it may have had a short proboscis like that of a tapir (Figs. 6.2, 6.3, 6.4A, B).

Figure 6.4. Primitive proboscideans. A. Skeleton of *Moeritherium*. B. Life-size model of *Moeritherium*. (See also next page.)

By the early Oligocene, the proboscideans had split into several lineages. One of these was a group called the deinotheres ("terrible beasts"). They are best known from their most famous fossils, those of *Deinotherium*, a huge beast bigger and longer-legged than any living elephant (Figs. 6.2, 6.3, 6.4C, D). *Deinotherium* had no upper tusks, but its downward- and backward-curved lower tusks are its most distinctive feature. Its molar teeth were fully developed into cross-crests, like those of a tapir, indicating a diet of pure leaves, which they could have stripped from the tops of trees with their great height and long trunk. The lower tusks were probably used to hook branches and help pull them down as they fed on the leaves. They evolved from *Prodeinotherium* from the early Miocene of Africa, but by the middle Miocene deinotheres were spread across Eurasia and were typically the largest animal wherever they occurred. The last *Deinotherium* died out during the early Pleistocene, and its role as the largest land mammal was taken over by its distant relatives, the mammoths.

The main lineages of proboscideans are represented by early Oligocene fossils, again from the Fayûm beds of Egypt but from rocks younger than late Eocene *Moeritherium*. The best known of these are *Palaeomastodon* and *Phiomia* (Fig. 6.3). About 2 m (6.5 ft) high at the shoulder and weighing as much as a small rhinoceros, *Palaeomastodon* had a long jaw, a flat forehead, and short tusks in both its upper and lower jaw. *Phiomia serridens* (whose name means "saw-toothed animal of the Fayûm") was a bit smaller, about 1.3 m (4 ft) at the shoulder, and had cylindrical upper tusks with an oval cross section but lower tusks that were flattened into a spatula-like shape. Thus, they are classic intermediate fossils, coming between *Moeritherium* and later proboscideans in size and in the lengths of their tusks and trunks (longer than those of *Moeritherium* but shorter than those of later mastodonts).

The next split in proboscidean evolution was between the mastodont family (Mammutidae) and the rest of the Proboscidea (Figs. 6.2, 6.3). Mammutids can be traced back to *Losodokodon*, a very primitive fossil from the late Oligocene (24–27 Ma) of Kenya, followed by the early Miocene *Eozygodon* from Kenya and Uganda. By the middle Miocene, they had evolved into *Zygolophodon*, which had spread beyond Africa by about 19 Ma, and by 18 Ma had spread across Eurasia and had even crossed the Bering Land Bridge to North America. But the best-known mammutid is the American mastodon, *Mammut americanum* (Fig. 6.4E). It was smaller than most living elephants, only about 3 m (10 ft) at the shoulder, and weighed about 4500 kg (9900 lb). *Mammut* lacked the steep forehead and the shoulder hump found in elephants and mammoths. It had a long, flat head, slightly curved tusks, and a deeper chest, broader hips, shorter legs, and a longer back than mammoths. Its teeth retained the primitive condition of rounded conical cusps that formed connected cross-crests as they were worn away. With these primitive

Figure 6.4. Primitive proboscideans. C. Skeleton of *Deinotherium*. D. Restoration of *Deinotherium*. E. Skeleton of an adult and juvenile of the American mastodon, *Mammut americanum*, at the Page Museum of La Brea tar pits.

teeth, mastodonts ate mostly leaves and pinecones (their diet has been confirmed by the gut contents of mummified specimens) in the forests they inhabited of the Miocene, Pliocene, and Pleistocene, their lifestyle in contrast to the mostly grazing habits of mammoths and other elephants. Mastodonts had a thick coat of shaggy hair to keep them warm during the Ice Ages, but because of their restriction to forested regions, they were not as common or widespread as mammoths, which could live anywhere, though they thrived mostly in open plains and steppe-tundra environments. Mastodonts were thought to have gone extinct with the rest of the Ice Age megamammals about 10,000 years ago, but there are legends of individual mammoths surviving into more recent times, and a few mastodonts are known to have survived until about 4,000 years ago.

D

E

The main lineage of Proboscidea can be traced from *Phiomia* of the early Oligocene to the gomphotheres of the early and middle Miocene (Figs. 6.3, 6.5A). Gomphotheres were widespread across both North America and Eurasia during this time span, performing the role that mammoths and elephants later performed as the largest herbivore. Gomphotheres were about 3 m (10 ft) at the shoulder and weighed about 4000–5000 kg (8800–11,000 lb). They had long, flat-topped skulls, with two well-developed tusks in both their upper and lower jaws, and probably a short trunk. The lower tusks were shaped like spatulas and are thought to have been useful for digging up roots and food, as well as stripping bark off trees.

From the gomphotheres evolved many different groups of proboscideans. One of these was the shovel-tuskers (subfamily Amebelodontinae, containing five genera), whose name refers to the fact that the lower tusks are shaped like a pair of broad shovels (Fig. 6.5B, C). Amebelodonts first appeared in the early Miocene of Africa; the advanced shovel-tuskers evolved in North America about 9 Ma and then spread to Asia in the late Miocene. Traditionally shovel-tuskers were thought to have used their lower tusks for scooping up water plants in swampy habitats, but detailed analysis of the wear on their "shovels" shows abrasion from scraping bark off trees, so they probably had a diet of leaves, twigs, and bark, like most mastodonts. Shovel-tuskers vanished at the end of the Miocene in North America, at the same time when rhinos, protoceratids, dromomerycines, musk deer, and many other groups typical of the American savanna also vanished.

Figure 6.5. More advanced proboscideans. A. Skeleton of a gomphothere. B. Skull and jaws of the shovel-tusked mastodon *Platybelodon*. C. Life-size model of *Platybelodon*. D. Life-size model of the anancine gomphothere *Anancus*.

C

D

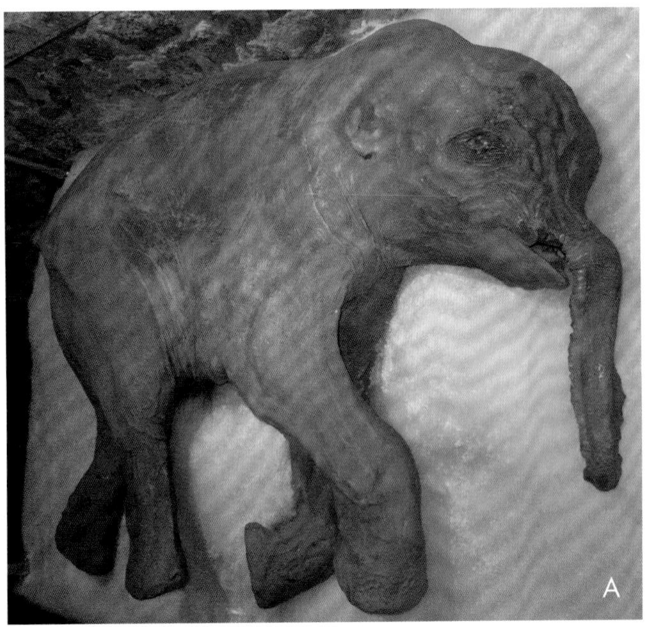

The only surviving lineage of Proboscidea is the Elephantoidea, containing the family Elephantidae, which includes the living elephants and the extinct mammoths (Fig. 6.2). Elephantids evolved in a new direction, undergoing a shortening of the face and lower jaw and a raising of the crest on the top of the skull, which allowed them to develop the huge pair of upper tusks and lose the lower tusks entirely (Fig. 6.3). Meanwhile, their molar teeth became composed of a big set of tightly folded enamel and dentin plates that make a large grinding surface on the top, so that they are well adapted for grinding tough vegetation like grasses. Eventually, these huge teeth dominated their short faces and jaws, so they had only one or two molars in each side of the upper and lower jaw at the same time.

Instead of having baby teeth that fall out when the adult teeth erupt beneath them (as in most mammals), proboscideans erupt their teeth from the back of the jaw. The teeth push forward in the jaw as they wear out, until the broken and worn oldest teeth are pushed out the front of the jaw. This is known as serial tooth displacement and also occurs in the sirenians.

Figure 6.6. Mammoths. A. Woolly mammoth mummy. B. Skeleton of pygmy mammoth from the Channel Islands, near Santa Barbara, California.

The evolutionary trend in Elephantoidea can be seen in anancine gomphotheres, which were very large gomphotheres with long flat skulls and only upper tusks that were long and straight, reaching up to 3 m (10 ft) in length (Figs. 6.2, 6.5D). The next stage is the stegodonts, whose upper tusks were also up to 3 m long but had a J-shaped curvature and were so close together at the base that the trunk could not fit between them but draped across the top (Fig. 6.2). The ensuing stage was the stegotetrabelodonts, which evolved from the advanced gomphothere *Tetralophodon*. They were the first to show the true shortening of the face and lower jaws and the loss of the lower tusks completely.

The final stage was the family Elephantidae, the modern family of elephants and mammoths. They can be traced back to the *Stegotetrabelodon* and then to the oldest fossil of the group, *Primelephas* ("first elephant" in Latin), a late Miocene genus that still had short upper and lower tusks like a gomphothere. During the Pliocene, this lineage evolved into *Loxodonta*, the African elephant, in the late Miocene; *Elephas*, the Asian elephant, first appearing in the earliest Pliocene of Africa; and *Mammuthus*, the mammoths, which spread from Eurasia across the Northern Hemisphere and Africa. As mammoths evolved, their molars became larger and more complex, with more and more folds of enamel and dentin. Mammoth evolution culminated with the huge Columbian mammoth, which had a naked hide like a modern elephant (not the long hairy coat found only on woolly mammoths) and is well known from temperate and tropical latitudes of North America. The Columbian mammoth reached 4 m (13.1 ft) high at the shoulder and weighed about 7–9 metric tons (7.7–9.9 short tons). The fringes of the Pleistocene glaciers were inhabited by woolly mammoths, which were slightly smaller and were covered by a thick coat of long fur that protected them in their cold habitat. These mammoths are known from a number of mummified specimens (Fig. 6.6A) frozen in the tundra of Siberia and Alaska, which show us not only what they looked like alive but even reveal what they ate. Some had a fondness for buttercups.

Finally, at the end of the last Ice Age, about 10,000 years ago, all the mammoths vanished from the continents, along with most of the rest of the large Ice Age mammals. However, populations of dwarfed mammoths managed to survive on the Arctic islands of the Aleutians and north of Siberia and persisted to only about 6,000–4,000 years ago, surviving until after the pyramids were built in Egypt. There were also populations of dwarfed mammoths on other islands, including several islands in the Mediterranean and the Channel Islands off the coast of Santa Barbara, California (Fig. 6.6B). They were no larger than sheep.

ORDER SIRENIA (MANATEES AND DUGONGS, OR SEA COWS)

The manatees and dugongs are both familiar creatures from tropical and subtropical waters around the world, with a fossil record that goes back to the middle Eocene. In 2013, a group of 10 scientists led by Julien Benoit published a report about new specimens from the early Eocene (about 50 Ma) in Tunisia. These included a number of skull bones with ear regions that are distinctly sirenian and some other fragments of the skeleton. The locality is known as Chambi, so the specimens are known as the "Chambi sea cow" (Fig. 6.1) for now, since they are yet too incomplete to merit a formal taxonomic name. Fragmentary though they are, the Chambi sea cow ear regions complete a puzzle by showing that the earliest sirenians, like their relatives the earliest proboscideans, hyraxes, and other tethytheres, first appeared in the Tethys region (primarily Africa). The sea cows were aquatic and soon spread all the way from Africa to the Caribbean to India.

Back in 1855 Richard Owen described the very first fossil sirenian, originally from the early Eocene deposits of Jamaica, and named it *Prorastomus sirenoides*. It had the slightly downturned snout, the nasal opening high on the skull, and many other features of the sirenians (Figs. 6.7A, B, 6.8). The skeletal parts of the animal were mere fragments, but they suggested a sheep-size mammal that could walk on four legs. The skull and skeletal fragments of *Prorastomus* were found with pieces of ribs that were thick and very dense, a diagnostic feature of all sirenians. These thick, heavy ribs provide ballast, preventing the animal from floating too high in the water, and the extremely dense bone, even of a single rib fragment, is unique and diagnostic for every sea cow.

The next step in sirenian evolution was *Pezosiren portelli* ("Portell's walking siren"), discovered in the late 1990s from the early Eocene of Jamaica (Figs. 6.7C, 6.8). *Pezosiren* was about the size of a large pig (about 2.1 m/6.5 ft long) and had a very primitive sirenian skull. In some ways (having a crest along the top of the skull), it was more primitive than later sirenians, but in most features (ear region, and the downward deflection of the tip of the lower jaw) *Pezosiren* was advanced. It also had the classic thick, solid rib bones of all sirenians, which were part of a long barrel-shaped trunk, and a short tail. Most important, *Pezosiren* had nearly complete front and hind limbs and hip bones. The limbs were short, but had perfectly normal fore and hind feet for walking on land, with no obvious specializations for swimming. The details of the limbs and spine reveal that *Pezosiren* apparently swam by paddling with its feet and propelling itself along the bottom in shallow water (as hippos do today), rather than by swimming with up and down motions of its tail, as all modern sirenians, whales, seals, sea lions, and otters do.

In 1904 Othenio Abel described a skull of a more advanced sirenian, *Protosiren fraasi* (Fig. 6.7D), from the middle Eocene (40–47 m.y. old) lower Building Stone Member of the Gebel Mokattam Formation in Egypt (the same rocks from which the Pyramids of Giza were built). The skull was much more like those of modern sirenians, with a more strongly downturned snout, the specialized nasal opening farther back on the skull, and other more advanced features. Later specimens were found in many different places—North Carolina, Hungary and France, Pakistan and India—indicating that *Protosiren* had an almost worldwide distribution in warm tropical and subtropical waters. When the skeletal remains were found, it turned out that *Protosiren* had tiny hind limbs. In addition, its hips were not strongly attached to its lower backbone, showing it was almost completely aquatic and could barely walk on land. Most of the sirenian fossils younger than *Protosiren* have shrunken hind limbs and could no longer walk, having become completely aquatic like all the living sea cows.

After *Protosiren*, there were many different fossil sirenians through the group's 50 m.y. history (Fig. 6.8). During the rest of

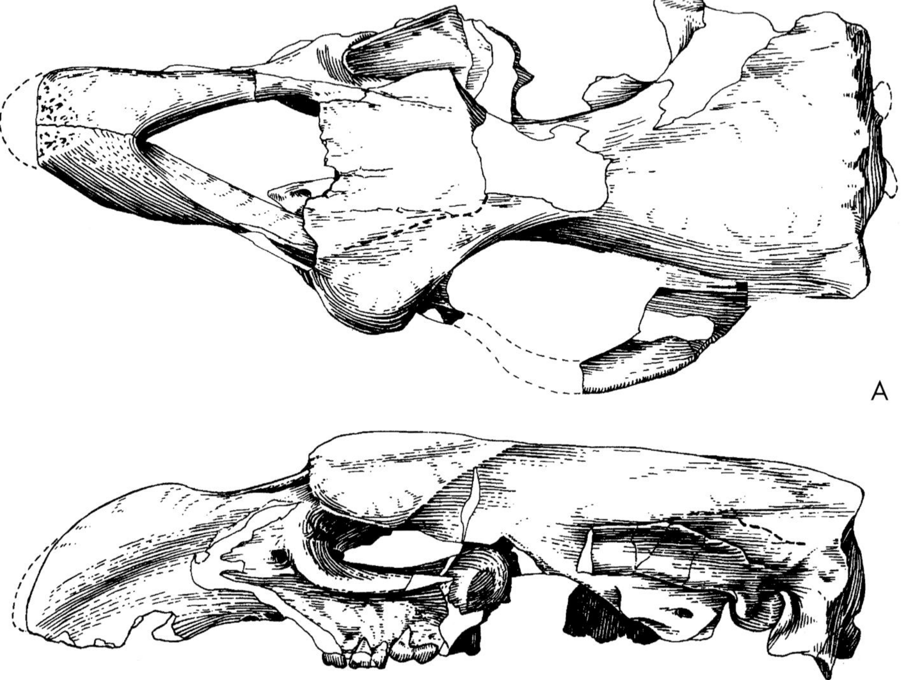

Figure 6.7. Sirenians. A. The skull of *Prorastomus sirenoides*. B. Restoration of *Prorastomus*. C. The skeleton of *Pezosiren portelli*, with sirenian expert Daryl Domning, who named and described it, standing next to it. D. The skull of *Protosiren fraasi*.

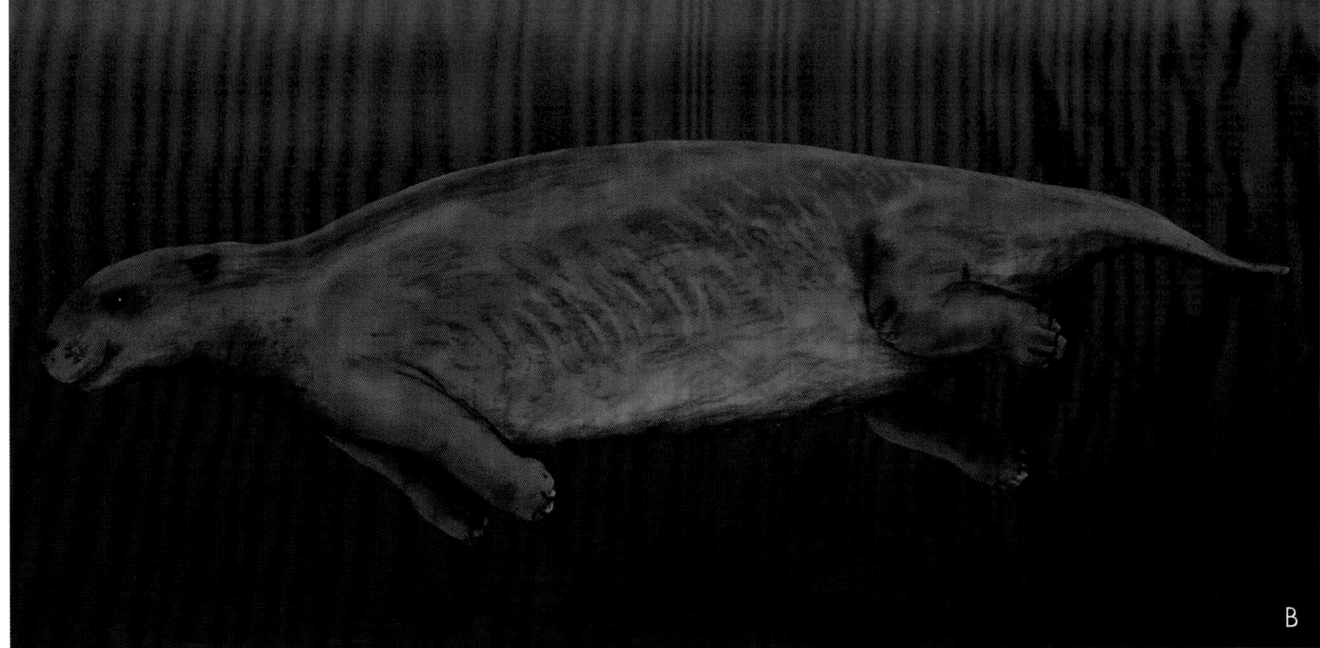

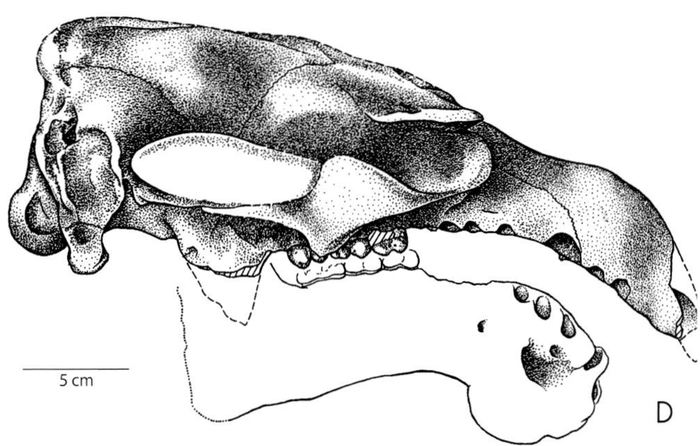

the Eocene and Oligocene, a primitive group of sirenians known as halitheres spread around the tropical oceans of the world. By the early Miocene, there were primitive members of both the manatee and dugong families. For most of the sirenians' history, their evolution can be tied to their need to live on large areas of sea grasses. Today, just the manatees and the dugong survive.

There was another, gigantic species of sirenian alive on this planet just a few hundred years ago. During Vitus Bering's expedition to discover and claim Alaska for the Russian Empire in 1741–42, the official expedition scientist Georg Steller named and described a huge creature quietly feeding on kelp in the Aleutian Islands (Fig. 6.9). As large as some whales, Steller's sea cow (*Hydrodamalis gigas*) grew to a length of 8–9 m (26–30 ft) and weighed about 8–10 metric tons (8.8–11 short tons). It was completely docile and unafraid of humans, since no Europeans had ever hunted it, and the limited population of a few thousand was easily slaughtered for meat, or just for sport, by Russian fur trappers hunting sea otter and seal pelts. By 1768, only 27 years after Steller had first seen it, the largest of all the known sirenians was extinct.

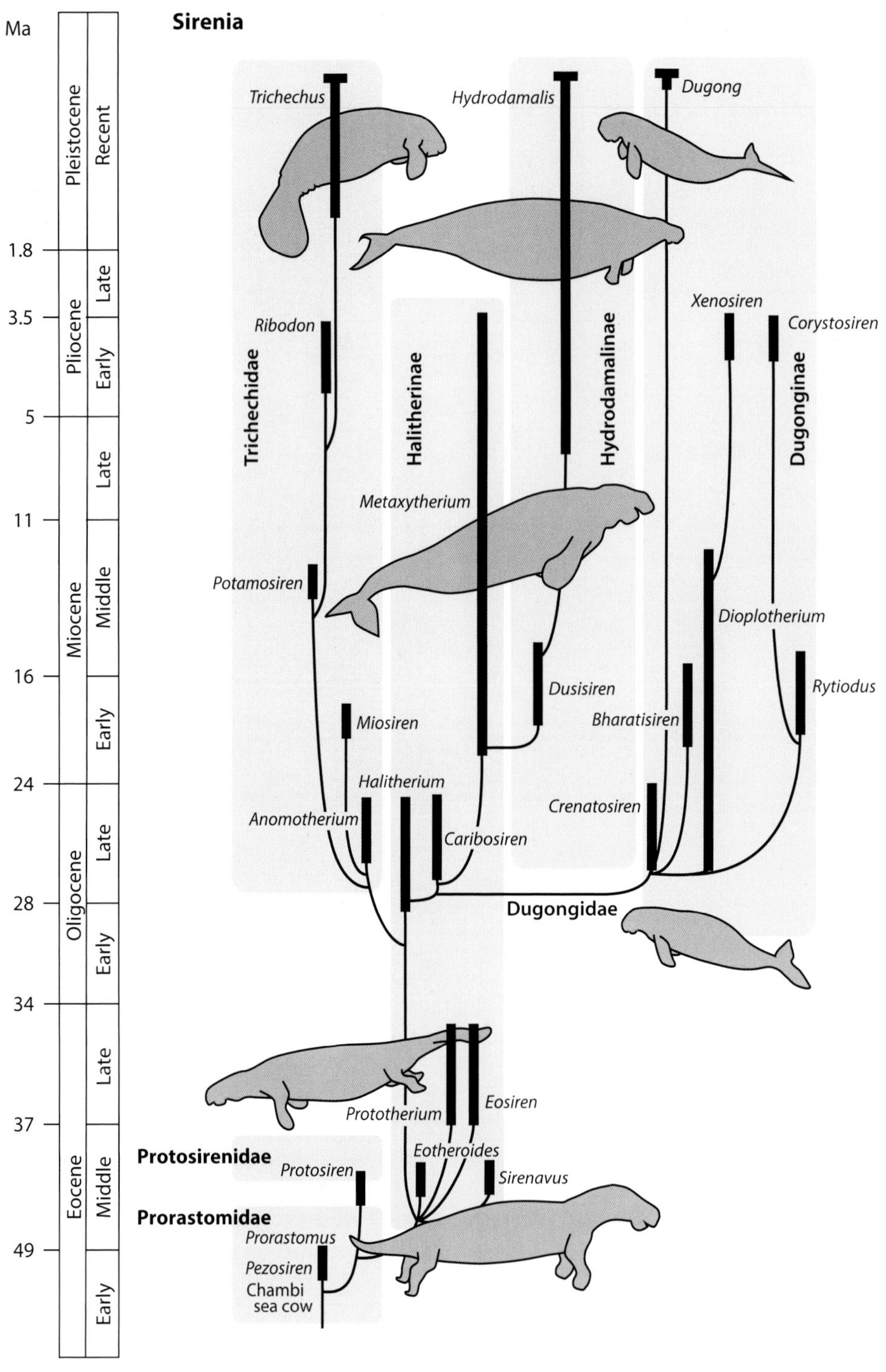
Ma
Sirenia
Pleistocene
Recent
1.8
Late
3.5
Pliocene
Early
5
Late
11
Miocene
Middle
16
Early
24
Late
Oligocene
28
Early
34
Late
37
Eocene
Middle
49
Early
Trichechus
Hydrodamalis
Dugong
Xenosiren
Corystosiren
Ribodon
Trichechidae
Halitherinae
Hydrodamalinae
Dugonginae
Metaxytherium
Potamosiren
Dioplotherium
Dusisiren
Rytiodus
Bharatisiren
Miosiren
Halitherium
Crenatosiren
Anomotherium
Caribosiren
Dugongidae
Prototherium
Eosiren
Protosirenidae
Eotheroides
Protosiren
Sirenavus
Prorastomidae
Prorastomus
Pezosiren
Chambi
sea cow

Figure 6.8. The evolutionary history of the sirenians.

Figure 6.9. Restoration of the extinct giant Steller's sea cow, *Hydrodamalis gigas*.

ORDER EMBRITHOPODA (ARSINOITHERES)

Among the finds that Andrews and Beadnell made in the late 1890s in the Fayûm beds was a truly amazing creature known as *Arsinoitherium* (Figs. 6.1, 6.2, 6.10). Named after the Hellenistic Egyptian queen Arsinoe, who once had a palace nearby, *Arsinoitherium* looked somewhat like a rhino, except that it had two massive conical curved bony horns on its forehead and a second pair of tiny bony knobs just behind the big pair. (True rhino horns are made of cemented hairs and have no bone in them.) The details of its skull and teeth, however, were truly distinctive, and *Arsinoitherium* looked like no other mammal ever found. It was about 1.75 m (6 ft) tall at the shoulders and over 3 m (10 ft) long, and had a heavy skeleton, with legs and hips that looked much like those of a hippopotamus; there are adaptations suggesting that it had a semiaquatic lifestyle as well, although its bone chemistry suggests that it was completely terrestrial.

But what was *Arsinoitherium* related to? Many ideas were proposed, but there were no convincing anatomical similarities with any known fossils at the time it was described in 1902 or for another 75 years afterward. *Arsinoitherium* remained a zoological puzzle, isolated from all other mammals in the classification. It was placed in its own order, Embrithopoda, but the name concealed the lack of any real information about its kinfolk.

Ironically, the first pieces of the puzzle were sitting in museum drawers, described but misunderstood. A partial lower jaw of another mysterious creature, *Phenacolophus*, had been found in upper Paleocene rocks in Mongolia in 1923 by American Museum paleontologists. Forty years later, Malcolm McKenna was visiting the collections in the Soviet Academy of Sciences in Moscow, when he found another jaw fragment of the same creature. It turned out that the two specimens were the front and

Figure 6.10. Skeleton of the huge embrithopod *Arsinoitherium*, from the Fayûm beds of Egypt.

back halves of the same jaw, collected from the same place by different expeditions 25 years apart!

The restoration of the complete jaw of *Phenacolophus*, plus the discoveries of Turkish Eocene fossils called *Hypsamasia* and *Palaeoamasia*, described in 1966, and of the teeth of a strange Eocene fossil known as *Crividiatherium* in Romania in 1976, led Malcolm McKenna and Earl Manning to publish a paper in 1977 that argued that all of these fossils were extinct embrithopods and that they have anatomical similarities with primitive proboscideans. Subsequently, Nick Court looked closely at the skull, brain, and inner ear anatomy of these fossils and found further evidence that arsinoitheres were tethytheres, distantly related to proboscideans and sirenians (Fig. 6.1). After almost 80 years of mystery, arsinoitheres and their relatives are now accepted as another branch of tethytheres that lived mostly in Africa and the Middle East in the Eocene (and in Mongolia, if *Phenacolophus* is truly an embrithopod). When *Arsinotherium* vanished from the jungles of Africa (Ethiopia, Kenya) and Saudi Arabia in the late Oligocene, the entire order vanished as well.

ORDER DESMOSTYLIA (DESMOSTYLIANS)

Another longstanding zoological puzzle was the mysterious group of creatures known as desmostylians. Their name means "bundle of columns" in Greek, because their peculiar cheek teeth look like a bunch of cylinders clumped together (Fig. 6.11A). When the tips are worn down into open "craters," they resemble a cluster of tubes or organ pipes or even volcanoes. The complete creature was just as strange and mysterious as its teeth (Fig. 6.11B). Most desmostylians were big beasts about the size of cows or hippos, reaching 1.8 m (6 ft) in length and weighing at least 220 kg (485 lb), with massive bodies and short tails. Their peculiar skulls had long snouts with forward-pointing tusks, thick cheekbones that flared out around the eye socket, heavy hoofed feet, and many other features that resemble primitive mastodonts.

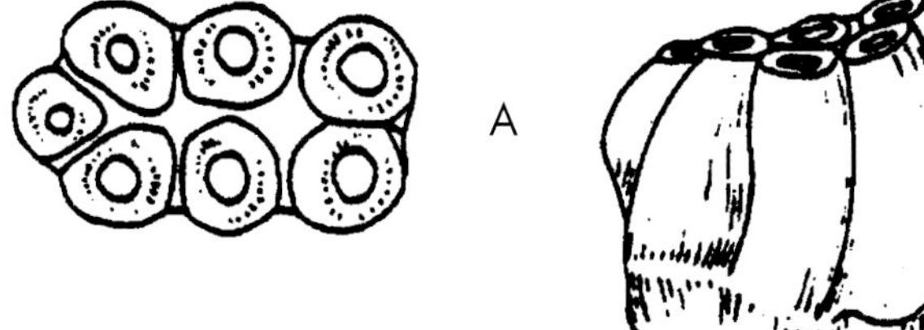

Figure 6.11. Desmostylians. A. A desmostylian molar, with its distinctive cusps shaped like a bundle of cylinders that look like volcanoes when they wear down. B. Skeleton of the recently described *Neoparadoxia*, from the late Miocene of Orange County, California. (See also next page.)

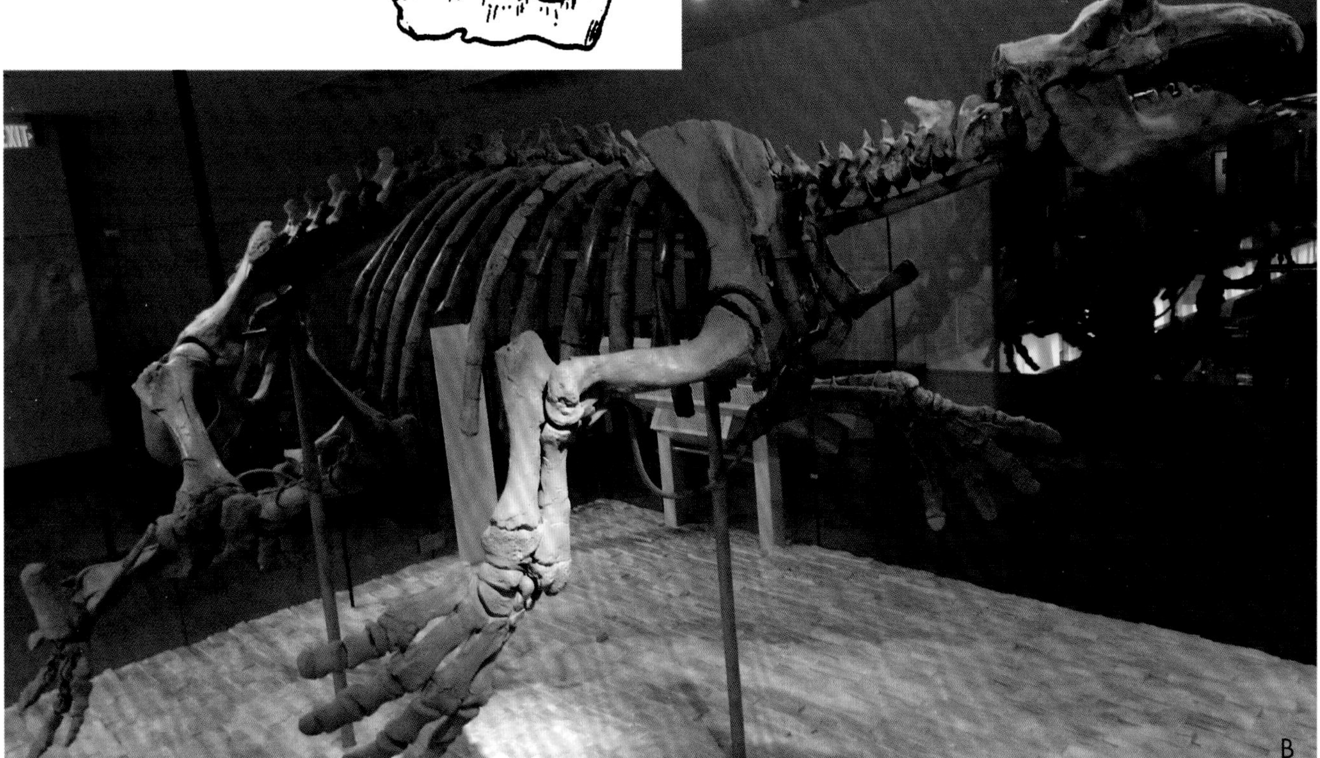

Figure 6.11. Desmostylians. C. Reconstruction of *Paleoparadoxia*. D. Reconstruction of the primitive Oligocene desmostylian *Behemotops* (swimming) and the advanced Miocene *Neoparadoxia* (on the shore).

Another peculiarity is their distribution: they are known only from areas around the North Pacific (Baja California to Alaska in the eastern Pacific, and Japan and Russia in the western Pacific) and only since the late Oligocene; no earlier fossils are known. Their fossils are always found in shallow marine rocks (especially their distinctive molars), and their bodies seem to be adapted for a hippo-like existence on the shoreline, mostly swimming, with occasional walking on the shallow bottom as hippos do. Their retracted nasal bones indicate they had a small proboscis, and their eyes are on the top of their heads so they can see while nearly immersed. Details of their bone structure show they could not walk on land much without the support of water. Chemical analysis of their teeth shows they had an aquatic diet of seaweed, kelp, and possibly some shellfish, and lived in both brackish waters of lagoons and river mouths, and marine waters. Thus, they could be thought of as nearshore hippo-mimics or maybe omnivorous sea lions with hooves.

There are eight genera and a dozen species of desmostylians known, divided into two families. One of the strangest, given the name *Paleoparadoxia* ("ancient paradox"), was uncovered during excavation of the site for the Stanford Linear Accelerator in California (Fig. 6.11C). But what were these strange beasts related to? Early paleontologists had only teeth and a few bones to go on, and usually considered them to be peculiar sirenians (or in one case, a giant marine platypus!). But when nearly complete skeletons were found in Sakhalin Island in 1941, it was clear they were not members of any known group, and so they were given their own order, Desmostylia. The breakthrough came in 1977 when the legendary collector Doug Emlong found a mysterious jaw from the late Oligocene in Oregon (more fossils were found later in British Columbia and Japan). After Emlong's death in 1980, Daryl Domning, Clayton Ray, and Malcolm McKenna finally described the fossil in 1986. They called it *Behemotops* ("Behemoth-like," after the hippo-like Behemoth in the Bible) and realized that it was not only a very primitive desmostylian, but also showed similarities to other primitive tethytheres such as mastodonts and sirenians (Fig. 6.11D). Thus we have another group of tethytheres, although one far from Tethys or Africa—yet, since we have no fossils older than the late Oligocene, there are undoubtedly more specimens still uncollected that will one day connect this group to the tethytheres of the Eocene of Africa.

ORDER HYRACOIDEA (HYRAXES)

Manatees and elephants are familiar to most of us, but another group of living afrotheres is less so. The hyraxes (Fig. 6.1) are small furry creatures from Africa and the Middle East that somewhat resemble a woodchuck, until you look closer at their anatomy or their behavior. The name hyrax means "shrew-mouse" in Greek, while the family name Procaviidae means "before the guinea pigs"—clearly there has long been zoological confusion about what the hyraxes really are. They are also known as conies or dassies, or "feeble folk" in the Bible. In Phoenician and Hebrew they are called *shaphan*, "hidden ones," because of their habit of hiding in rock outcrops. Three thousand years ago Phoenician sailors talked of a land in the western Mediterranean they thought was inhabited by hyraxes, which they called Ishaphan. The Romans modified this name to Hispania. Thus, Spain got its name from the Phoenician legends of hyraxes there (which were probably rabbits and not hyraxes at all).

The zoological confusion continued until the great French zoologist Georges Cuvier dissected specimens in 1800 and realized they had hoof-like nails and many other features not found in rodents like woodchucks or guinea pigs. He and British zoologist Richard Owen both noticed their similarities to the "odd-toed" hoofed mammals, and when Owen erected the order Perissodactyla in 1848, he included hyraxes with the horses and rhinos and tapirs. In the 1980s, Martin Fischer dissected more specimens and noticed that hyraxes had large inflated sacs in their Eustachian tubes (connecting the middle ear with the throat), a feature found elsewhere only in perissodactyls. In a paper published in 1986, Fischer, Earl Manning, and I argued that Owen had been right in 1848. By the late 1990s, however, the molecular evidence (plus additional anatomical features) made it clear that hyraxes were clustered with the afrotheres and that hyraxes were most closely related to the tethytheres.

Despite their woodchuck-like appearance, hyraxes do not behave like rodents. Most live in rocky outcrops in the African savanna and in the Middle East, where they can hide in crevices and caves from predators such as eagles, jackals, rock pythons, and leopards; one genus of hyrax lives in trees. Their feet have soft pads so the animals can move very fast in the rocks and even cling to steep surfaces as they climb. When cornered, hyraxes will stand and fight with their sharp incisors, but they have

Figure 6.12. Living and extinct hyraxes. In the foreground is the living genus *Procavia*, the rock hyrax. On the right is the extinct *Pliohyrax*. On the left foreground is *Kvabebihyrax*, and behind it is *Megalohyrax*.

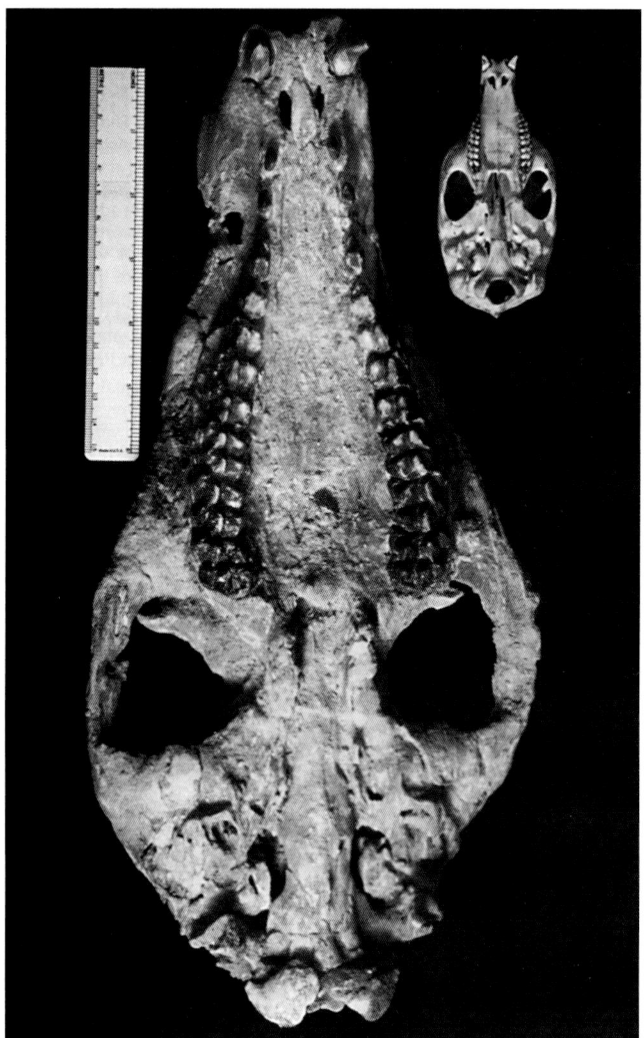

no real defensive weapons, and hiding is their primary strategy. They feed largely on grass, with some leaves and shoots, but they do not have the chisel-like cropping teeth in the front of the mouth like most grazers, so they must nip off vegetation with the side of the mouth. They move their jaws in a manner that resembles cud chewing in cows. Although they are not ruminants and cannot regurgitate their food as a cud to chew again, their stomach is nearly as efficient in digesting cellulose and plant material. Hyraxes make a variety of strange noises as their groups move around, so they are nicknamed "pig-crickets." The nocturnal tree hyraxes make an eerie, penetrating frog-like croak that unnerves explorers in the African jungle, who may imagine a much bigger and scarier animal than a tree hyrax.

There are only four living species in three genera today, but the hyraxes have an extensive record of 19 extinct genera (Fig. 6.12) going back to the Eocene of Africa, and were once widespread across Eurasia as well. The earliest fossils come from the early Eocene of Algeria, and include the very primitive form *Seggeurius* and the rabbit-size *Microhyrax*. In the upper Eocene–lower Oligocene beds of the Fayûm Depression in Egypt, there are eight different genera of hyraxes, usually placed in the subfamily Saghatheriinae. They were performing some of the ecological roles that hoofed mammals occupied on other continents, and they make up almost half the mammal fossils known from these beds. They include the rhino-size fossil *Titanohyrax* (Fig. 6.13), which had a skull over 60 cm (24 in) long, larger than any mastodont skull from the Fayûm; and the tapir-size

Figure 6.13. Comparison of the skulls (bottom view) of a modern hyrax and the gigantic extinct *Titanohyrax*.

Megalohyrax, which had crests on its teeth suitable for leaf browsing (Fig. 6.12). There were many hyraxes with different types of teeth adapted for different types of vegetation, from *Bunohyrax*, *Saghatherium*, *Pachyhyrax*, and pig-size *Geniohyus*, all with low rounded cusps on their teeth for browsing and omnivory, to *Thyrohyrax*, with teeth whose cusps are partially connected by cross-crests, to *Selenohyrax*, presumably a grazer, with crests on its teeth resembling those of ruminant artiodactyls. Also found were *Dimaitherium*, the oldest member of the modern hyrax family Procaviidae, and *Antilohyrax*, a fast runner with moderately long legs like those of an antelope (hence its name).

In the late Oligocene and Miocene, this great diversity of hyraxes vanished, replaced by invaders from Eurasia, especially the diversity of artiodactyls, both pigs and ruminants. Only the giant *Megalohyrax* and the pig-size *Meroehyrax* remained from Africa to Pakistan (Fig. 6.12). During the middle and late Miocene in Eurasia and Africa, there was a second great radiation of hyracoids, two genera and four species in the subfamily Pliohyracinae. These were horse-size creatures with teeth that resembled those of the chalicotheres (see Chapter 15), so some were confused with those creatures for 37 years. Pliohyracines included *Prohyrax* and *Parapliohyrax* in Africa and the hippo-like grazing Eurasian hyraxes *Pliohyrax*, *Kvabebihyrax* (Fig. 6.12), *Sogdohyrax*, and *Postschizotherium* (once mistaken for a chalicothere). Pliohyracines had short legs, barrel-shaped bodies, eyes high on top of their short-snouted skulls, and very high-crowned teeth for grazing, so apparently pliohyracines were performing the hippo role in Eurasia long before hippos ever left Africa. The pliohyracines vanished in the late Pliocene, to be replaced by members of the modern family Procaviidae, including not only the living rock hyrax genus *Procavia*, but also the huge rhino-size African fossil *Gigantohyrax*, from the Ice Age caves in Africa that have yielded early hominids such as *Australopithecus africanus*.

The Hyracoidea today, with just four species, is but a tiny remnant of the group's once great number and variability. In places such as Africa during the Eocene and Oligocene and Asia in the Miocene, hyraxes once occupied niches that were later filled by hippos, rhinos, elephants, pigs, and many other animals.

ORDER TUBULIDENTATA (AARDVARKS)

Nearly everyone has heard of aardvarks, famous for being the first entry in many dictionaries. Today found only in sub-Saharan Africa, the aardvark (Fig. 6.1, 6.14) is a solitary pig-size mammal. Its name means "earth pig" in Afrikaans. It has strong forelimbs and sharp claws that it uses to rip open termite nests. It is mostly nocturnal and digs deep burrows where it nests. Many other non-digging African mammals depend on old aardvark burrows for shelter. The aardvark has a tough hide with a thin coat of fur (since it is always digging and getting covered in dirt) and large ears for hearing and radiating excess body heat. It has no natural defenses, so if threatened it must dig a hole quickly—and it can dig 1 m (3 ft) of tunnel in a few minutes.

In some ways, aardvarks resemble the South America anteater and the scaly pangolins, but these similarities are due to convergent evolution because of their similar ant-eating diets. Instead, the genus and species *Orycteropus afer* is the sole surviving member of its own order, the Tubulidentata. This Latin name translates as "tubule toothed," in reference to the short peg-like teeth made only of dense tubules of dentin bundled together,

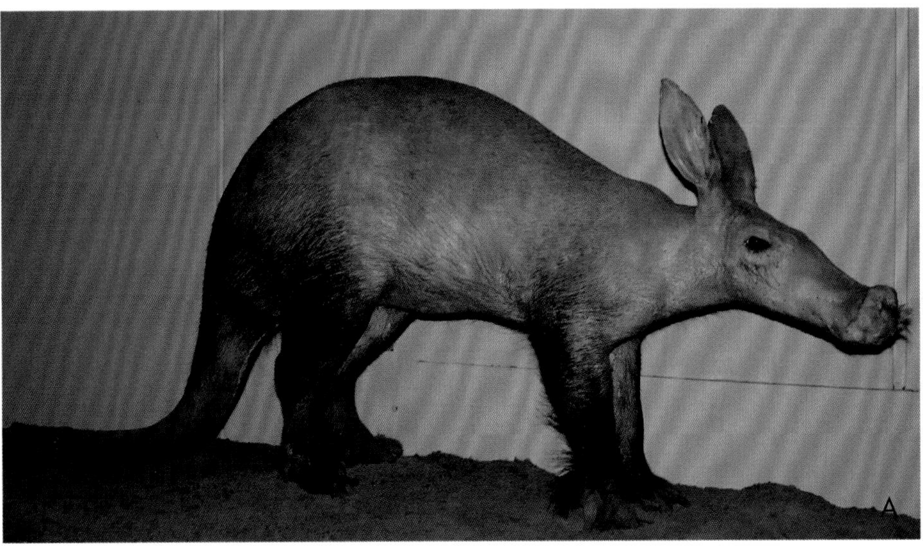

Figure 6.14. Aardvarks. A. The living aardvark, *Orycteropus afer*. (See also next page.)

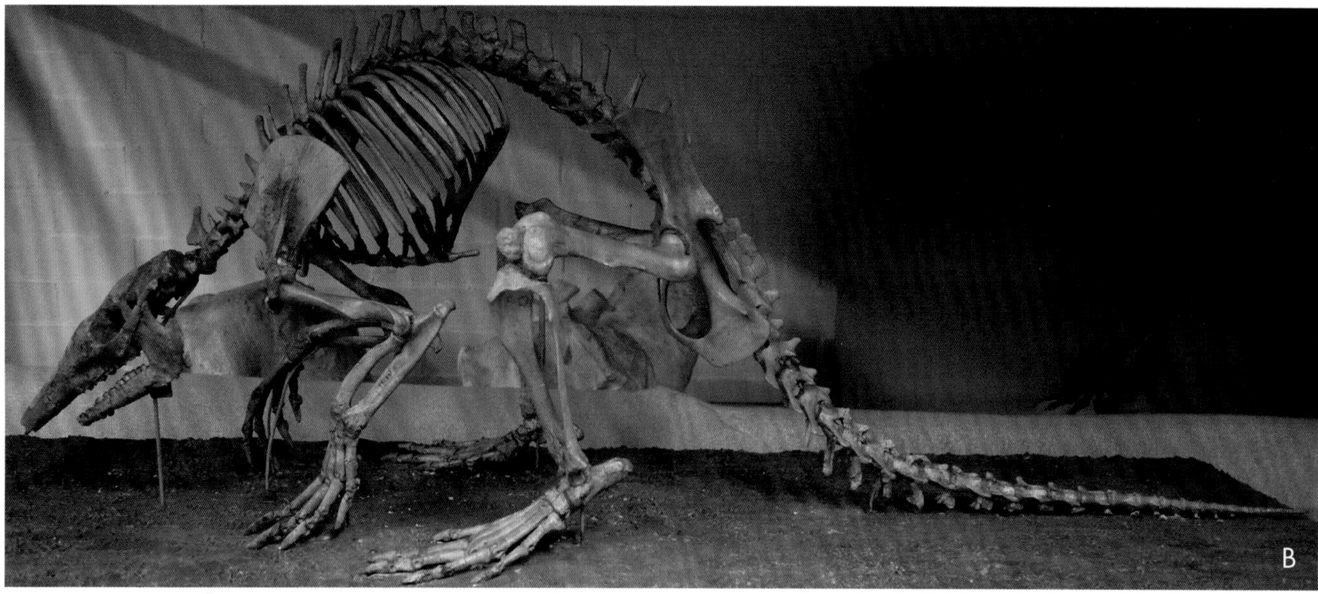

Figure 6.14. Aardvarks. B. The complete skeleton of the Miocene aardvark *Orycteropus gaudryi*, found on the Greek island of Samos.

with no hard enamel coating. Like other ant-eating mammals, aardvarks have a long sticky tongue to lap up termites with, and most of the time they eat their insects whole, without much chewing with their simple peg-like teeth. But the zoological affinities of aardvarks have long been controversial. The aardvark was thought to be a relict of the ungulate radiation until molecular data showed that it was an afrothere, related to the tenrec-golden mole–elephant shrew group (Fig. 6.1).

The fossil record of aardvarks is relatively poor. The oldest fossil species, *Myorcteropus africanus*, comes from the early Miocene of Kenya, so there is no fossil record of how the group diverged from other afrotheres in the Paleocene and Eocene. By the middle Miocene aardvarks were found over much of Europe and the Middle East (Russia, Ukraine, Moldova, Turkey, and Pakistan), including the Greek island of Samos, where the beautifully preserved skeleton of *Orycteropus gaudryi* was unearthed (Fig. 6.14B). Aardvark fossils are occasionally found in the Pliocene and Pleistocene beds of East Africa (where our earliest human ancestors were found). There was even an aardvark on Madagascar during the late Ice Ages.

ORDER MACROSCELIDIA (ELEPHANT SHREWS)

Elephant shrews are small insectivorous mammals found only in Africa, where they are known as "sengis." As their English name suggests, they look much like shrews, except for the long flexible snout reminiscent of the trunk of an elephant (Fig. 6.1). They are adept at running beneath dense foliage and hiding in the leaf litter, seeking out insects, spiders, centipedes, millipedes, and worms to eat. Four different genera and 13 species still roam Africa today, but their fossil record yields at least 14 genera, going back to the early Eocene fossils from Tunisia and Algeria known as *Chambius*. Some of these extinct elephant shrews, like *Myohyrax*, converged with hyraxes on teeth and diet, while others, such as *Mylomygale*, looked more like a rodent.

The relationships of elephant shrews were long a mystery. Traditionally, they were lumped into a wastebasket taxon of insectivorous mammals, along with many other groups with insect-eating diets but little else in common. But under molecular analyses it became clear that they were afrotheres, related to the rest of the afrothere groups rather than to other insectivores.

ORDER AFROSORICIDA

Family Tenrecidae (Tenrecs)

Tenrecs are a small family (30 species in 10 genera), mostly restricted to Madagascar, where they are the primary mammalian insect eaters (moles, shrews, and rodents never made it to the island). Thanks to convergent evolution, different species have evolved to resemble not only shrews (Figs. 6.1, 6.15A, B) but also hedgehogs, opossums, and mice. Some even resemble otters, such as the genus *Potamogale*, the giant otter shrew (Fig. 6.15C), which lives in the West African jungles.

The fossil record of tenrecs goes back to three extinct genera from the early Miocene of eastern and southern Africa. Thus, tenrecs have a long history on that continent (which once included Madagascar) and have never lived anywhere else. Yet, as with the elephant shrews, their zoological affinities were long a mystery. They were often tossed in the wastebasket group "Insectivora," based solely on their common diet, but no anatomical evidence supported this group (other than their specialized insect-cutting teeth). But molecular analyses have shown that tenrecs are in fact afrotheres, closely related to the other groups, such as elephant shrews and golden moles, and have nothing in common with Northern Hemisphere insectivores like shrews, moles, and hedgehogs.

Family Chrysochloridae (Golden Moles)

The golden moles (Fig. 6.1) are another uniquely African group of about 21 species in eight genera. They resemble the Australian marsupial "moles" in their burrowing adaptations for "sand swimming," including their powerful digging forelimbs, tiny eyes completely covered by skin, dense fur to repel the dirt and moisture, and snouts adapted to pushing through the soil. Like true moles, they feed on insects and other arthropods that fall into their burrows, which they detect by sensing vibrations in their snout.

Golden moles are known from fossils as old as the early Miocene, found in eastern Africa. They were once lumped into the "Insectivora" based on their similarity to true moles of the Northern Hemisphere. But like elephant shrews and tenrecs, they are now considered afrotheres because they have strong molecular similarity to these other groups.

Figure 6.15. Tenrecs are insectivorous mammals from Africa and Madagascar related to the Afrotheria. A. The common tenrec, *Tenrec ecaudatus*. B. The lowland streaked tenrec *Hemicentetes*. C. The giant otter shrew *Potamogale*.

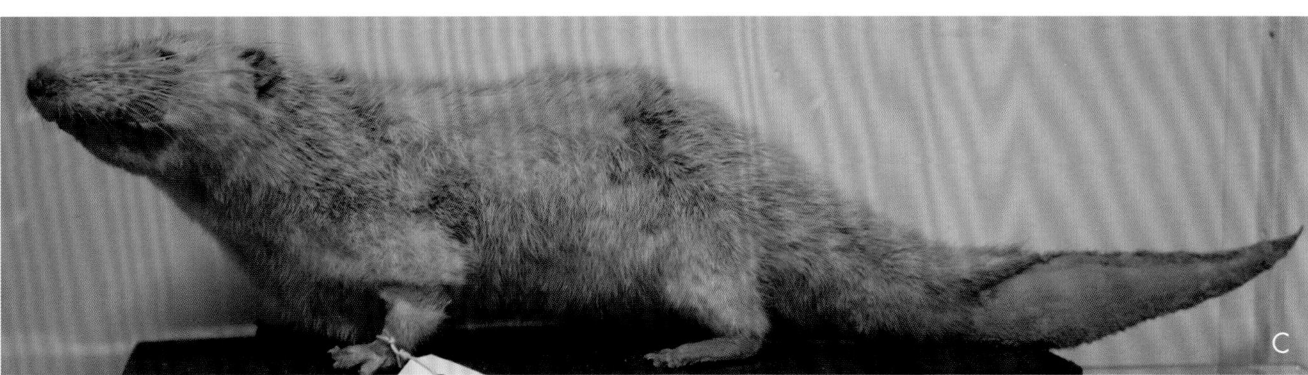

CHAPTER 7

EUARCHONTOGLIRES: EUARCHONTA

Primates, Tree Shrews, and Colugos

ARCHONTANS

In 1758 the founder of modern classification, Linnaeus, recognized the order Primates (Latin *primus*, "first rank") for the group that includes lemurs, monkeys, apes, and humans and their relatives (Fig. 7.1). In 1910 William King Gregory suggested that primates might be closely related to some other groups of mammals, including tree shrews, colugos, and possibly bats (Fig. 7.2). He coined the term "Archonta" ("ruling creatures") for this group. In both cases, the chosen name referred to a particular species of primate, *Homo sapiens*, which they considered of first rank and ruler of the world.

Figure 7.1. The tarsier-like plesiadapid *Carpolestes* has traits typical of many primates, such as grasping hands and arboreal habitat.

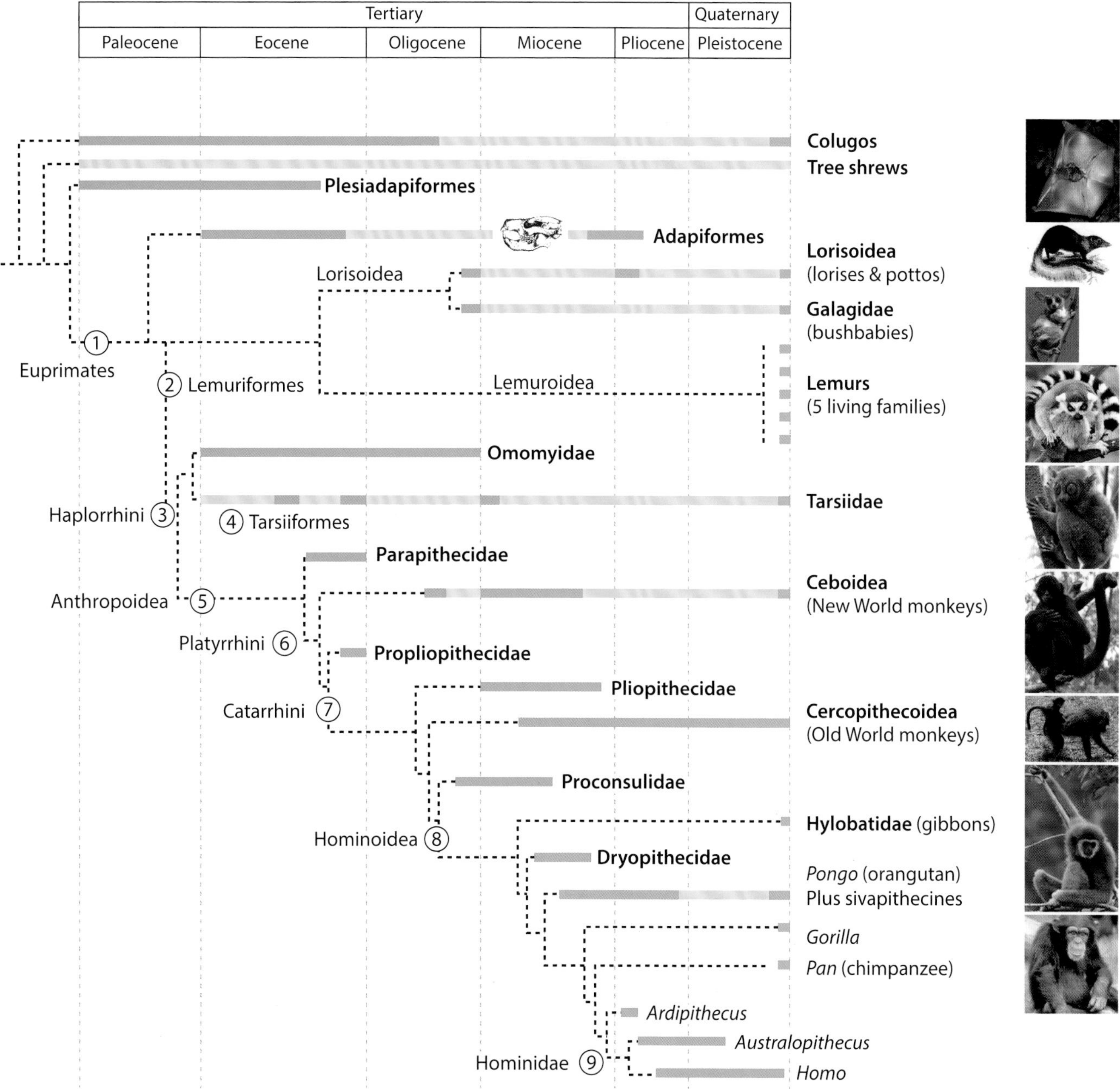

Figure 7.2. The family tree of the different groups of archontans. The green bars indicate the known fossil record of the group; the yellow bars indicate the inferred range of the group where fossils are absent.

The idea of the archontans was well supported in anatomical-based analyses of the 1980s and 1990s. By the 2000s the molecular evidence clearly supported a grouping of primates plus tree shrews and colugos, but placed the bats with other groups of mammals. Recent classifications have coined the term "Euarchonta" to represent this concept of archontans without bats. Molecular data also cluster the Euarchonta with the Glires (rodents and lagomorphs) into the "Euarchontoglires," while bats are closer to other mammals within the Laurasiatheria (Fig. 4.3).

ORDER SCANDENTIA (TREE SHREWS)

The tree shrews (families Tupaiidae and Ptilocercidae) are "living fossils" that demonstrate the transition between primates and their insectivorous ancestors (Fig. 7.2). Despite their name, tree shrews do not live exclusively in trees and they are not true shrews (which belong to the family Soricidae), although they vaguely resemble them. Today there are 20 species of tree shrews in five genera in two families found entirely in the jungles of Southeast Asia. They are shrew-like in build (although usually larger and longer-limbed), live in trees and on the ground during the daytime, and feed on a wide variety of foods (insects, small vertebrates, fruits, nuts, and seeds). Originally tree shrews were thrown into the taxonomic wastebasket "Insectivora" with other insect-eating mammals. But throughout 100-plus years of study, scientists had noticed their similarities to the most primitive fossil primates, and their relationship was confirmed by molecular analyses in the 1990s.

Tree shrews have a sparse fossil record. The oldest known is *Eodendrogale*, from the middle Eocene of China, which consists of just a few teeth. They are represented by more complete fossils from the Miocene and Pliocene of China, Pakistan, India, and Thailand. Today, tree shrews are found in nearly every book about primate evolution as the example of the type of creatures from which primates evolved.

ORDER DERMOPTERA (COLUGOS, OR "FLYING LEMURS")

The most unusual group of living archontans is the Dermoptera ("skin wing" in Greek), or colugos (Figs. 7.2, 7.3). They are sometimes mistakenly called "flying lemurs," although they are not lemurs nor do they fly. Only two species survive today, in the jungles of the Philippines and Southeast Asia. Vaguely resembling the largest species of flying squirrels in size, they have a large furry membrane (**patagium**) stretching between their front and hind limbs and toes, and between the hind limbs and tail. They use this to glide from tree to tree. However, their heads are more like those of primates, with large forward-facing eyes for good stereovision, especially at night when they feed on leaves, sap, flowers, fruits and seeds. Their cheek teeth bear some resemblance to those of the most primitive primates, and each of their lower front incisors is comb-like for grooming their fur, convergently evolved with the "tooth combs" formed of the lower incisors in some primates.

The fossil record of colugos is relatively poor, since they apparently were restricted to the jungles of Asia, where fossilization is rare. The oldest fossil colugo is *Dermotherium* from the Eocene and Oligocene of Thailand and Pakistan.

Figure 7.3. The living colugo, *Cynocephalus*. A. Flying. B. Clinging to a tree trunk.

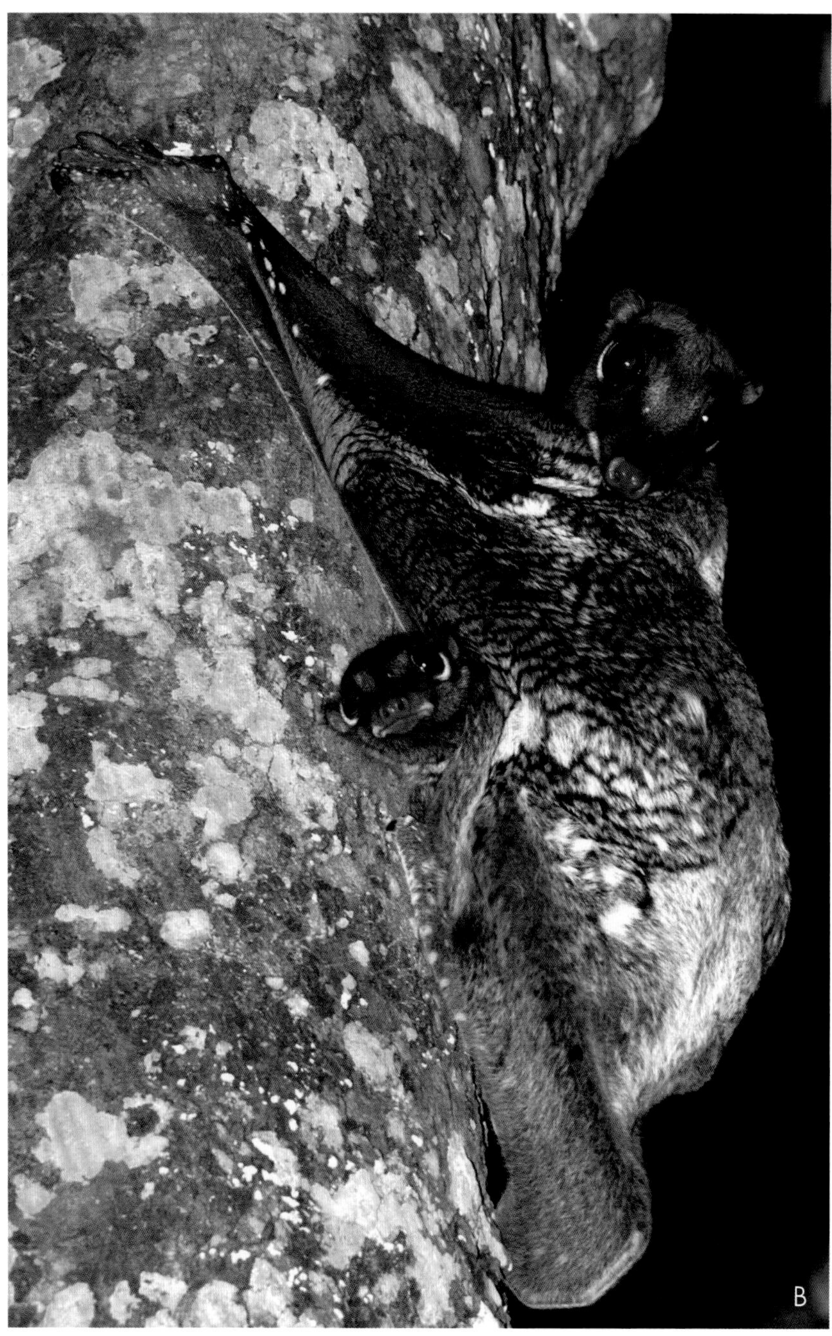

However, two extinct groups of mammals that were abundantly preserved in the Paleocene and Eocene of North America and Eurasia are thought to be related to the colugos. One of these, the family Mixodectidae, is known from a small number of teeth and jaws of three genera from the early Paleocene to early Eocene of North America. Even though they are poorly known, mixodectids seem to be transitional from insectivorous ancestors to the archontans. The other group, the family Plagiomenidae, is known from seven North American genera ranging from the late Paleocene to the late Eocene, mostly found in upper elevations of the Rockies or in the Canadian Arctic.

ORDER PLESIADAPIFORMES (PLESIADAPIDS)

Among the most common mammals in the Paleocene of North America and Eurasia were the plesiadapids, a group of mammals closely related to primates but usually placed just outside the order rather than within it (Fig. 7.2). Plesiadapids had some similarities to true primates in their cheek teeth, and a few of them had nails instead of claws, opposable thumbs, and bodies like lemurs or tarsiers (for example, *Carpolestes*; Fig. 7.1). Otherwise they lacked primate specializations such as forward-facing eyes for binocular vision. Many, such as *Plesiadapis* (Fig. 7.4A, B), were built like squirrels, while the biggest of them (*Platychoerops*; Fig. 7.4C) had the build and size of a woodchuck and were probably ground dwellers. *Craseops*, of the late middle Eocene, may have weighed up to 4 kg (almost 9 lb). Plesiadapids also had large forward-pointing incisors and a big gap (diastema) between the incisors and the cheek teeth (Fig. 7.4B), somewhat like the condition found in rodents, although plesiadapids did not develop the full chisel mechanism or ever-growing incisors found in all rodents and rabbits. Their huge diversity and abundant fossils indicate that plesiadapids apparently thrived in the Paleocene jungles of Montana, Wyoming, and Europe. They probably ate a variety of fruits, flowers, and seeds.

Dozens of species of plesiadapids in 38 genera in nine families are known from the fossil record. The earliest, from the early Paleocene, is *Purgatorius*, a shrew-like creature that was very primitive and insectivore-like but still had primate-like features in its teeth and ankles. It gets its name from the locality where it was first found, Purgatory Hill in the Tullock Formation of Montana. Plesiadapids were most abundant in the Paleocene of North America and Europe, where they were so common and rapidly evolving that their fossils are used to tell relative ages of the deposits in this time interval. Most plesiadapids vanished in the early Eocene with the origin of true primates (adapids and omomyids) and rodents, but a few genera straggled on into the middle Eocene, and a few even lasted until the end of the middle Eocene (40 Ma). Since most were tree dwellers, their extinction was probably due to a combination of the vanishing of the tropical forests, which occurred through the middle and late Eocene, and the competition from true primates.

Figure 7.4. Plesiadapids. A. Reconstruction of the squirrel-like *Plesiadapis*. B. Skull of *Plesiadapis*, showing the distinctive chisel-like front incisors and gap between the front teeth and cheek teeth, which evolved convergently with those of the rodents and rabbits. C. The woodchuck-like *Platychoerops*.

Figure 7.4. Plesiadapids. C. The woodchuck-like *Platychoerops*.

ORDER PRIMATES (EUPRIMATES)

The plesiadapids were very similar to primates, but today they are usually considered the primates' closest extinct relatives. True primates, or euprimates, have a number of specializations, such as forward-pointing eyes for stereoscopic vision with a bar of bone behind the eye socket, nails on most of their fingers and toes instead of claws, relatively large brains, and many other features of the teeth and the skull region.

The living primates are very diverse, comprising 13 families, 71 genera, and 424 species of lemurs, lorises, pottos, tarsiers, monkeys, apes, and humans (Fig. 7.2). Many hundred more extinct species and genera are known from dozens of families. Most are specialized tree dwellers, living largely on fruits, seeds, and maybe leaves, although some (including baboons, gorillas, chimps, and humans) spend most or all of their time on the ground. Primates are typically split into two larger groups, the strepsirhines (once called "prosimians"), which include lemurs, lorises, and galagos; and the haplorhines, which include the tarsiers plus the anthropoids (or Simiiformes), which encompass the monkeys, apes, and humans.

Suborder Strepsirhini (Lemurs, Lorises, Bush Babies, and Their Extinct Relatives)

The earliest fossil euprimate, known as *Altiatlasius*, is from the late Paleocene of Morocco, and by the early Eocene euprimates had spread widely to Eurasia and North America. The early Eocene *Teilhardina* first appears in Asia, then in Europe, and finally in North America over a few thousand years. Two main families dominate the early Eocene of the northern continents (Figs. 7.2, 7.5): the larger, more lemur-like, long-snouted adapids and the smaller, more tarsier-like omomyids, which had short snouts and large forward-facing eyes. Both groups were abundant and diverse through the early and middle Eocene of North America but appeared to vanish completely by the end of the middle Eocene (40 Ma). However, if the late Oligocene *Ekgmowechashala* (pronounced *iggy-moo-we-SHAH-she-lah,* "little cat man" in Lakota) from Oregon and South Dakota is really a primate, then strepsirhines lasted until the late Oligocene in North America. In Asia, the much larger sivaladapids, which weighed up to 4 kg (8.8 lb), lasted to the Miocene (as late as 9 Ma) and were the last survivors of the Adapidae.

Figure 7.5. Primitive euprimates. A. The nearly complete, articulated adapid skeleton from the Messel beds known as *Darwinius*. B. Reconstruction of the omomyid *Tetonius*.

Figure 7.6. The diversity of living and extinct lemurs. Left to right, from top to bottom: Ring-tailed lemur (*Lemur catta*); diademed sifaka (*Propithecus diadema*); aye-aye (*Daubentonia madagascariensis*); reconstruction of the extinct giant sloth lemur (*Archaeoindris fontoynonti*); gray mouse lemur (*Microcebus murinus*); red-tailed sportive lemur (*Lepilemur ruficaudatus*); red-fronted brown lemur (*Eulemur rufifrons*); and black-and-white ruffed lemur (*Varecia variegata*).

The survivors of this early radiation of strepsirhines are the living lemurs, found entirely in Madagascar, plus the lorises and galagos of Asia and Africa, totaling more than 100 species in 15 genera and five families. Lemurs apparently evolved from adapid relatives that reached Madagascar in the early Eocene (54 Ma), and there are fossils of close relatives of lemurs from the Eocene of Algeria and Tunisia. Lemurs have lived in isolation on Madagascar since about 54 Ma. They developed into many different sizes and shapes, from the tiny mouse lemur (only 30 g, or 1.1 oz) to the gorilla-size, ground-dwelling *Archaeoindris* (Fig. 7.6),

Figure 7.7. Lemurs. A. Skeleton of the gigantic extinct lemur *Megaladapis*. (See also next page.)

Figure 7.7. Lemurs. B. The aye-aye, a bizarre nocturnal lemur with an elongate middle finger.

which weighed 200 kg (440 lb). The Ice Age lemur *Megaladapis* (Fig. 7.7A), which was 1.5 m (5 ft) tall, and may have reached 50 kg (110 lb), was still roaming the forests of Madagascar until humans wiped it out in about 1420. Another group, the long-armed sloth lemurs (family Paleopropithecidae), which hung upside down from branches with their long, curved fingers, like modern sloths, vanished around 1620. Today many species of lemurs are endangered, their decline due to the human population explosion in Madagascar and the widespread deforestation, as well as to poachers who kill them for bushmeat.

Among living lemurs many are climbers, while some (such as indris and sifakas) hop along the ground on two legs as if they were skipping. The most peculiar of all is the nocturnal aye-aye (Figs. 7.6, 7.7B), which has black fur, huge eyes for seeing in the dark, and big ears that provide for excellent hearing. It taps on trees to locate grubs by sound, then uses its long rodent-like incisors to gnaw holes to access the grubs, and then employs its extremely long middle finger to fish the grubs out of the hole.

Suborder Haplorhini (Tarsiers, Monkeys, Apes, and Humans)

Haplorhines include the tarsiers plus the anthropoids (Fig. 7.2). The earliest haplorhine fossils include the early Eocene (55 Ma) *Archicebus* and the middle Eocene (45 Ma) *Xanthorhysis*, both from China. Today tarsiers are restricted to Southeast Asia, but their fossil record goes back to the Eocene of China, Thailand, and even Africa. Tarsiers have huge eyes that dominate their face and are strictly nocturnal. They have very short snouts, small ears, large hands and feet with long fingers and toes bearing nails, and long tails. They are the only entirely carnivorous primates; they catch insects by leaping from branches and also eat snakes, lizards, bats, and birds.

Other than tarsiers, the remaining primates are the anthropoids, or simian primates (Simiiformes). These include the Platyrrhini (New World monkeys) and the Catarrhini, which includes the Old World monkeys (Cercopithecidae), apes, and humans. These two large groups are named for the shapes of their snouts and noses. Platyrrhine means "flat-nosed," and the platyrrhine primates have relatively flat noses compared to the narrower noses (catarrhine means "narrow-nosed") of Old World group.

Like the oldest fossil tarsiers, the oldest known anthropoids come from the early Eocene; fossils include *Eosimias* ("dawn monkey") from China and *Afrasia* from Myanmar, as well as the larger Amphipithecidae from China and Myanmar. By late in the middle Eocene, anthropoids were well established in Africa, as shown by fossils such as *Afrotarsius*. They are much better known in the late Eocene, especially in the Fayûm beds of Egypt. These include the squirrel-size forms such as *Apidium* and the dog-size *Aegyptopithecus*, *Propliopithecus*, and *Parapithecus*, which might be considered the most primitive members of the catarrhines.

By the early Oligocene, Old World primates had all but vanished from Eurasia (a few fossils have been found in the Oligocene of Pakistan), and anthropoid evolution then occurred only in Africa. The oldest advanced catarrhine fossils are the late Oligocene *Nsungwepithecus*, from beds 26 m.y. old in Tanzania, and the gibbon-size *Saadanius*, from beds about 29 m.y. old near Mecca in Saudi Arabia. The next most recent fossil is *Kamoyapithecus*, from beds dated to 24 Ma in Kenya. By the Miocene there were many primitive catarrhines, such as *Victoriapithecus*, the earliest Old World monkey (from 20 Ma), and *Prohylobates*, dated to about 17 Ma. There was a big evolutionary radiation in the middle Miocene of Old World monkeys, not only in Africa but also in Eurasia, where they are known from hundreds of sites. Today Old World monkeys are spread across this entire region, with mainly baboons and colobus monkeys in Africa, and macaques and rhesus monkeys more common in Eurasia and North Africa.

Today the New World monkeys, or Platyrrhini, are found only in Central and South America. They include more than 64 species and 17 genera in five families, including the marmosets and tamarins, spider monkeys, squirrel monkeys, howler monkeys, capuchins and uakaris, woolly monkeys, and sakis. Many have prehensile tails. In fact, the group includes the only primates that can grab branches and similar supports with their tails.

Platyrrhini has a long but sparse fossil record in South America, starting with the earliest fossil New World monkeys, reported from isolated teeth in beds from about 36 Ma (late Eocene) in the Peruvian Amazon. The oldest relatively complete New World monkey fossil is *Branisella*, from late Oligocene beds of about 26 Ma in Salla, Bolivia. It is very similar to the late Eocene African fossil *Proteopithecus*, suggesting that platyrrhines traveled from Africa to South America about 36 Ma (Fig. 7.8), rafting on floating vegetation at a time when the Atlantic was 1000 km (about 600 mi) narrower than it is now. (The New World rodents, or caviomorphs, did the same thing, and at about the same time; see Chapter 8). By the Miocene, the radiation of New World monkeys was in full swing, and at least 20 different genera are known from Argentina to Bolivia to Colombia.

Hominoidea (Apes and Humans)

In addition to the Old World monkeys, the catarrhines include the apes and humans (all now lumped into the family Hominidae; see Fig. 1.6). Apes differ from monkeys in that they have lost their tails and they usually have a wider degree of motion in the shoulder joint, which allows them to hang down from branches. Some of them, especially the gibbons, swing through branches hand over hand, a mode of travel known as **brachiation**.

Although there are only a few types of living apes (at least two species each of chimpanzees, gorillas, and orangutans, and perhaps 15 species of gibbons), this group had a much greater diversity in the geologic past. The oldest known ape fossil is *Rukwapithecus*, from upper Oligocene beds about 25 m.y. old in

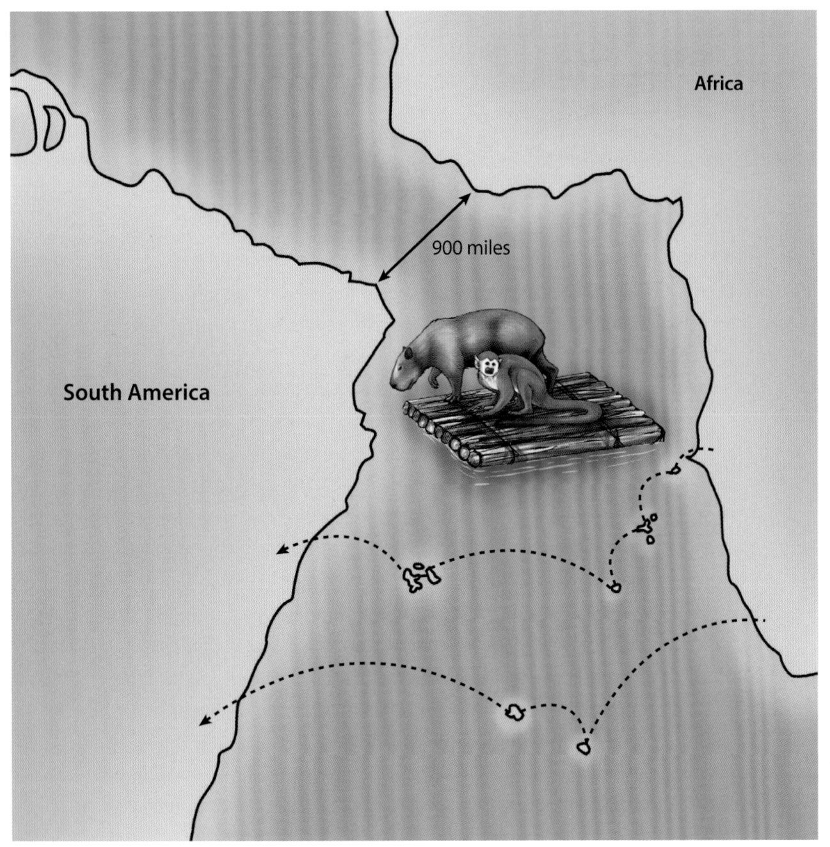

Figure 7.8. Both New World monkeys (Platyrrhini) and the caviomorph rodents (capybaras, agoutis, guinea pigs, chinchillas) are restricted to Central and South America, but their nearest relatives come from the Eocene of Africa. They must have rafted across a much narrower South Atlantic sometime in the middle or late Eocene, since their oldest South American fossils are from about 36–41 Ma.

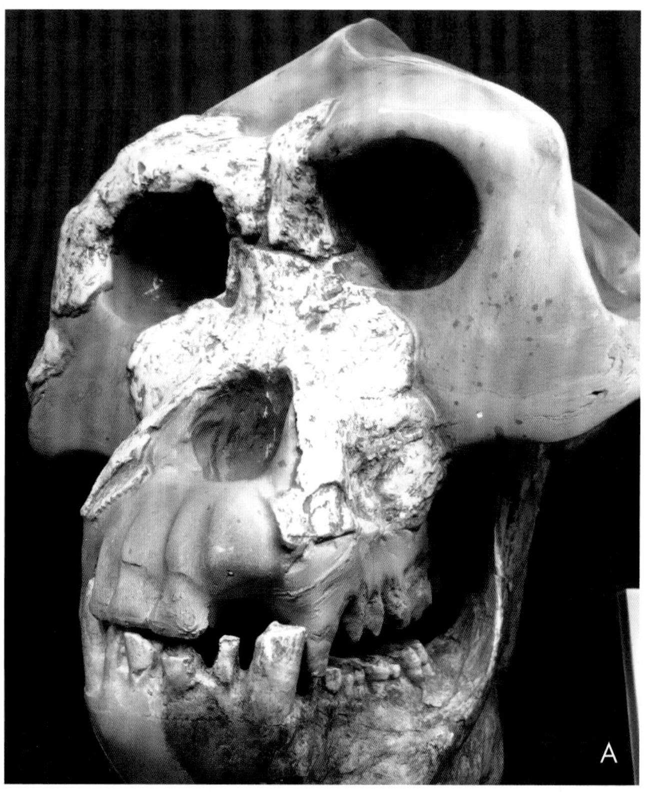

Tanzania. During the Miocene, apes underwent a spectacular evolutionary radiation in Africa and Eurasia and were much more common and diverse than monkeys. More than 40 fossil genera and over 100 species are known—14 genera in just the Miocene of Africa alone. They ranged from the size of a house cat (3 kg/6.6 lb) to the size of a gorilla (80 kg/176 lb), and ate a wide variety of foods, from leaves and fruit to a more omnivorous array. At about 16.5 Ma, *Afropithecus* escaped Africa, and its descendants spread across Eurasia, leading to a whole new evolutionary explosion of apes. Some are particularly well known, such as *Sivapithecus* (which includes *Ramapithecus*) from 12 Ma in Pakistan (Fig. 7.9A), once confused for the human ancestor but now recognized as a primitive orangutan. A group of apes called the dryopithecines was widespread across much of Eurasia. One of them, *Oreopithecus*, was first described in 1872 from a fossil found in Europe (Fig. 7.9B). Living about 7 Ma, *Oreopithecus* was much more specialized for leaf eating than most apes. By the end of the Miocene the climate had changed, and as the environment began to dry up the great ape radiation was decimated. Only a few species survived into the Pliocene, at which time the Old World monkeys diversified instead.

Figure 7.9. Fossil apes. A. The skull of the primitive ape from Pakistan *Sivapithecus*. B. Fossil skeleton of the primitive Miocene ape *Oreopithecus* next to the skeleton of a modern orangutan.

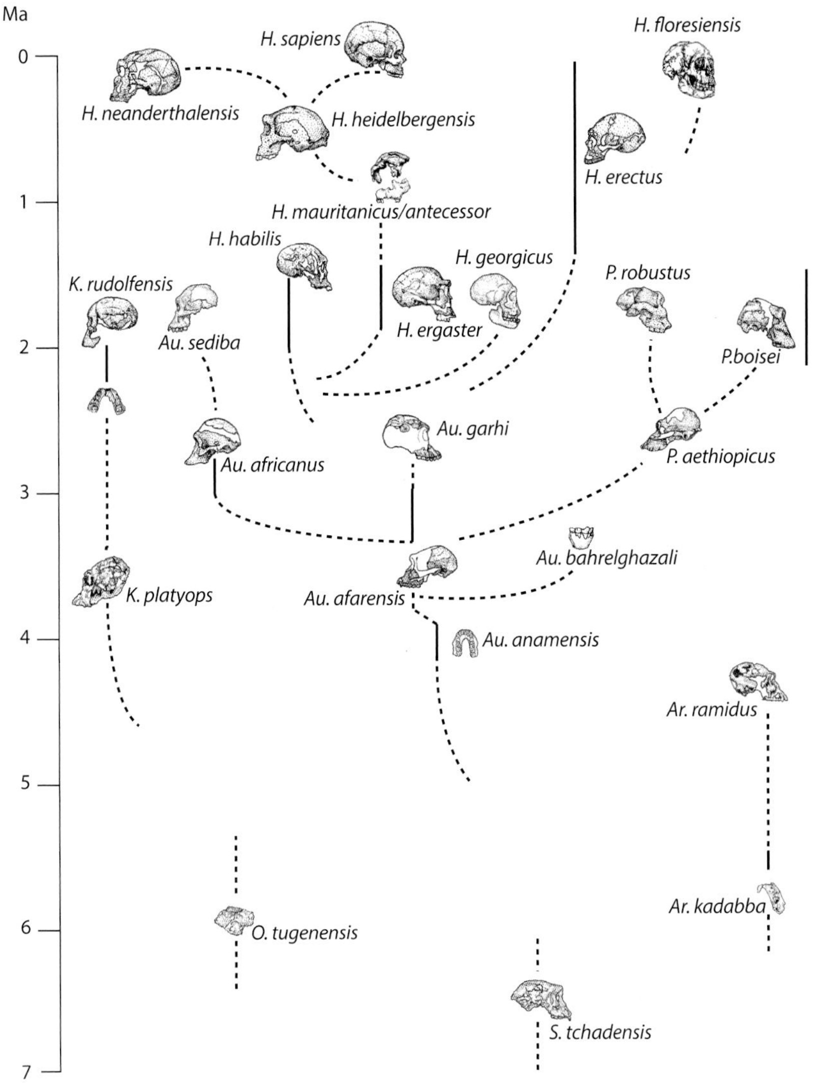

Figure 7.10. Evolutionary history of hominins.

HUMAN EVOLUTION

Finally, let us look at our own lineage, the humans or Hominini (hominins). The current oldest known human fossil is *Sahelanthropus*, from beds deposited about 7 Ma in Chad (Fig. 7.11A). About the size of a chimp, *Sahelanthropus* had a flat, humanlike face, small canines, and had a fully upright posture. In fact, upright posture occurs in the very beginning of human evolution, while large brain size evolved only at the very end. This is contrary to the early ideas about human evolution, which proposed that it was driven by evolution of large brains and bipedalism. More than a dozen different species of fossil hominins are known from the late Miocene and Pliocene of East Africa (Fig. 7.10), including *Orrorin* of 6 Ma, *Ardipithecus* from 5.8-4.4 Ma (Fig. 7.11B), and the first species of *Australopithecus* at about 4.3 Ma. Between 4.3 and 2.5 Ma, there were multiple species of *Australopithecus*, including *A. afarensis* from Ethiopia (one specimen of which is well known as Lucy, 4.0–3.0 Ma; Fig. 7.11C), and *A. africanus* from South Africa (3.0–2.0 Ma).

There were also three species of robust hominins now placed in the genus *Paranthropus*, including Louis and Mary Leakey's hyper-robust "Nutcracker Man" from Olduvai Gorge (*P. boisei*, 2.2–1.1 Ma; Fig. 7.11D), and the first robust hominin, *P. robustus*, to be discovered in South Africa (2.0–1.4 Ma). These three *Paranthropus* species coexisted in southern and eastern Africa around 2.0–1.8 Ma, yet somehow they did not push one another out. They lived with the first species of our genus, *Homo habilis* ("handy man"), which comes from beds from 2.5–1.4 Ma in Kenya (Fig. 7.11E). Recently a new species of human, *Homo naledi*, has been recovered from a large population sample in a cave in South Africa.

By 1.9 Ma the first member of the modern human lineage had evolved and eventually escaped Africa. Known as *Homo erectus*, it was originally made famous by fossils from Indonesia ("Java man") and China ("Peking man"), although it lived around much of Eurasia. *Homo erectus* was the longest-lived human species ever, lasting from 1.8 Ma to as recently as 143,000 years ago.

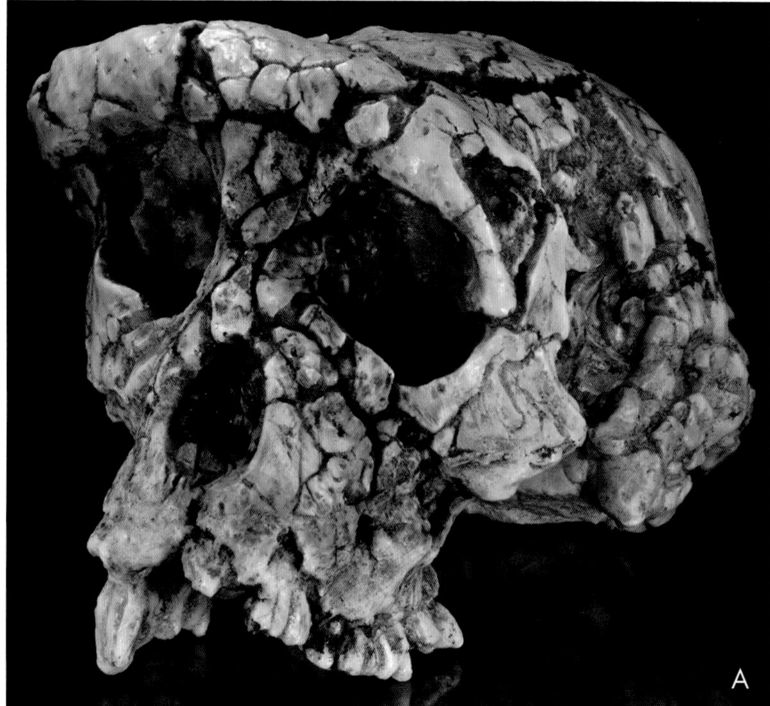

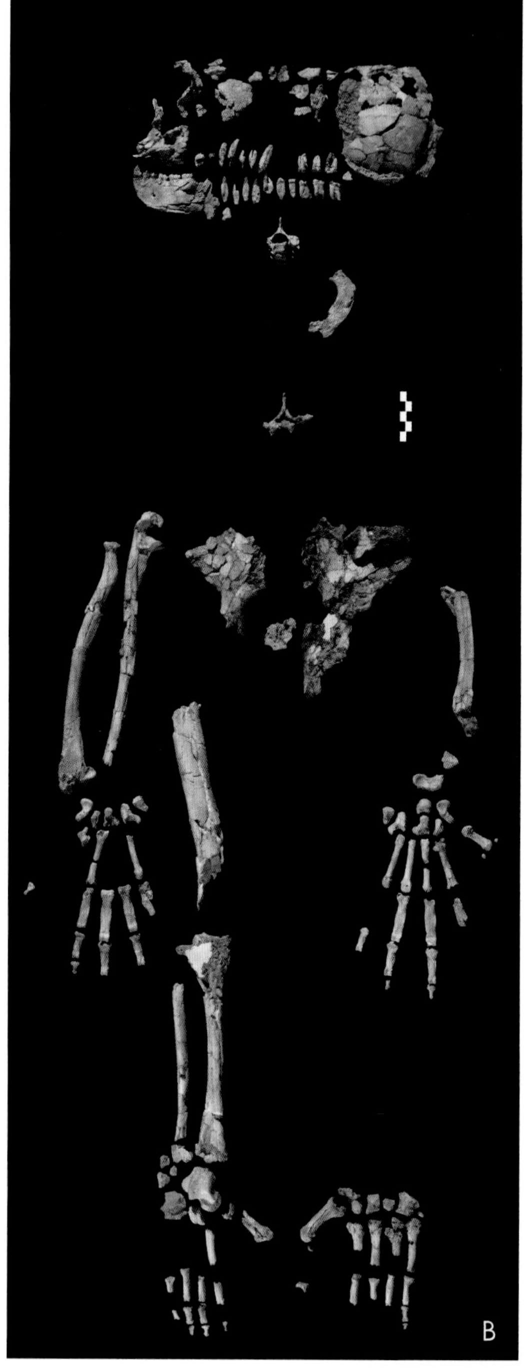

Figure 7.11. Fossils of some of the key species of hominins. A. The skull of the earliest known hominin, *Sahelanthropus*. B. The oldest partial hominin skeleton, *Ardipithecus*.

During this period a number of other human lineages evolved in Africa and Eurasia.

Fossils show that the first members of our own species, *Homo sapiens*, appeared in Africa about 100,000 years ago and migrated to Eurasia in several waves. There they encountered the more robust, cold-adapted Neanderthal peoples, *Homo neanderthalensis*, who lived on the glacial margins of Europe between 200,000 years ago and 40,000 years ago. DNA analysis has shown that modern humans interbred with Neanderthals on occasion, although otherwise they are considered distinct species. Since the extinction of Neanderthals, however, there has been only one species of hominin on this planet. This is one of the few times when our species diversity has been so low—even as human population has passed 7 billion, and we are threatening to destroy the habitability of the planet that gave us birth.

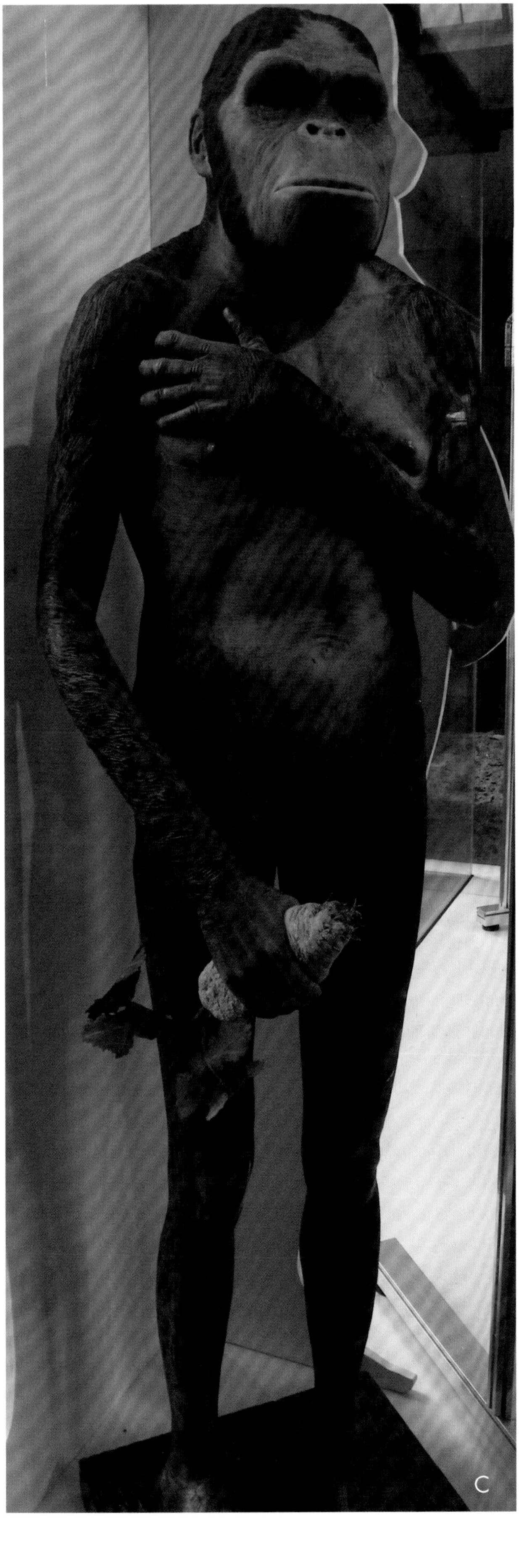

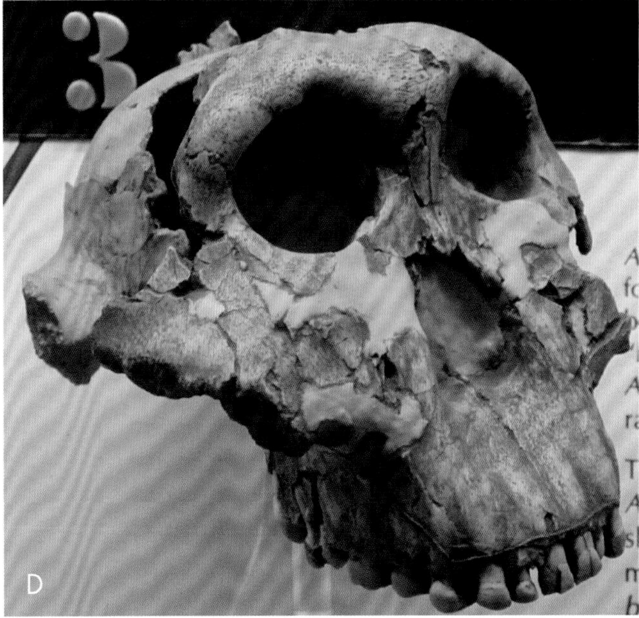

Figure 7.11. Fossils of some of the key species of hominins. C. Life-size model of *Australopithecus afarensis* ("Lucy"). D. Skull of the robust hominin *Paranthropus boisei* (known as "Dear Boy" to Louis and Mary Leakey, who found it). E. Two fossil skulls of *Homo habilis* (the one on the left is sometimes called *H. rudolfensis*).

CHAPTER 8

EUARCHONTOGLIRES: GLIRES

Rodents and Lagomorphs

CHISEL TEETH

One of the best-established mammalian groups is the Glires (Fig. 8.1), the rodents plus the lagomorphs (pikas and rabbits). It was first recognized and named in 1910 by William King Gregory, based on anatomical features, and its standing has been confirmed by recent analyses and by fossils from the Paleocene and Eocene of Mongolia and China. The name "Glires" refers to the chisel-like gnawing front teeth, or incisors, a configuration known as "gliriform." Nearly all the molecular analyses done in the last 20 years have also shown that rodents and rabbits are closely related to one another, although a few studies have not.

Most people are surprised to learn that rabbits are *not* rodents. But the differences between these groups were noticed as far back as the 1850s, when zoologists placed them in separate orders. Although both groups are small-bodied herbivores or omnivores with ever-growing chisel-like front incisors for gnawing, they are very different in anatomical details (Fig. 8.2). In particular, rodents have only one pair of upper chisel-like incisors, while rabbits have two pairs. The snout region of the skull in

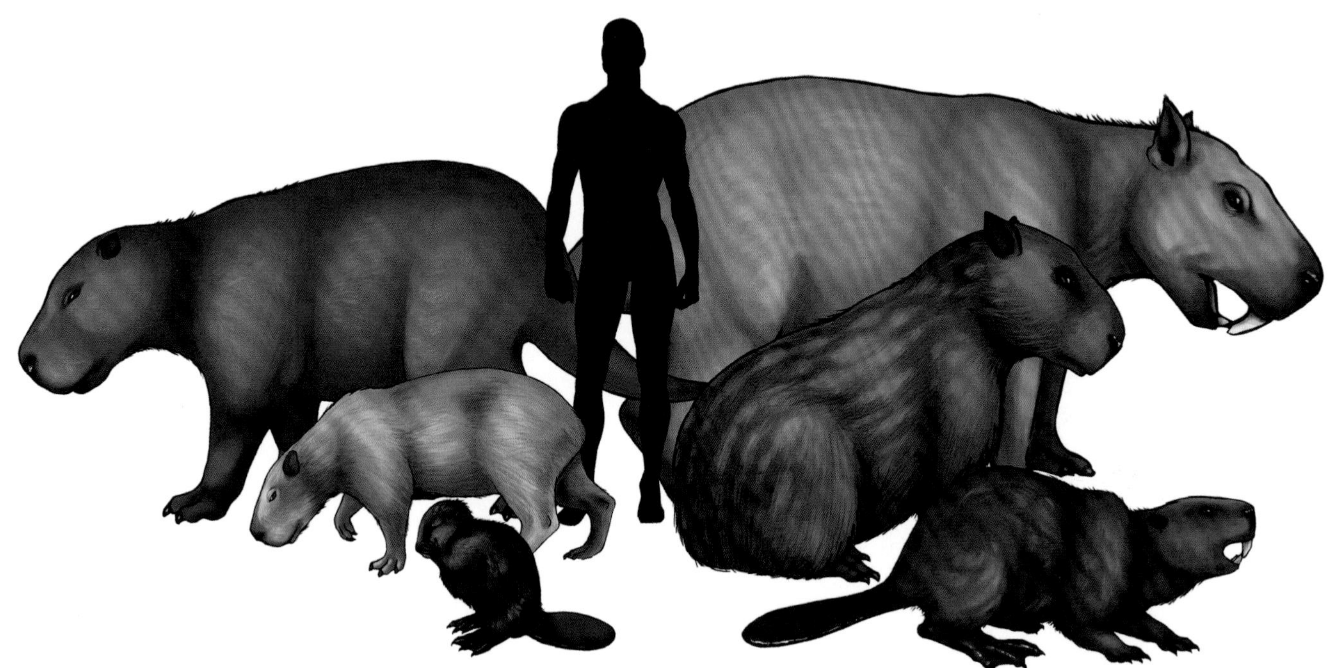

Figure 8.1. Rodents come in many shapes and sizes, and not all of them small; some extinct rodents were huge. In the left foreground are the beaver and the capybara, the largest living rodents. In the right foreground is the bear-size Ice Age beaver *Castoroides*. The rodent in the right background is *Josephoartigasia*, an immense pacarana from South America. In front of it on the right is the giant capybara *Neochoerus*, and on the left is *Telicomys*.

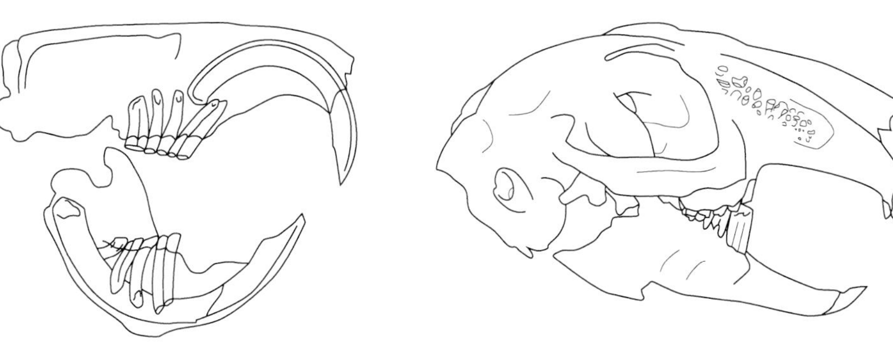

Figure 8.2. Left: Skull of a typical rodent, showing the deeply rooted ever-growing upper and lower chisel-like incisors and the diastema, or the gap between the front teeth and the cheek teeth. Most rodents have just a few specialized grinding cheek teeth, which are ever-growing from deep roots. Right: Skull of a rabbit, showing the two pairs of front chisel-like incisors, the porous bone of the snout region, and the other key differences in the skull.

rabbits is made of porous spongy bone. Functional morphology studies have suggested that rabbit evolution has minimized the amount of dense bone needed in the snout, since the stresses on the skull are taken up by the solid bones in the rim of the snout. Some species of rabbits have a joint in their skull that serves as a shock absorber when leaping.

There are also numerous detailed differences in the skull and skeleton. Rabbits tend to have short tails, and longer hind legs because they move by jumping, while rodents are mostly runners and climbers (although kangaroo mice, kangaroo rats, gerbils, jerboas, spring hares, and a few other rodents have convergently evolved to perform the jumping locomotion used by rabbits). For much of the 20th century, biologists and paleontologists regarded rodents and lagomorphs as unrelated groups that had independently developed the gnawing lifestyle in different ways. Only since the 1980s has opinion swung the opposite way, as new anatomical analyses, new fossils, and finally molecular data have changed the weight of scientific evidence.

The best lines of evidence for the close relationship between the two orders are fossils of primitive ancestral Glires from the early Cenozoic of Asia that were closely related to both rodents and lagomorphs but not a member of either group. These fossils form their own wastebasket groups, often known by the names "anagalids" and "eurymylids," and some of them (such as *Tribosphenomys* and *Rhombomylus*) were closer to rodents, while others (such as *Mimotona* and *Gomphos*) were closer to rabbits. Some paleontologists trace these animals back to a well-known group of Cretaceous Mongolian fossils known as zalambdalestids. By the Paleocene of China, there are numerous fossils of a group called pseudictopids, which are increasingly similar to both rodents and lagomorphs. These fossils are as big as most rodents and rabbits, and show many of the specializations of the skeleton as well, but they do not yet have the fully developed ever-growing gnawing incisors. *Mimotona* from the early Paleocene of China has the lagomorph-like double pair of upper chisel incisors, but the crown pattern of the molar teeth is very primitive and not yet truly lagomorph. *Eurymylus*, *Rhombomylus* (Fig. 8.3), and other fossils from the Paleocene and early Eocene of China have the characteristic single pair of gnawing incisors like rodents but do not yet have the cheek tooth specializations that separate rodents from more primitive forms. Thus we have the classic transitional fossils that not only link rodents and lagomorphs but also allow us to trace their origins back into the Paleocene and even the Late Cretaceous.

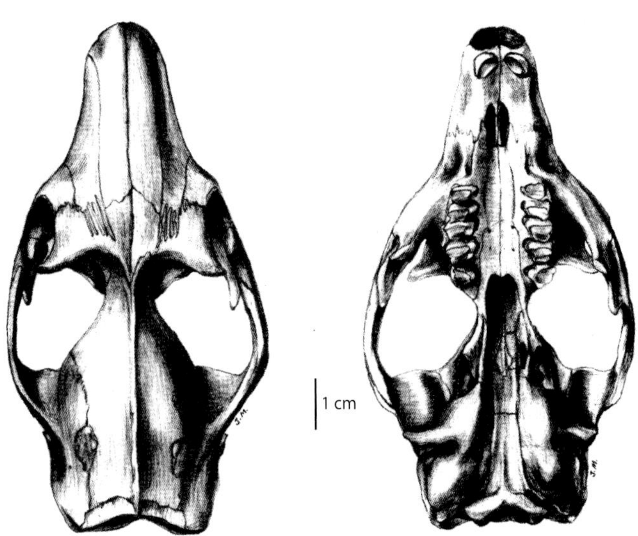

Figure 8.3. Skull of the primitive relative of rodents and rabbits from the Paleocene of China known as *Rhombomylus*.

ORDER RODENTIA (RODENTS)

The rodents are by far the most numerous and diverse mammals on the planet, with some animals (such as certain rats and mice) having populations in the millions all around the world. The latest classification schemes recognize 33 families, 481 genera, and 2,277 species of living rodents. They make up about 40% of the living mammal species, and there are least five times as many extinct rodent genera and species as there are living genera and species. They range from the small rats, mice, hamsters, voles, gophers, chinchillas, guinea pigs, and chipmunks to medium-size creatures like squirrels, and to large species such as beavers, muskrats, porcupines, and capybaras, which are dwarfed by some enormous extinct representatives (Fig. 8.1).

The interrelationships among rodents remain a contentious issue, and many different classification schemes and family trees have come and gone over the years. Most of them are based on features of the jaws, the skulls, and especially the muscles on the cheekbone and snout (Fig. 8.4). Even the latest molecular analyses (Fig. 8.5) have not resolved some of the questions and problems, so there is still a lot of work to do. There are some general things about rodents that can be pointed out, but there is no room in a field guide to discuss every living species or family of living rodent, let alone the thousands of fossils species, most of which are distinguished only by small differences in their teeth.

As noted above, the most striking feature of the rodents is their chisel-like pair of gnawing front incisors in both upper and lower jaws (Fig. 8.2). Rodents use them not only to gnaw on their food but also to gnaw on wood, to create holes or cut down trees, and even to dig burrows. Rootless and constantly growing, these incisors curve deeply back into the skull and lower jaw. The tips are continuously worn down by use as the animal gnaws, and the incisors grow continuously to compensate for this wear. The incisors are made of soft dentin with a band of hard enamel only on the front surface, so as the teeth gnaw, the enamel forms a self-sharpening edge in front of the easily worn dentin. Rodents often sharpen their teeth by grinding them against each other, and they must continually keep these teeth sharp or they will starve. In addition, if the incisors don't occlude and wear properly, the unworn incisors can grow out of control, curve around, and eventually pierce the skull. Behind the incisors is a toothless gap (**diastema**), then a small number of premolars and molars adapted for whatever the rodent eats (Figs. 1.8, 8.2).

In addition to gnawing incisors, there are other features of the chewing mechanism that are important to rodent classification (Figs. 8.4, 8.5). The most primitive rodent skulls are those of the **protrogomorphs**, in which the masseter muscles that close the jaw are attached to a limited area of the skull. This is the primitive condition found in most mammals. From this starting point, three different lineages with different jaw-muscle configurations evolved. The **sciuromorphs**, or squirrel-like rodents, are just slightly more advanced than protrogomorphs in that the masseter muscles of the jaw extend up the front of the cheekbones alongside the snout. Sciuromorphs include squirrels, chipmunks, woodchucks, marmots, and beavers. Another skull and muscle configuration has a large opening (the **infraorbital foramen**) in the front of the cheekbones on each side of the snout. In the **hystricomorphs** (porcupines plus the South American and some African rodents), the masseter muscle passes through the infraorbital foramen to attach the jaw to the front of the snout. Finally, in the **myomorphs** (rats, mice, and most other rodents), one branch of the masseter muscle passes through the infraorbital foramen (as in hystricomorphs), and the other branch curves around the front (as in sciuromorphs), a combination of both anatomies. Although there are other anatomical features used in classification, these distinctive jaw-muscle conditions seem to define natural groups within the Rodentia and are partially confirmed by most of the recent molecular work as well. They are also significant because the shift in the position of these jaw muscles gives rodents their ability to slide the jaw front to back rather than side to side when they chew, making it possible to grind up tough seeds and plants. (Large herbivores have only a side-to-side chewing motion).

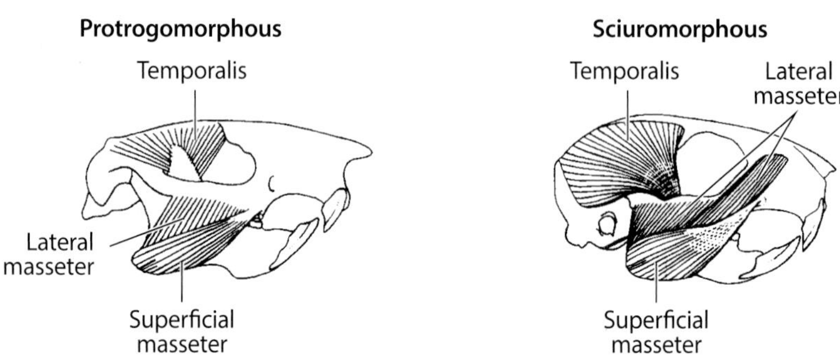

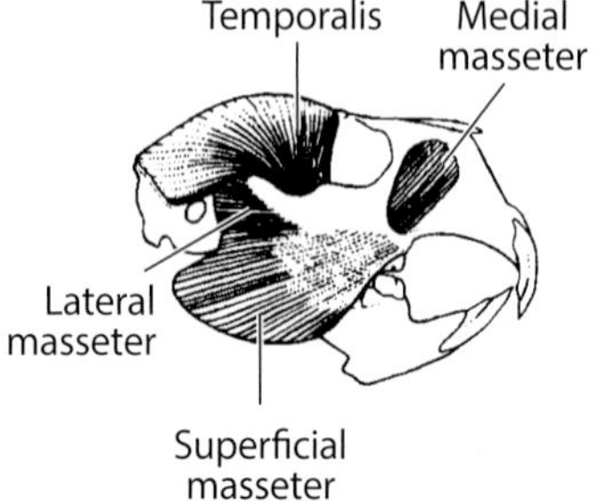

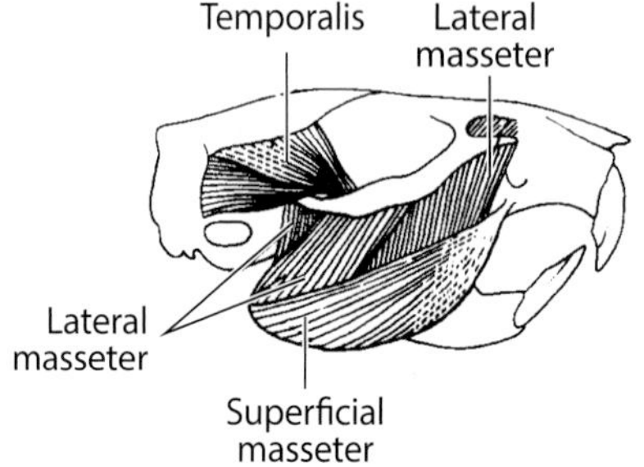

Figure 8.4. Skull muscles of rodents. Protrogomorphous is the primitive condition. Sciuromorphous skulls have masseter muscle passing across the front of the cheekbone and attaching to the snout. Hystricomorphous skulls have a strand of the medial masseter muscle passing through a hole in the cheekbone, the infraorbital foramen. Myomorphous skulls have both muscle configurations.

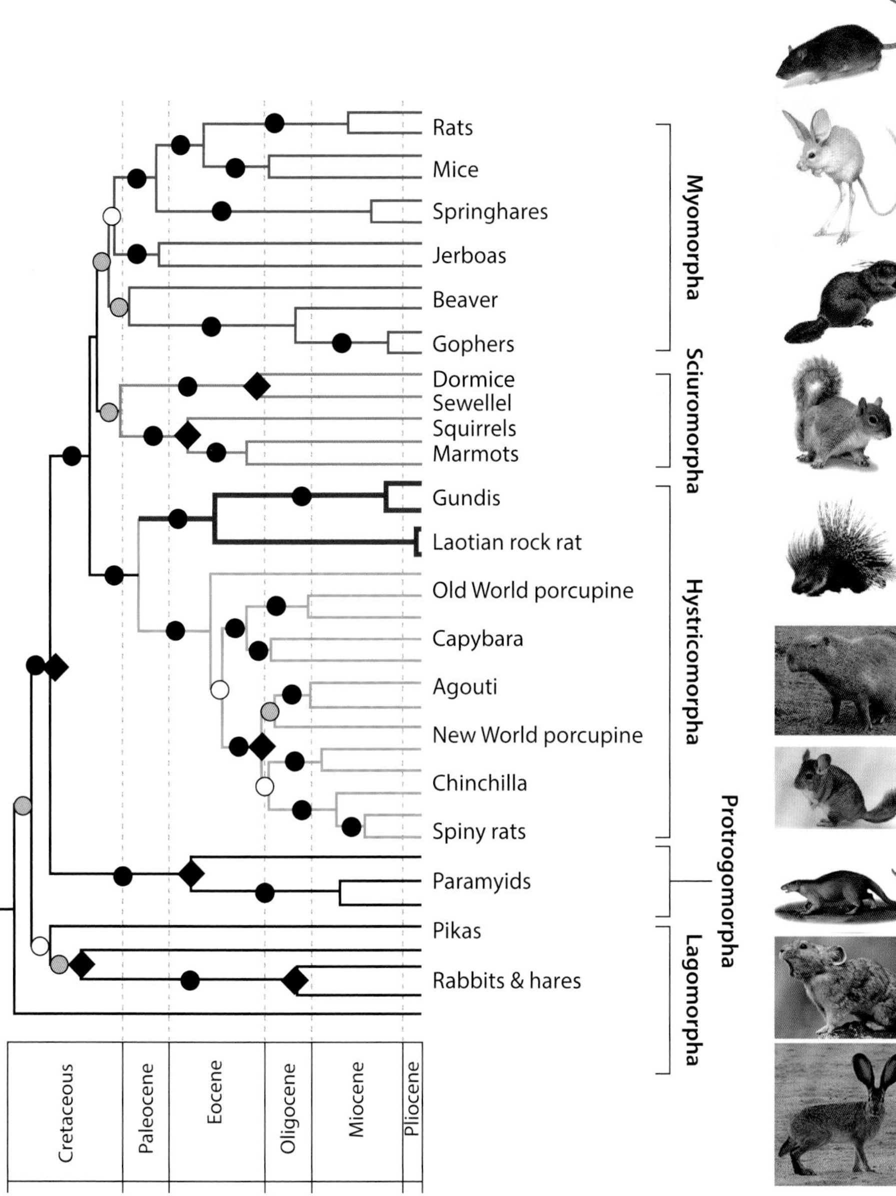

Figure 8.5. Molecular tree of rodent and lagomorph phylogeny. Note that the time scale is based on molecular clock distances. The fossil record indicates that the oldest rodents appeared no earlier than the late Paleocene.

Suborder Protrogomorpha

As discussed already, a number of primitive fossils (such as *Tribosphenomys*) from the Paleocene of China and Mongolia seem to be very close to the ancestral lineage of the rodents. In the latest Paleocene, rodents escaped from their East Asian homeland, and show up in fossils in North America and Europe for the first time. The best known of these, from the early to the middle Eocene, is *Paramys*, which was a common squirrel-size rodent that had the primitive protrogomorphous jaw condition (Fig. 8.6A). Paramyid rodents were quite diverse in the early and middle Eocene of North America and Europe, but by the late Eocene, a number of other rodent groups had appeared that drove these

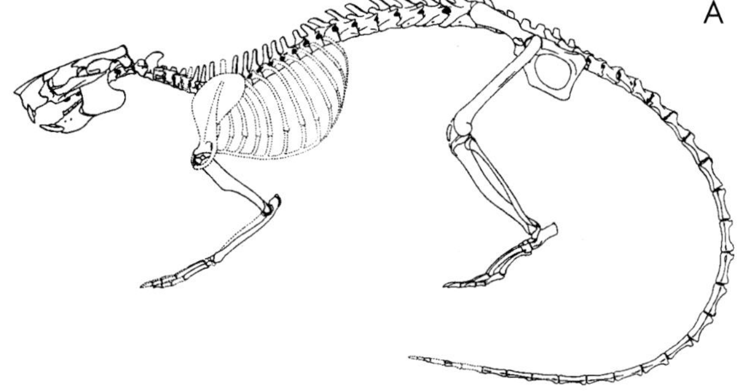

Figure 8.6. Some extinct protrogomorph rodents. A. The earliest and most primitive North American rodent, *Paramys*. (See also next page.)

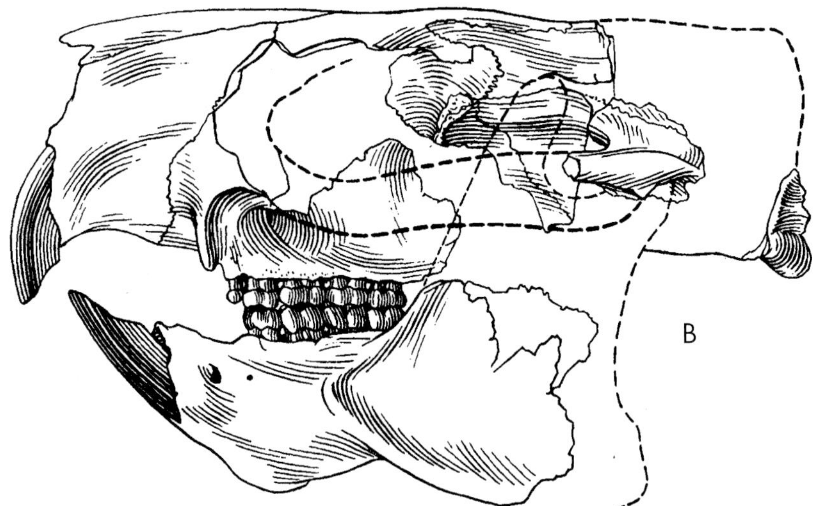

Figure 8.6. Some extinct protrogomorph rodents. B. The skull of *Manitsha tanka*, the biggest and last of the paramyid rodents.

archaic gnawers to extinction. The last of the paramyids was a beaver-size rodent (Fig. 8.6B), weighing almost 9 kg (20 lb), from the early Oligocene of northwestern South Dakota called *Manitsha tanka*, which means "giant ground squirrel" in the Lakota Sioux tongue. It was the largest rodent to evolve for a long time.

Suborder Hystricomorpha

The latest molecular and anatomical evidence (Fig. 8.5) shows that the first branching point of advanced rodents from the protrogomorph paramyid ancestors occurred with the hystricomorphs, whose masseter muscles pass through the infraorbital foramen (Fig. 8.4). Except for some Asian rodents known as gundis, all the rest of the living and extinct hystricomorphs also have hystricognath jaws, characterized by large flanges that stick out sideways from the back corner of the jaw for anchoring the chewing muscles.

The most primitive hystricognaths are some African rodents, including a big radiation of fossils from the Eocene–Oligocene Fayûm beds of Egypt known as Phiomorpha, plus the living African porcupines, dassie rats, and cane rats (Fig. 8.5). Hystricognaths also include the bizarre naked mole rats, which live their entire lives in burrows underground, using their teeth to dig burrows and gnaw roots and tubers. Naked mole rats have a social structure much like an ant or termite colony, with one dominant female who rules the nest and many subordinate daughters who carry out the daily tasks.

The most amazing of all the Hystricomorpha, however, is the huge radiation of South American rodents known as caviomorphs. This group includes 14 families, dozens of genera, and hundreds of species, from chinchillas, pacas, agoutis, and guinea pigs to the largest living rodents of all, capybaras (Fig. 8.5). They are all descended from an African hystricognath ancestor that must have rafted across the much narrower Atlantic Ocean during the middle Eocene (Fig. 7.8), which we know because the earliest fossil rodents in South America date back to 41 Ma. Since South America had no native mammals competing in the rodent niche, the caviomorphs underwent a huge evolutionary radiation there through the rest of the Cenozoic, resulting in the gigantic diversity they exhibit today. They have evolved into gopher-like forms (the tuco tuco), tree-dwelling forms (the porcupine and spiny rats), fast grassland runners (the maras), jungle runners (pacas and agoutis), mountain dwellers (chinchillas), and large grazing, swimming herbivores (the nutrias and capybaras). These rodents evolved to eat grasses long before grazing ruminants ever reached South America across Panama.

A few hystricomorphs successfully went northward when the Panama land bridge formed in the late Miocene (Fig. 5.4), including ancestors of the North American porcupine and a gigantic Ice Age capybara, *Neochoerus*, which at 113 kg (250 lb) was the size of a large boar (Fig. 8.1). Meanwhile, South America developed even larger giant rodents, including the rhinoceros-size Plio-Pleistocene pacarana *Josephoartigasia* (Fig. 8.1), which was 3 m (10 ft) long and weighed up to 1500 kg (3300 lb). This giant was discovered only a few years ago; the previous record holders were the bison-size Miocene pacarana *Phoberomys* ("terror mouse"), which was 3 m (10 ft) long but weighed only about 700 kg (1540 lb), and the slightly smaller late Miocene–Pleistocene cow-size fossil *Telicomys*, which was about 2.7 m (6.5 ft) long.

Suborders Sciuromorpha and Castorimorpha

Some recent molecular data (Fig. 8.5) suggest that the beaver group (Castorimorpha) is closely related to the Myomorpha. Other molecular phylogenies indicate that the two remaining rodent groups, the sciuromorphs and myomorphs, have a common ancestor that evolved after the hystricomorphs split off.

The rodents with sciuromorphous jaw muscles are now split into two suborders: Sciuromorpha (squirrels, chipmunks, flying squirrels, ground squirrels and prairie dogs, marmots, and dormice, with 307 living species spread among 61 genera in seven families, and many more fossil species), and the Castorimorpha (beavers, kangaroo rats, and pocket gophers, plus many extinct groups). Sciuromorphs were abundant in the late Eocene, as they apparently drove the archaic paramyids to extinction, and they have continued to diversify in the Oligocene and ever since.

Among the most bizarre sciuromorphs were the horned burrowing rodents known as mylagaulids (Fig. 8.7B). Their fossils have been discovered in the bottoms of huge corkscrew-shaped burrows known as *Daemonelix*, or "devil's corkscrews" (Fig. 8.7A), common in the late Oligocene to early Miocene rocks of Nebraska and also China. The function of these strange horns on their snout has been the source of much speculation, but recent research has ruled out digging, and display (as with the antlers and horns of ruminants). This leaves the simple idea that mylagaulids defended themselves from larger predators (such as the burrowing bear dogs *Daphoenodon*, found in the same beds near Agate Springs, Nebraska) with the fierce show of short sharp horns to any intruder chasing them down their burrows.

The Castorimorpha also features a diverse radiation of rodents, especially in North America, where the pocket gophers (Geomyidae) and kangaroo rats and mice (Heteromyidae) have long fossil records. The beavers (Castoridae) also have an extensive fossil

Figure 8.7. Fossil sciuromorphs and their traces. A. The corkscrew burrows dug by horned rodents, known as "devil's corkscrews," or *Daemonelix*. B. Skeleton of the Miocene horned mylagaulid rodent *Epigaulus hatcheri*. The skeleton was found in the bottom of the burrow.

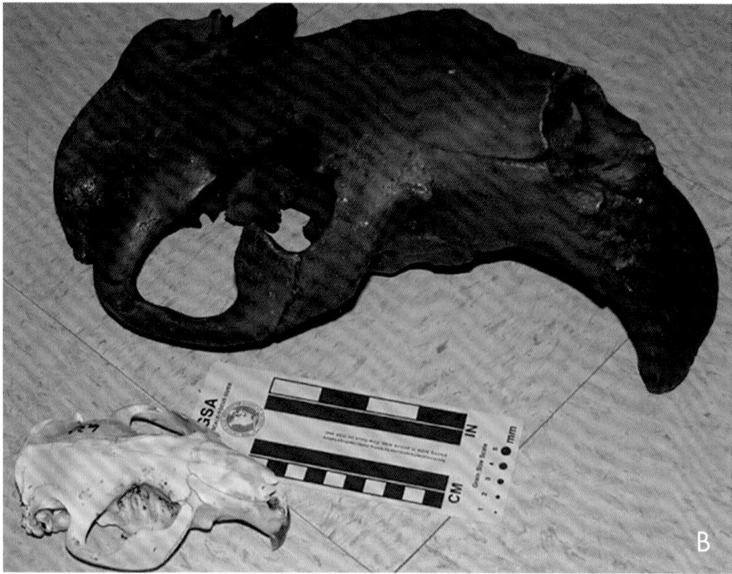

Figure 8.8. The bear-size Ice Age beaver *Castoroides*. A. Complete skeleton. B. Comparison of the gigantic skull of *Castoroides* with that of a modern beaver.

record, from the earliest castorids of the late Eocene, which were adapted for more generalized gnawing than for the specialized wood gnawing that came later in their evolution. The living beaver is the largest rodent in North America today, but there were even bigger beavers in the Ice Ages in both North America (*Castoroides*) and Europe (*Trogontherium*). *Castoroides*, which roamed widely in the eastern forests of North America during the Ice Ages, was the size of a small bear (Figs. 8.1, 8.8), about 2 m (6.5 ft) long and weighing up to 125 kg (275 lb). It had a proportionally narrower tail than a modern beaver. Its huge incisors were more than 15 cm (6 in) long; however, their shape suggests that they were not as specialized for wood gnawing as those of modern beavers. It seems likely that *Castoroides* was not a dam builder, as modern beavers are.

Suborder Myodonta

The latest research shows that myomorph rodents share a common squirrel-like ancestor with the Castorimorpha (Fig. 8.5). The Myodonta are incredibly diverse, and include 1,137 living species in seven families. One branch of the Myodonta is the Dipodoidea, which includes the living jumping mice and jerboas of northern Africa and Asia. These creatures evolved the long hind limbs and tail for jumping completely independently of the American kangaroo rats and mice, which are in the Castorimorpha (family Heteromyidae).

The other branch, the Muroidea, includes the familiar rats, mice, muskrats, packrats, voles, hamsters, lemmings, dormice, gerbils, climbing rats, crested rats, spiny mice, bamboo rats, and hundreds of other species from the Northern Hemisphere and Africa known mostly to specialists. The muroid rodents can be traced back to the Oligocene, but they underwent an explosive evolutionary radiation in the Miocene, at which time they spread to nearly every continent except Antarctica. Today, rats and mice are universal in every part of world, especially where humans have created habitat for them to live.

ORDER LAGOMORPHA (RABBITS, HARES, AND PIKAS)

The other main branch of the Glires is the order Lagomorpha, the rabbits, hares, and pikas. The living members include about 58 species of rabbits and hares in 12 genera and 29 species of pika in one genus. More than 75 genera and 230 species are known from fossils. As discussed earlier, lagomorphs are recognized by their two pairs of gnawing incisors in the upper jaw, compared to only one pair in rodents (Fig. 8.2). Rabbits and hares have short tails compared to most rodents, long ears that help them hear predators and also radiate excess body heat, and long hind legs for their jumping gait.

They are also unusual in that they eat large quantities of vegetation though they lack a foregut fermentation chamber, as ruminants have, instead relying on cellulose-digesting bacteria in the caecum near the end of their digestive tract. To compensate, most rabbits will eat their feces and thus run the food through their gut a second time, so the bacteria from the first passage will have time to break down the cellulose and perform more efficient digestion.

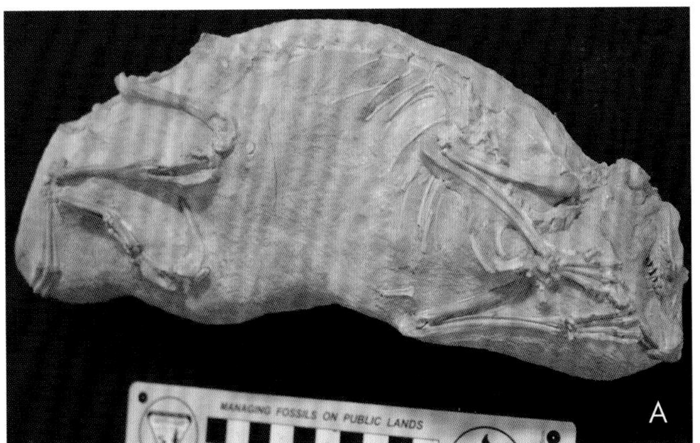

Lagomorphs are also famous for their fast breeding rates. Typically, the female can bear a large litter of babies, leave them in the burrow for safety while she feeds, wean them in about a month, and then immediately become pregnant again. This allows lagomorphs to have several large litters in a year, and in good times with abundant food or few predators, they can multiply at incredible rates.

Only two families of lagomorphs survive today (Fig. 8.5): the Leporidae (hares and rabbits), and the Ochotonidae (pikas). Leporids can be traced back to the early Eocene of Asia, where fossils such as *Shamolagus*, *Lushilagus*, *Dituberolagus*, and *Strenulalagus* represent extremely primitive rabbits just slightly more advanced than archaic Glires such as *Gomphos* and *Mimotona* from the Paleocene and early Eocene. In the late middle Eocene, leporids crossed the Bering Land Bridge and began to evolve in North America as well. Some of them, such as *Palaeolagus* from the Big Badlands of South Dakota (Fig. 8.9A), are extremely common in the fossil record, known from many complete skeletons as well as thousands of jaws. Leporids continued in low diversity during the Miocene, apparently outcompeted by ochotonids. When the spread of cold grasslands occurred in the Pliocene and Pleistocene, leporids underwent a huge, explosive radiation in diversity. They also spread to Africa in the late Miocene and were among the immigrants from North America to South America during the Pliocene, when they crossed the Panama land bridge and evolved new genera there (Fig. 5.4).

There are also some remarkable extinct rabbits. The most amazing is *Nuralagus rex* (Fig. 8.9B), found in deposits from about 3–5 Ma on the Mediterranean island of Minorca, near Spain. It weighed about 12 kg (26 lb), about six times as much as a modern rabbit! It was bulky and heavy, not slim and fast like modern rabbits. This body form is typical on many islands

Figure 8.9. Fossil lagomorphs. A. An articulated skeleton of *Palaeolagus*, a typical late Eocene rabbit from the Big Badlands of South Dakota. B. Comparison of a modern rabbit to *Nuralagus*, the gigantic rabbit from the Pliocene of Minorca.

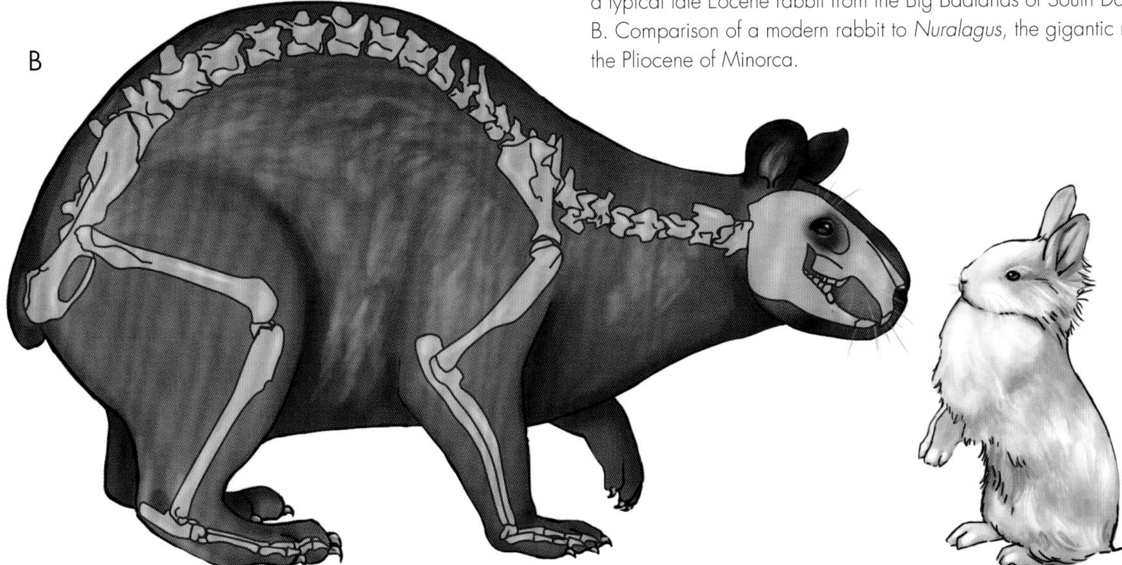

where there are few predators, allowing normally small and fast animals, like rabbits and shrews, to grow gigantic in the absence of strong selection pressure by predation.

Pikas (Fig. 8.10) or ochotonids are unusual for lagomorphs, in that they have smaller ears and lack the long hind legs for running on open ground, instead finding safety by living in crevices in rocks. To the untrained eye, these small lagomorphs look a bit like lemmings or voles. Most live on the tops of mountains above the tree line in Asia and North America, where the vegetation is sparse grasses and flowers. They spend their time harvesting food and storing hay piles in their burrows for the winter months, since they do not hibernate. Pikas emit a series of distinctive whistles to warn others of predators (mainly raptors and foxes) and to guard their territory against other pikas that might poach their food or hay pile.

The earliest fossil ochotonid is *Desmatolagus* from the middle Eocene (42.5 Ma) of Asia; by the late Eocene ochotonids had reached North America as well. By the late Oligocene and middle Miocene they had a great evolutionary radiation, and dozens of fossil genera have been found in North America and Eurasia, although most of these fossil species were larger than today's pikas, were built more like rabbits, and lived in the lowlands, where they outcompeted rabbits. For example, in the early Miocene there were 16 genera of ochotonids in the world and only three genera of leporids. By the Pliocene, there was a major extinction of the pikas, and the earlier ratio had reversed, leaving 18 genera of leporids and only four of ochotonids. Some speculate that it was due to the cooling and drying climate at the end of the Miocene, which decimated the savannas and their plants, and spread the modern grassy plains vegetation that rabbits and hares favor. Only late in their evolution did pikas diminish to one genus, *Ochotona*, and become restricted to the tops of mountains, after the glaciers retreated at the end of the last Ice Age.

Figure 8.10. *Ochotona princeps*, the North American pika, a specialist in the alpine meadows of the Rocky Mountains.

CHAPTER 9

LAURASIATHERIA: INSECTIVORES

Order Eulipotyphla and Other Insectivorous Mammals

Most Mesozoic mammals had teeth suitable for chopping up insects and other arthropods with a hard cuticle, which have always been an abundant food source for small mammals (Figs. 1.9, 1.10). Only later in their evolution did some mammals develop rounded blunt cusps for eating soft vegetation or sharp blade-like teeth for eating meat. Thus, almost all mammals are descended from "insectivorous mammals," a broad category for all mammals with an insectivorous diet.

Mammals that focus on tiny prey such as insects tend to be tiny themselves, and the smallest mammals known are insectivores, including the living Etruscan shrew, which weighs 2 g (0.07 oz) and reaches only 3.5 cm (less than 1.5 in) in length. Even smaller was the extinct *Batodonoides vanhouteni* (Fig. 9.1) from the early Eocene (53 Ma) of Wyoming. At 1.3 g (0.05 oz), it was the smallest mammal that ever lived.

ORDER EULIPOTYPHLA

Unfortunately, this primitive characteristic of an insectivorous diet was used as early as 1821 to define a wastebasket group, the order "Insectivora." It included not only the four closely related groups that make up the Eulipotyphla—the shrews, moles, hedgehogs, and solenodons (Fig. 9.2)—but also nearly every other primitive Cretaceous and early Cenozoic mammal group that had not modified its primitive insectivorous dentition for some other kind of diet. Indeed, there are many families and dozens of genera of extinct mammals that had insectivorous teeth but did not show any other anatomical specializations of the Eulipotyphla. For many years, paleontologists recognized the futility of "Insectivora" but then replaced it with other groups that were also wastebaskets for different assemblages of unrelated

Figure 9.1. Reconstruction of *Batodonoides vanhouteni*, the smallest mammal (living or fossil) ever discovered.

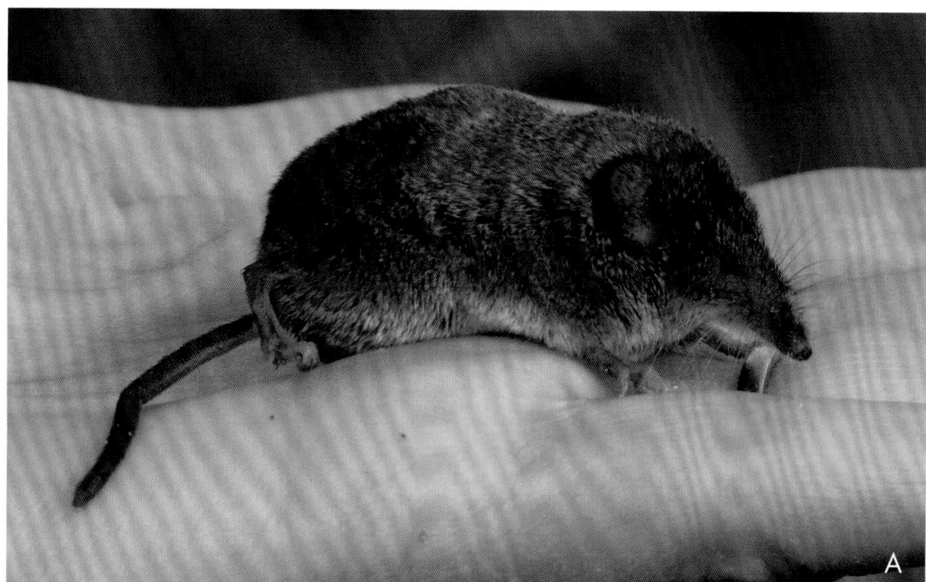

Figure 9.2. The four living groups of eulipotyphlans. A. Shrew.

insectivorous mammals, such as "Proteutheria," "Palaeoryctoidea," and the "Cimolesta." The evidence clustering these groups of extinct insectivores was just as tenuous as that for the original wastebasket "Insectivora."

In addition to extinct insectivorous mammals, there are a number of insectivorous living mammals that have been thrown into the "Insectivora" wastebasket. These included the tree shrews (families Tupaiidae and Ptilocercidae) and the colugos (family Cynocephalidae), which are archontans, related to the primates (Chapter 7); and the golden moles (family Chrysochloridae), tenrecs (family Tenrecidae) and elephant shrews (family Macroscelididae), which molecular evidence suggests are afrotheres (Chapter 6). For many years scientists used another designation, "Lipotyphla," for the closely related shrews, moles, and hedgehogs but also the golden moles and tenrecs, but the "Lipotyphla" are not closely related, either. The wastebasket names "Insectivora," "Proteutheria," "Lipotyphla," "Palaeoryctoidea," and "Cimolesta" are now considered obsolete. The only natural group of insectivorous mammals is the Eulipotyphla. There are many extinct groups of insectivorous mammals that are not in Eulipotyphla, but whose relationships are uncertain.

Family Soricidae (Shrews)

One of the most familiar insectivorous mammals is the shrew (Fig. 9.2A). The label "shrew" is applied to someone (particularly a woman) who nags, carps, and scolds people, as the lead character in Shakespeare's *Taming of the Shrew*, as if in imitation of real shrews, which constantly attack prey and even much larger animals. Although shrews are tiny and look somewhat like mice, they are very active and voracious predators with distinctive features of the teeth for eating arthropods and worms. In fact, shrews are active almost 24 hours a day hunting prey, because their tiny size means they are always losing body heat. They have a very high metabolic rate to compensate for the heat loss and must eat constantly (80–90% of their body weight every day) or they will die within a few days. During lean times with little food they may starve to death quickly or go into a form of torpor that resembles hibernation.

Shrews are famous for being ferocious for such tiny animals, often driving off much larger animals with their fearless attacks. Some even have venom in their teeth (among the few mammals that do) for subduing prey and attacking predators. They have poor eyesight but excellent hearing and smell, appropriate to a predator that hunts in the dark leaf litter, in dense underbrush, or in burrows.

The largest species is the Asian house shrew, which is about 15 cm (6 in) long and weighs 100 g (3.5 oz), while the smallest is the Etruscan shrew mentioned earlier. The shrews are a diverse group, with over 385 living species in 26 genera grouped into three subfamilies. This makes them the fourth largest family of mammals, after two rodent families and one family of bats.

The fossil record of shrews, however, is relatively poor because of their tiny size. Most are known only from isolated teeth and occasionally jaws recovered from screenwashing lots of sediment. True soricids first appear in the middle Eocene (43 Ma) of North America (*Domnina*). They spread to Eurasia in the middle Oligocene and eventually to Africa by the middle Miocene. Relatively few genera lived until the middle Miocene, when there was a big radiation of all three modern subfamilies in North America and a lesser diversification in Eurasia. Shrews have never been native to Australia, New Zealand, or Antarctica, and only one genus occurs in South America.

Family Talpidae (Moles)

Moles (Fig. 9.2B) are a familiar part of Western culture, well known as the blind furry little creatures that dig tunnels

Figure 9.2. The four living groups of eulipotyphlans. B. Mole. (See also next page.)

beneath our lawns, as a slang term for an embedded traitorous spy, and from Mr. Mole in *Wind in the Willows*. Contrary to popular myth, they do not eat the roots in their tunnel. Moles use the tunnel as a trap. Whenever an insect, a larva, or a worm falls in, the sound attracts the mole, which races down the tunnel to kill and eat its prey.

Moles have small weak eyes and most are nearly or completely blind. They spend almost their entire lives in darkness in their tunnels, and rely on sound, smell, and touch to sense their world. Their short thick fur protects them in their burrows, since it sheds dirt, and they have highly sensitive flexible snouts, often with strange star-like structures on them. Their most striking feature is their large powerful forelimbs with digging claws, which they use to create the burrows on which they depend.

There are 46 living species of moles, divided into 17 genera and three subfamilies. As discussed in Chapter 6, the golden moles of Africa are not moles at all; they are afrotheres that have converged on the mole-like body shape. There is a group of Australian marsupials, the notoryctids, that has converged on the mole lifestyle as well.

The fossil record of moles is sparse because their tiny teeth and jaws are hard to find, but they can be traced back to the late Eocene of Europe and North America. Some of the genera from the middle Miocene are still alive today, so their genera and species have remarkably long durations.

Family Erinaceidae (Hedgehogs)

Since ancient times hedgehogs (Fig. 9.2C) have been familiar creatures of the undergrowth in Eurasia and Africa. Today they are familiar to some because of the video-game character Sonic the Hedgehog. They are famous for rolling up into a ball and exposing their spiky hairs as protection, as well as burrowing or hiding in undergrowth. They are predators, catching insects, worms, and other small invertebrates, but occasionally they eat carrion, seeds, or fruit.

Hedgehogs are the largest members of the Eulipotyphla; the greater moonrat (which despite its name is actually a hedgehog, not a rodent) reaches 45 cm (18 in) long and weighs up to 1.4 kg (3 lb). In the geologic past there have been some surprisingly large hedgehogs, including the Miocene fossil *Deinogalerix* from Gargano Island in the Italian part of the Mediterranean (Fig. 9.3). This giant hedgehog was the size of a badger, with a skull as big as that of a large dog, and probably caught vertebrate prey. Its large size is probably due to the fact that it had little competition on Gargano Island from any other similar-size predatory mammals and so became large to fill this vacant niche.

There are 43 living species of hedgehogs and their relatives in 12 genera and two subfamilies. Their fossil record goes back to the Eocene of Eurasia, and they spread to Africa during the Miocene. Although none live in the New World today, there

Figure 9.2. The four living groups of eulipotyphlans. C. Hedgehog.

Figure 9.2. The four living groups of eulipotyphlans. D. *Solenodon*.

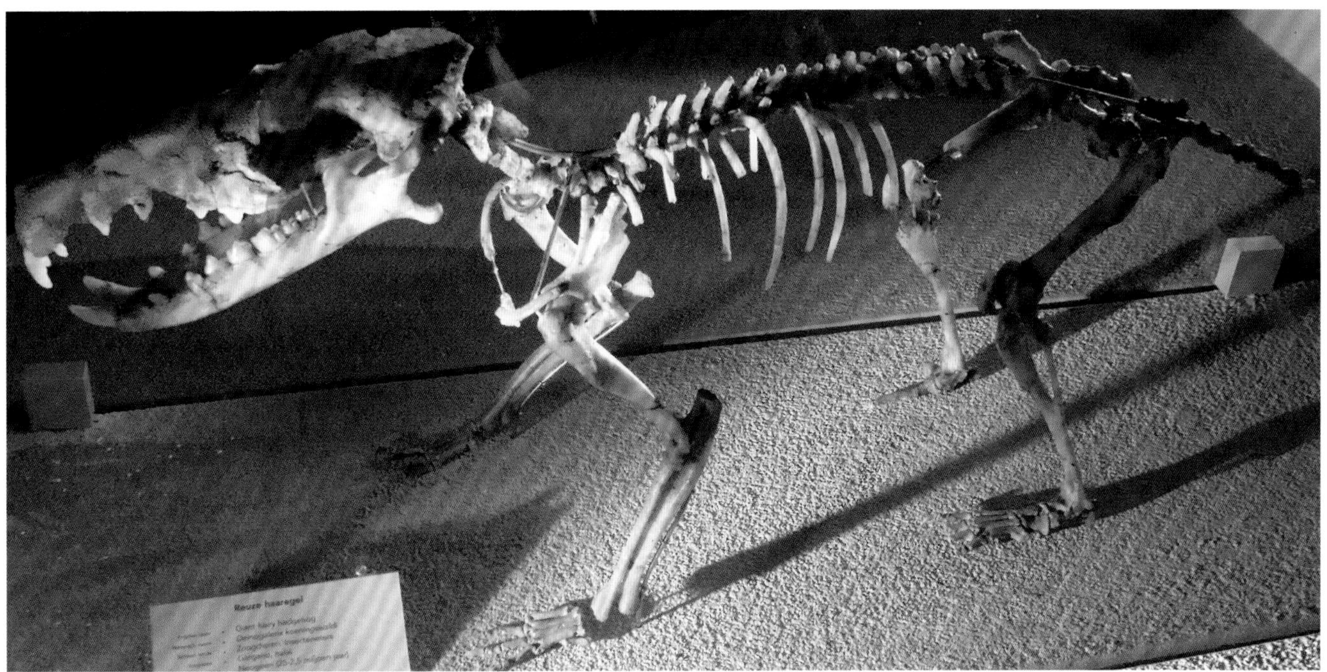

Figure 9.3. Skeleton of *Deinogalerix*, the hedgehog that reached the size of a dog or a badger. It evolved in isolation on Gargano Island in the Mediterranean, in a place that had no large predators to compete with it.

are hedgehog fossils going all the way back to the Paleocene in North America, and the group became quite diverse there in the late Oligocene and the Miocene.

Family Solenodontidae (Solenodons)

Solenodons are often called "living fossils" because they are primitive insectivorous mammals that resemble those that survived the Cretaceous extinctions unchanged (Fig. 9.2D). They have been allied with several groups, including tenrecs and even with rodents, but the most recent studies of their DNA suggest they are the most primitive group of the eulipotyphlans, not closely related to moles, shrews, or hedgehogs, and that they diverged from other mammals in the Late Cretaceous about 76 Ma.

Solenodons look like large and long-legged shrews, 32 cm (13 in) from nose to rump, weighing about 1 kg (2.2 lb), and sporting a long rat-like naked tail. They are strictly nocturnal and very secretive, living in dense brush and in burrows and emit a musky odor when threatened. Like some shrews, they have grooved teeth that can inject venom in their prey. They have a long flexible snout, which they use to sniff out prey, usually insects, worms, and small invertebrates. Today they survive on the islands of Hispaniola (Haiti and Dominican Republic) and Cuba, where they are endangered due to human disruption of their habitat and predation from introduced species such as the mongoose. There was also another, related genus, the chipmunk-size *Nesophontes*, on these same islands, but it went extinct in the 1930s due to human activity.

EXTINCT INSECTIVOROUS GROUPS

As mentioned already, there are many families of extinct mammals with an insectivorous diet, as indicated by their teeth, but their relationships to the living Eulipotyphla are not clear. Most fossil species are known only from isolated teeth and jaws, so not much can be said of their body forms or mode of life, nor can these be reliably reconstructed to look much different from a modern shrew.

Family Leptictidae (Leptictids)

One of the oldest groups of Cretaceous insectivorous mammals to survive into the Cenozoic was the Leptictidae. Evidence based on a number of Late Cretaceous fossils, such as the genus *Gypsonictops*, indicates that they were one of the first groups to branch off from the earliest placental radiation. This is apparent in their extremely primitive skulls and skeletons, with the typical insectivorous teeth and skull bones. In other ways, they were specialized, as is demonstrated by their having a peculiar pair of ridges along the top of the skull. The largest species reached almost 90 cm (35 in) long, including their long tails, the size of a large cat. Some are known from beautifully preserved complete skeletons, such as those of *Leptictis* from the Eocene and Oligocene beds of the Big Badlands (Fig. 9.4A). Although 21 leptictid genera are known from the Paleocene

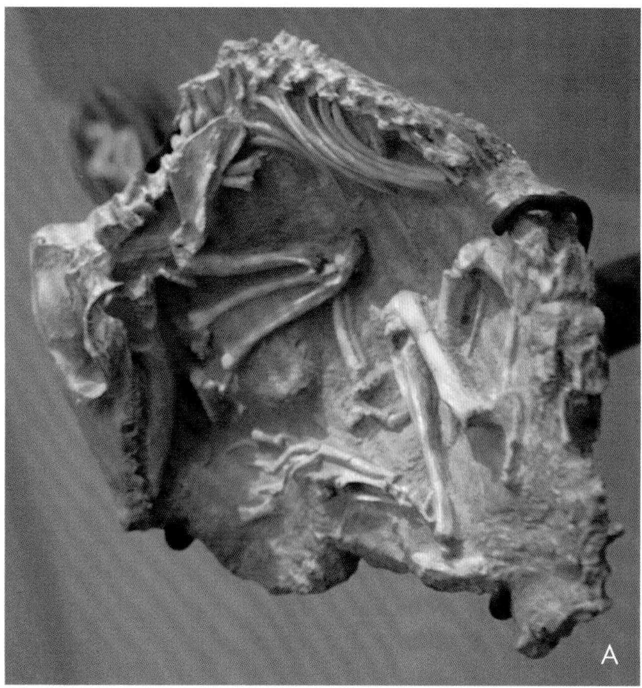

through Oligocene, *Leptictis* was the last of the family line, vanishing about 30 Ma.

One remarkable leptictid from the middle Eocene lake shales of Messel, Germany, is *Leptictidium* (Fig. 9.4B, C). Fossils of several different species preserve the entire articulated skeleton and even the body outline and hair impressions. This creature had a long slender snout and short proboscis for feeling around in the undergrowth for insect prey. Its relatively long hind legs would have given it a hopping, rabbit-like gait. *Leptictidium* had a remarkably long tail, probably for balance when hopping.

Family Apatemyidae (Apatemyids)

Although most apatemyids were mouse- to rat-size and insectivorous, there are a number of remarkable fossils that show some bizarre features. Known mainly from the Paleocene and Eocene of North America (but new specimens from Florida show they survived until the late Oligocene), they also reached Europe during the Eocene. Some of these apatemyids, such as *Apatemys* and *Labidolemur* from the middle Eocene lake beds of the Green River Formation in Wyoming and *Heterohyus* from the Messel deposits of Germany (Fig. 9.5A, B), are built like the living aye-aye,

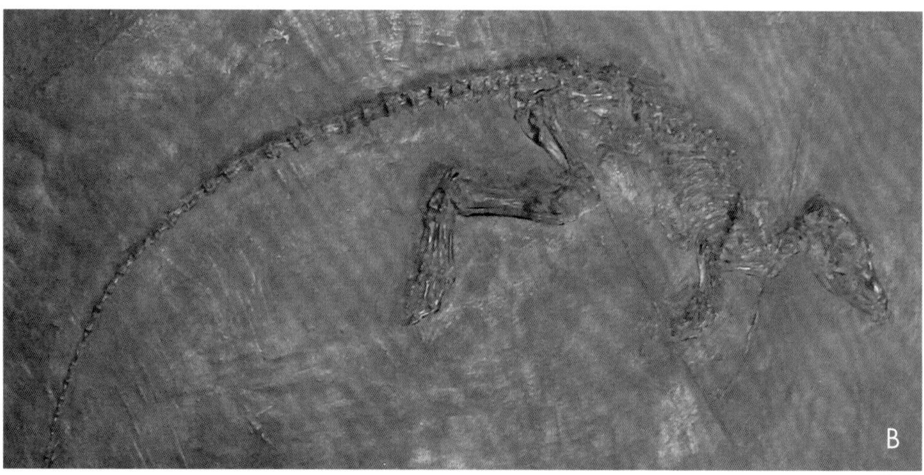

Figure 9.4. Leptictids were common insectivores during the early Cenozoic. A. A complete articulated skeleton of *Leptictis*, the common Oligocene fossil from the Big Badlands. B. *Leptictidium* was a long-legged hopping leptictid from the middle Eocene lake beds of Messel, Germany. This is a complete articulated skeleton preserved in death pose in the fine lake shales. C. A reconstruction of *Leptictidium* in life.

Figure 9.5. Apatemyids. A. A complete articulated skeleton of *Heterohyus* from the Eocene lake shales of Messel, Germany. It had an elongate middle finger, like the living aye-aye (see Fig. 7.7B). B. Reconstruction of *Heterohyus* probing a tree for grubs with its middle finger. C. The strangely shaped skull, with its peculiar front teeth, of the bizarre apatemyid known as *Sinclairella*.

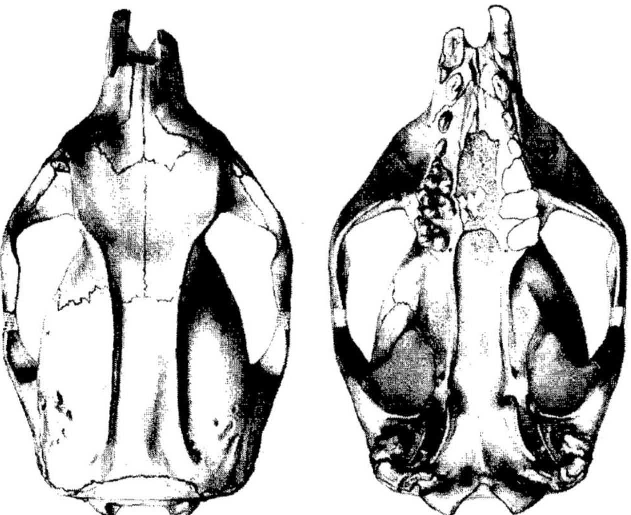

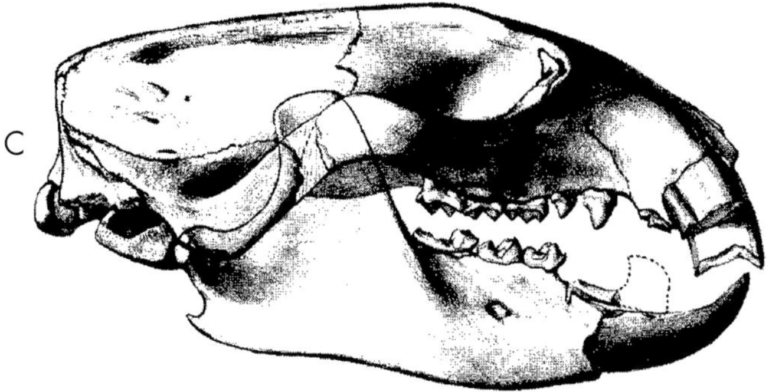

a nocturnal lemur with long front teeth for gnawing off bark and an extremely long middle finger for retrieving grubs (Fig. 7.7B). This is convergent evolution at work, since the aye-aye is a primate unrelated to *Heterohyus* and its kin. Convergence also produced a similar long middle finger adaptation in the marsupial *Dactyliopsis*.

An even more bizarre fossil is *Sinclairella*, one of the last of the apatemyids, found in the upper Eocene–lower Oligocene beds of the Big Badlands (Fig. 9.5C). It was described from a single fossil of the skull and jaws (now lost) that suggested it looked like a saber-toothed shrew, with a skull about 7.5 cm (3 in) long. Its long saber-like upper and lower incisors protruded forward from the mouth, and the upper incisors showed signs of wearing down to form chisels on their tips. What these peculiar creatures did for a living is speculative, since no skeleton has been found.

Family Pantolestidae (Pantolestids)

The family Pantolestidae first appeared in the early Paleocene of North America, where there were a number of genera that were only subtly different in their teeth from other insectivorous mammals. Soon, however, they evolved into a series of fossils that converged with raccoons or otters in their body form, from *Bessoecetor* in the middle Paleocene to *Paleosinopa* in the middle Eocene of the Green River Formation of Wyoming and *Kopidodon* and *Buxolestes* in the middle Eocene of Europe (Fig. 9.6). *Paleosinopa* is known from a complete skeleton from the middle Eocene Green River Shale of Wyoming (Fig. 9.6E). *Kopidodon* and *Buxolestes* are known from complete articulated skeletons from the Messel lake deposits. These were relatively large mammals for that time. They had a body length of 50 cm (20 in) and a powerful broad tail 35 cm (14 in) long and weighed up to 1.4 kg (3 lb). *Kopidodon* (Fig. 9.6A, B) was built much like a raccoon or mongoose, with a long bushy tail (the fur impressions are preserved in some specimens) and delicate fingers for tree climbing; its teeth suggest that it had a generalized omnivorous diet. *Buxolestes* (Fig. 9.6C, D) had long canines and crushing teeth that would have helped with eating clams and snails as well as fish; the crushing cheek teeth had thick enamel. *Buxolestes* also had strong, long-clawed forelimbs, built for swimming, just as living otters do. The long tail aided in swimming as well, and the strong neck muscles helped hold the head above water. Convergent evolution developed the otter-like body form in two different groups unrelated to otters: in these extinct pantolestids and in the living otter shrews.

OTHER INSECTIVOROUS GROUPS

Many other insectivorous fossil mammals of the Paleocene and Eocene (especially in North America) are known from isolated jaws and teeth, but there is little else to tell us much about their body shape or life habits. These include the lemur-like mixodectids (which have sometimes been allied with primates; see

A

Chapter 7), the very primitive palaeoryctids, the poorly known simidectids, and many others. There are some that are almost certainly eulipotyphlans, but again they are known largely from teeth and jaws and few skeletal parts, so most of them would look shrew-like if reconstructed. These include the geolabidids (such as the smallest mammal ever known, *Batodonoides vanhouteni*; Fig. 9.1). In this family there is one genus, *Centetodon*, that lasted for 34 m.y., one of the longest durations of any mammal genus known. Other groups include the archontan-like nyctitheres, the mole-like apternodontids, and many others.

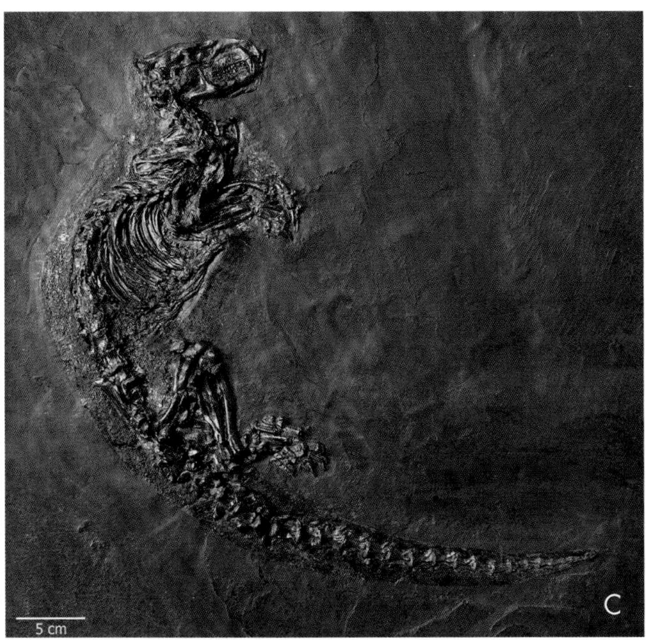

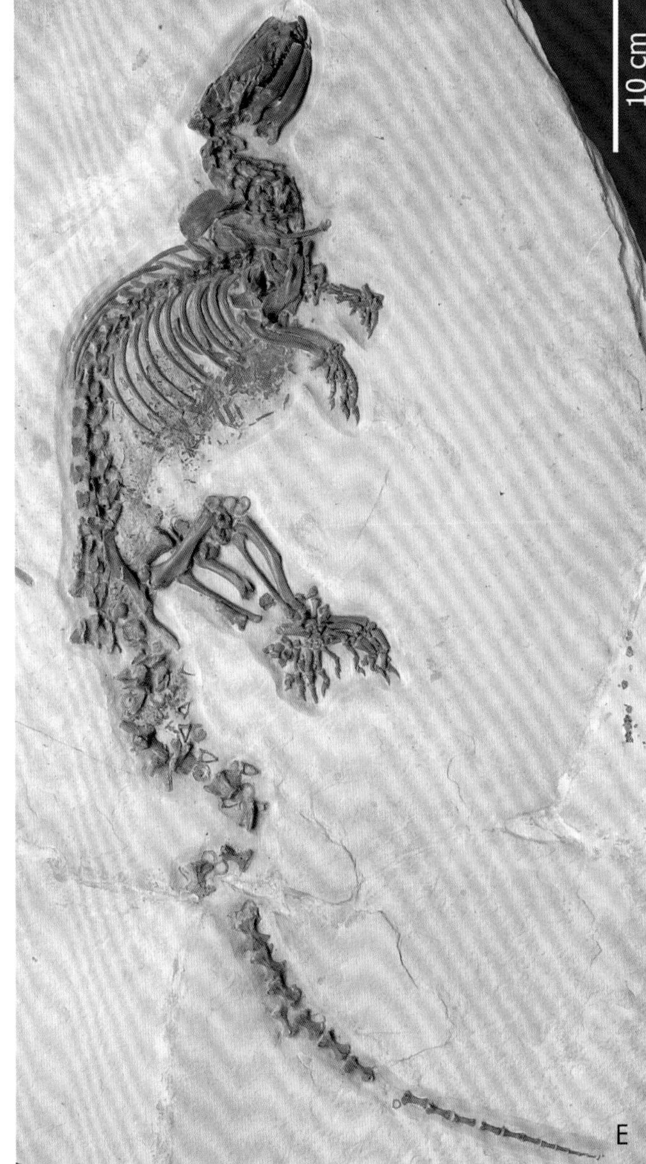

Figure 9.6. Pantolestids. A. Skeleton of *Kopidodon* from the middle Eocene Messel lake beds of Germany. B. Reconstruction of *Kopidodon*. C. Skeleton of *Buxolestes* from the Messel lake beds. D. Reconstruction of *Buxolestes*, which was otter-like in its size and shape. E. The middle Eocene pantolestid *Paleosinopa*, from the Green River lake shales of Wyoming

CHAPTER 10

LAURASIATHERIA: CHIROPTERA

Bats

Everyone is familiar with bats, since these winged mammals are found in every part of the world except Antarctica. We tend to associate bats with dark, creepy things and places (such as caves, Halloween, the legends of vampires like Dracula, or the "Dark Knight" Batman), or with insanity ("bats in the belfry" or "batty"). Many people are unjustifiably afraid of bats flying into their hair or attacking them, even though bats do not do so (except by accident when people block them from escaping a confined space). Vampire bats strike horror in many humans, even though they almost never feed upon us and live only in the tropical parts of Central America. Anywhere else in the world, bats are harmless to humans.

Putting aside these cultural issues, bats are actually amazing and highly successful mammals. They are the only mammals that

Figure 10.1. The skeleton of a modern fruit bat. Note how all the finger bones are highly elongated to support the wing membrane. Also apparent are the long toes for hanging upside down when resting and the portions of the feet and tail that attach to the wings.

truly fly. Most so-called flying mammals, such as "flying squirrels" and "flying lemurs," are actually gliders. Life has evolved multiple forms that have solved the problem of flight, including flying insects (which have thin, membranous wings with veins, unlike the wings of any vertebrate), pterosaurs (which had an extended fourth finger to support their wing membrane), and birds (which have fused finger bones and support their wings with feather shafts). By contrast, bat wings are an extension of the animals' hands, which have all four fingers (but not the thumb) elongated to support the wing membrane, or **patagium** (Fig. 10.1). The order Chiroptera literally means "hand wing" in Greek.

Because the "hand wing" is a unique solution to the problem of flight, this feature demonstrates that bats are a natural group. Evolution can come up with multiple solutions to the same problem (flight), depending upon the anatomical building blocks already available in the skeletons of the ancestors, as we see with each of these four groups. Like all fliers bats have very lightweight skeletons that are hollow and full of air, and they have many other specialized adaptations for flight as well.

Bats are also extremely diverse, with 1,240 living species in 186 genera and 18 families. Among living mammals, they are second in diversity only to rodents. Bats make up about 20% of all mammal species on earth. Because their skeletons are so small and fragile, bats are rarely fossilized. All but a few fossil bats are known from just their teeth and jaws, and scientists estimate that 60% to 90% of bat species have never left any fossil evidence. Bats are far more abundant in the tropics, and the conditions in these regions are notoriously bad for preserving small, delicate bones. It is only in Pleistocene caves that we find lots of preserved cave-dwelling bats, but this fossil yield represents a fraction of the diversity that once existed.

The living bats of the order Chiroptera are divided into two suborders, the Megachiroptera (fruit bats, "flying foxes," or megabats) and the Microchiroptera (the smaller bats or microbats, most of which are insect eating and echolocating). However, there are a number of bat fossils that do not below to either of these living orders but are more primitive relatives of both.

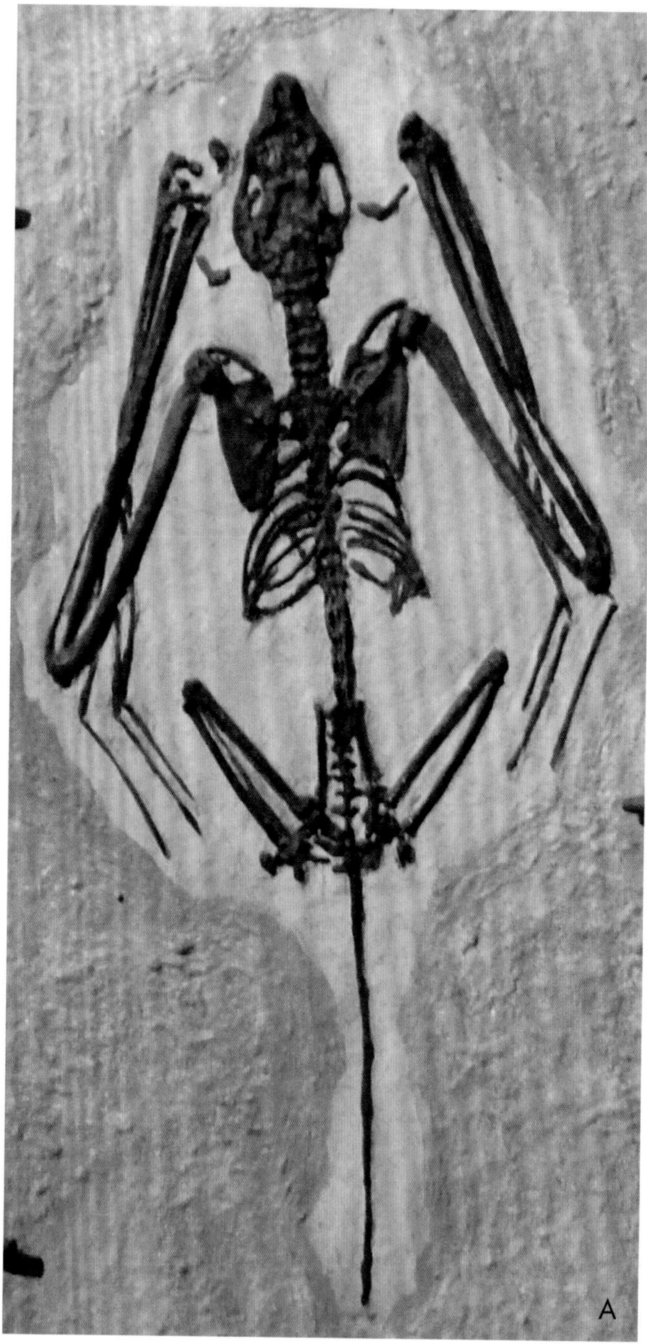

Figure 10.2. Some Eocene fossil bats. A. The most famous specimen of *Icaronycteris*, from the middle Eocene Green River Shale of Wyoming. (See also next page.)

Primitive Bats

The oldest fossil bats (*Wyonycteris*) come from the Paleocene and earliest Eocene, but they are just tooth fossils, since there are no deposits capable of preserving their delicate skeletons. The best fossils of early bats (Fig. 10.2) come from lake beds with extraordinary preservation, such as the middle Eocene of the Green River Formation of Wyoming (*Icaronycteris*) and the Messel deposits of Germany (*Archaeonycteris, Palaeochiropteryx, Hassianycteris*). Thus, a poor early fossil record hampers our understanding of bat evolution, for when good complete bat fossils finally appear they already have bat-like wings.

When you look closer, however, you find that these fossils are very primitive, and except for the wings they are not very bat-like. These specimens have primitive tooth crowns and relatively unspecialized skulls and brains, and they do not have the inner ear features necessary for echolocation. Their hands are not fully incorporated into the wing, and they have a long tail and lack the membrane between the tail and legs (**uropatagium**)

Figure 10.2. Some Eocene fossil bats. B. One of several beautifully preserved skeletons of different bats from the Messel lake beds in Germany. This one is *Palaeochiropteryx*.

seen in living bats. The fusion of the vertebrae and hip bones is much less developed than in modern bats, and there is no large keel on the sternum for strong flight muscles. Another, recently discovered Green River bat, *Onychonycteris*, which is an even less advanced transitional fossil, has many other features that are more primitive than those of any fossil or living bat. It had claws on all five fingers (living bats have them on only one or two), short broad wings more suited to a mixture of flapping and gliding rather than extended flights, and long hind legs and short forearms, suggesting that it evolved flight from a climbing way of life. Contrary to claims of creationists, these fossils are not "just bats," but are much more primitive, and without the wings they might not be identified as bats at all.

Suborder Megachiroptera

The megachiropterans, or fruit bats (Figs. 10.1, 10.3), range from species smaller than some microchiropterans to the largest "flying foxes," which have wingspans to 1.7 m (5.6 ft) and weigh up to 1.6 kg (3.5 lb). The latter are called "flying foxes" because their large heads have a long snout and ears much like those of a living fox. Unlike microbats, megachiropterans are daytime animals with large eyes and excellent sight that roost in trees, not in caves. Except for one species, they do not use echolocation to locate prey, since they feet on fruits, nectar from flowers, and nuts. They have no tail, while microbats may have a tail incorporated into their uropatagium. Most megabats land on a branch and crawl along with their hooked fingers to reach fruit, but smaller ones can hover in flight to feed on flower nectar with their long tongues. There are about 120 living species of fruit bats, and most live in the tropical jungle regions of Africa, southern Asia, and the eastern Pacific from Japan to Australia. Many of the tropical islands of the Pacific have their own endemic species of fruit bat, because all it took was one founding population to reach the isolated land, and then they diverged from their mainland ancestors.

Suborder Microchiroptera

The microchiropterans, or microbats (Fig. 10.4), are the more familiar group of bats. Most roost hanging upside down in caves or sheltered overhangs. They feed on flying insects at night and have poor eyesight, so they depend on echolocation to find their prey. Most are very small, just 4–16 cm (1.5–6 in) long. Nearly all feed on insects, except for three species of vampire bats and a few others that are active fishers or hunt birds, frogs, or other vertebrates. There are about 1,100 living species of microchiropterans, so they are by far the most diverse group of mammals on the planet other than rodents.

BAT ORIGINS

We have discussed the fossil record in terms of the oldest bat fossils, but to what other mammals are bats related? Lots of people call them "flying rats" (Germans call a bat a *Fledermaus*, "flying mouse"), but it is clear they are not related to rodents. For decades, bats were connected to archontans (primates, tree shrews, and colugos). In the 1980s, a few studies suggested that megachiropterans were more closely related to primates than they were to microchiropterans, implying that the entire specialized wing structure in both bat groups evolved convergently. But this only spurred much more detailed anatomical research, which confirmed the idea that the bat flight anatomy is unique and evolved only once.

Then molecular studies were done on most of the living bats, and their unique common ancestry was confirmed. Molecular data also showed that they are not close to primates at all, but are members of the Laurasiatheria (not the Euarchotoglires, which includes rodents and primates). The molecules suggest that they are distantly related to the ferungulate group, which includes the Ferae (carnivorans and pangolins) and the Ungulata (perissodactyls and artiodactyls).

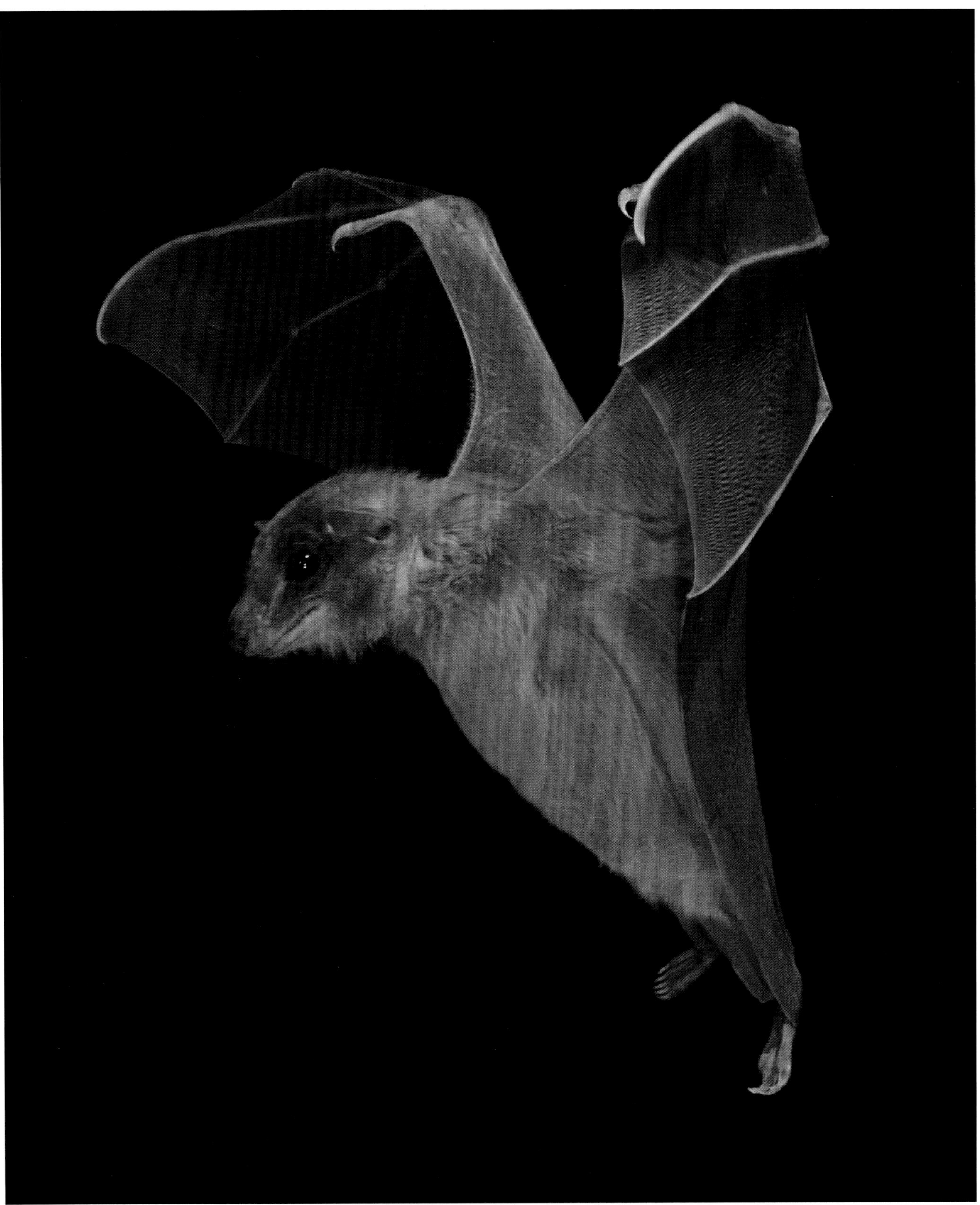

Figure 10.3. The large fruit bats (Megachiroptera) are often called "flying foxes" because of their fox-like head and ears. This is the Egyptian fruit bat, *Rousettus aegypticus*.

Figure 10.4. A microchiropteran bat grabbing prey in flight.

CHAPTER 11

LAURASIATHERIA: PHOLIDOTA

Pangolins, or "Scaly Anteaters"

When people first see a pangolin, they often mistake it for an anteater or a reptile, or are even reminded of a pinecone (Figs. 11.1, 11.2). Although a pangolin has an anteater-like body (and feeds on ants and termites), it is covered not with fur but with rows of plate-like scales, hence the nickname "scaly anteater." The scales are made of the same kind of keratin as hair and fingernails, and their edges are sharp. Pangolins curl up in a ball when threatened, and the scales form a bristling, forbidding defense against predators. Pangolins have a long tube-like snout and a sticky tongue almost 40 cm (16 in) long for snagging ants

Figure 11.1. The Chinese pangolin, *Manis pentadactyla*.

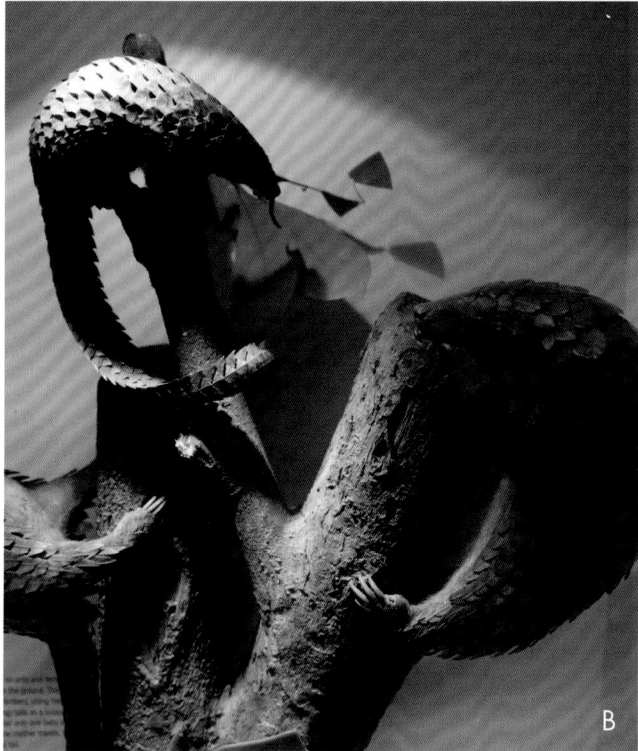

Figure 11.2. Pangolins. A. The giant pangolin, *Manis gigantea*. B. The tree pangolin, *Manis tricuspis*.

and termites and pulling them out of their nests. Their long front claws and powerful limbs are used for digging into those nests, as well as for climbing trees and for defense when necessary. Some species are strictly arboreal (Fig. 11.2B), using their prehensile tails to hang on while they rip open tree bark and nests for insects. Others, such as the giant pangolin (Fig. 11.2A), live on the ground and dig burrows for protection. Pangolins are mostly nocturnal, and they have poor eyesight, relying on sound and smell for getting around instead.

ORDER PHOLIDOTA (PANGOLINS)

The order Pholidota contains only one living family, the Manidae, and only one living genus, *Manis*, with eight species. These eight species are distributed all over Africa and southern tropical Asia. Sadly, their defenses are useless against humans, and they are hunted for bushmeat in Africa. In Asia they are near extinction because Chinese and Vietnamese cultures consider them a delicacy. In some Asian folk-medicine traditions, their body parts are considered to have curative value. (This is false, of course.) In 2013 one of many poaching operations was seized with over 10,000 kg (22,000 lb) of pangolin meat in its possession, representing thousands of unfortunate animals. Even strict protections for these amazing "living fossils" are not enough to stop the huge pressures of poachers and smugglers, and these animals are likely to be extinct in the wild before 2050.

The relationship of pangolins to other mammals has long been puzzling and controversial. Their order, Pholidota, has

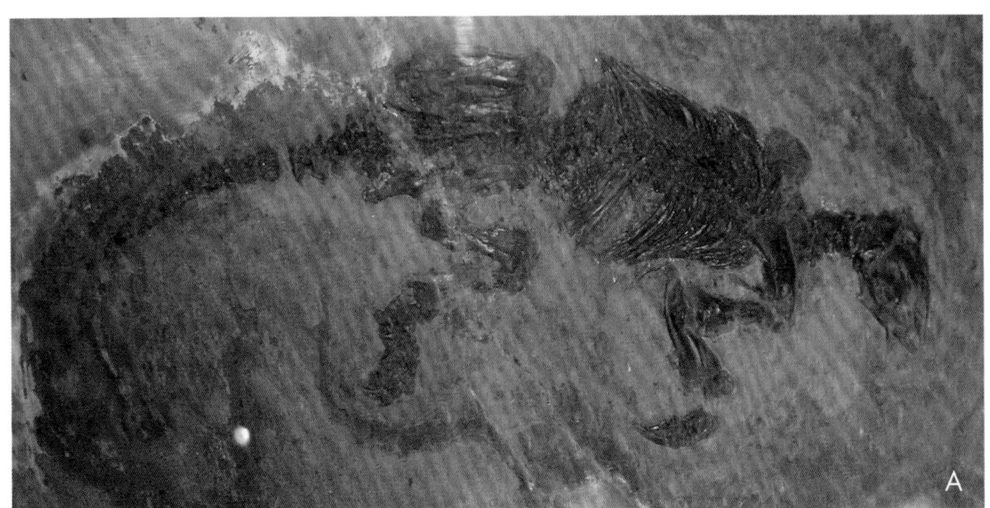

Figure 11.3. *Eomanis*, a middle Eocene pangolin from the Messel lake beds of Germany. LEFT: A. A complete articulated fossil. CENTRE LEFT: B. Restoration of *Eomanis* in life.

often stood in isolation in mammal classifications, or in some cases has been clustered with anteaters and aardvarks in the "Edentata" (see Chapter 5). This was their classification in the late 1990s when phylogenies were being produced from anatomical features. Then molecular phylogenies showed that pangolins are actually closely related to Carnivora, and not to Xenarthra or aardvarks or any other ant-eating group of mammals. Their anteater-like anatomy is purely evolutionary convergence.

Rare creatures like pangolins have only a sparse fossil record, and very little was known of their evolutionary history until the 1970s. *Neomanis* from the Oligocene and Miocene of France was known, but it was very incomplete. Then in 1970 a fossil found in the upper Eocene beds of Flagstaff Rim, Wyoming (misidentified as a juvenile carnivore skull), was identified and described by Robert Emry of the Smithsonian Institution as the pangolin *Patriomanis*. This discovery proved that pangolins had once lived in the Americas. In 1978 the beautifully preserved skeleton of a complete pangolin (even including the scales) from the middle Eocene Messel lake beds of Germany, named *Eomanis*, showed that the pangolin body plan was already established 50 Ma (Fig. 11.3). Another fossil from the same deposits, *Eurotamandua*, was once considered to be related to the tree-dwelling tamandua anteaters (Xenarthra), but is now recognized as a scaleless pangolin, since it does not have xenarthrous vertebrae (Fig. 11.4). These fossils, representatives of the family Manidae, tell us that pangolins were spread across the northern continents by the middle Eocene.

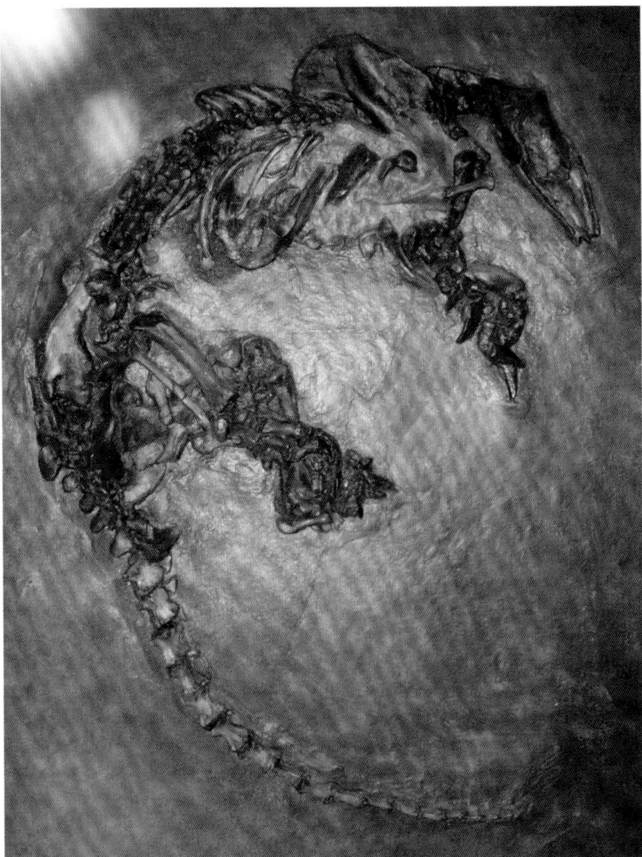

Figure 11.4. A complete skeleton of *Eurotamandua*, a pangolin with no scales but a furry skin instead, from the middle Eocene lake beds of Germany.

PALAEANODONTS

The pangolins, based on both their unique, bizarre anatomical features and molecular evidence, are clearly closely related to one another, and are not part of the Xenarthra (sloths, armadillos, and anteaters). However, there are numerous other fossils that seem similar to anteaters, armadillos, or pangolins in having developed the specialized snout and teeth for insect eating, and sometimes the enlarged front claws as well. They have been placed in the Xenarthra, the Pholidota, or in their own order, and their relationships have been controversial and confusing for a long time. None of them has the specialized xenarthrous vertebrae, so they are almost certainly not related to modern anteaters or armadillos, no matter how similar their ecology. The latest research suggests that they are primitive relatives of the pangolins, which were widespread across Eurasia and North America by the middle Eocene.

The oldest and strangest of these fossils is the bizarre *Ernanodon*, from the late Paleocene of China (Fig. 11.5). When it was first published in 1979, it was considered to be an anteater, and scientists thought that it was evidence that Xenarthra had originated in Asia before reaching South America. It had a deep skull and jaws with large tusks, yet relatively small cheek teeth. Its front feet bore large claws, suggesting affinities with the diggers in the Xenarthra. Yet it lacked the specialized xenarthrous vertebrae, present in all sloths, armadillos, and anteaters, and was missing other key skeletal features as well.

Another long-mysterious group was known as the Palaeanodonta, known from the Paleocene to the Oligocene of Eurasia and North America. One family, the Metacheiromyidae, includes armadillo-size fossils such as the Paleocene *Escavadodon* and *Propalaeanodon*, the early Eocene *Palaeanodon*, and the

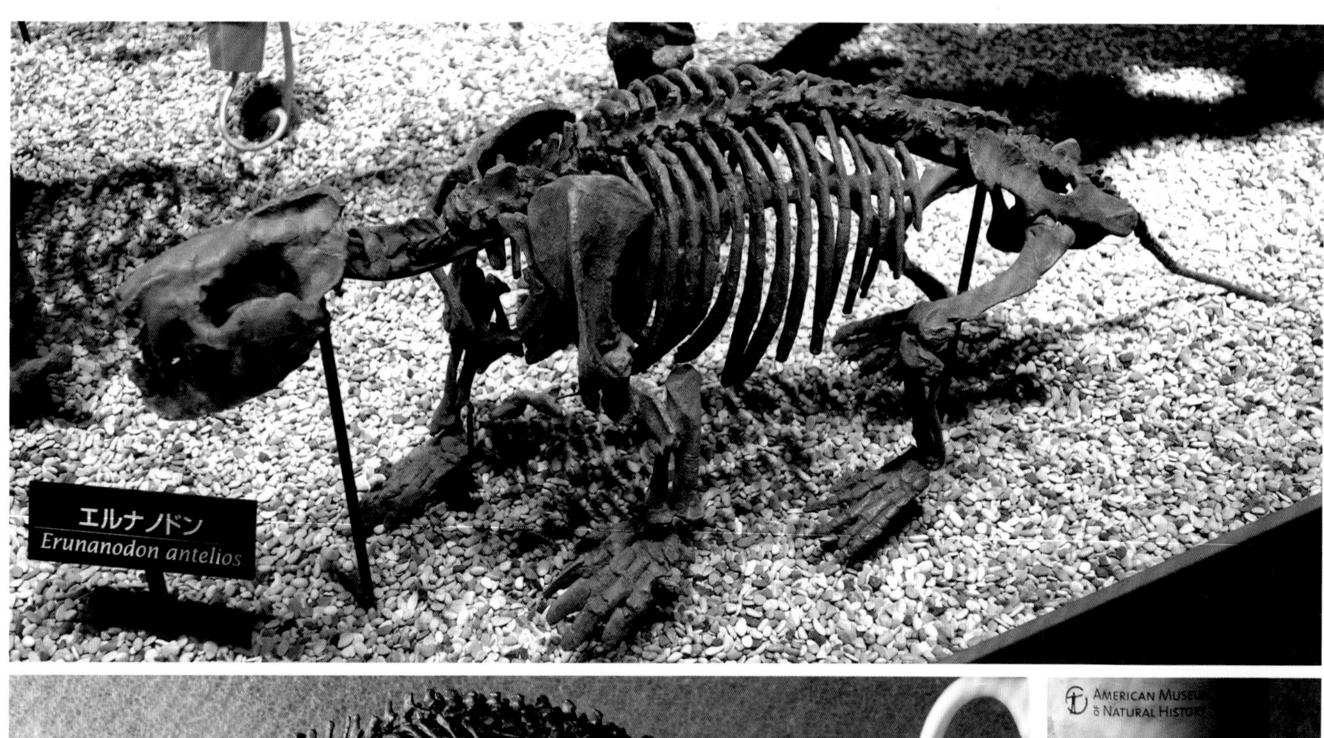

TOP: Figure 11.5. *Ernanodon*, the peculiar Paleocene Chinese fossil thought to be distantly related to palaeanodonts and epoicotheres.
ABOVE: Figure 11.6. Complete skeleton of the middle Eocene palaeanodont *Metacheiromys*.

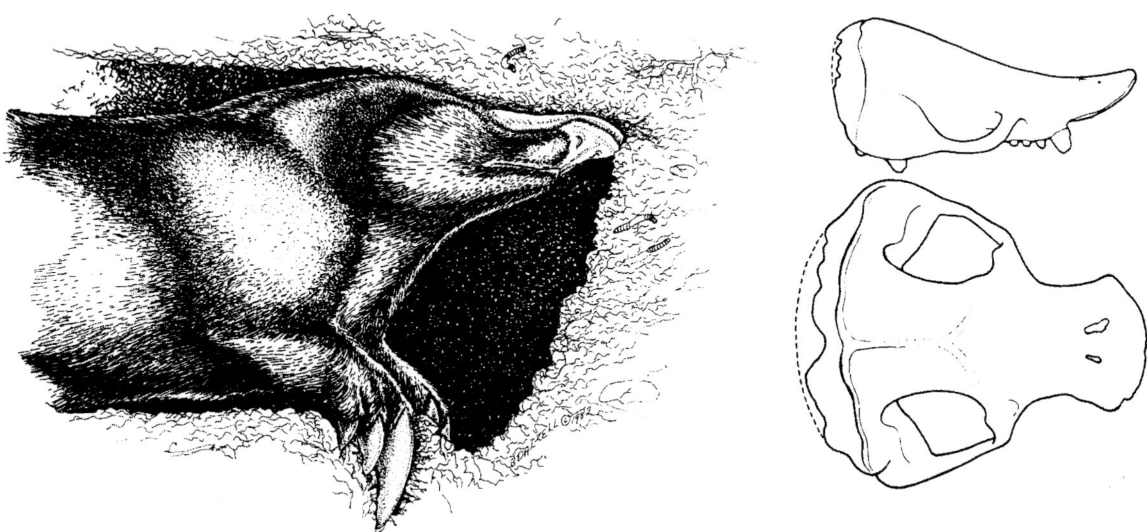

Figure 11.7. The peculiar burrowing epoicothere *Xenocranium* from the upper Eocene beds of the Big Badlands. Left: Reconstruction of the animal digging, using its spoon-like snout. Right: Skull.

middle Eocene *Metacheiromys* (Fig. 11.6). Like armadillos and anteaters, palaeanodonts had a long narrow snout and highly reduced peg-like cheek teeth, suggesting an ant-eating diet, and strong fore claws for digging up insects and other food. Unlike anteaters, however, they had large canine tusks, and the fossils suggest they used horny pads at the tip of the snout to crush their food, rather than chewing with their tiny cheek teeth. Their skeletons seem to be suited for digging and burrowing as well as insect eating.

The other family of the Palaeanodonta is the Epoicotheriidae. These tended to have shorter, wider skulls and retained more teeth, and their skeletons are almost like those of moles in their extreme specializations for burrowing. Not only were their forelimbs extremely strong, holding powerful digging muscles, but the tip of the snout on the most bizarre representative, *Xenocranium*, had a broad scoop on it, suggesting it scraped away soil in its burrow with its nose (Fig. 11.7)

All of these early anteater-like creatures from North America and Eurasia vanished by the end of the Eocene, and were not replaced with any other anteaters and diggers for a long time. In North America, for the most part, the anteater niche was never filled again. In Eurasia, the niche had few occupants until the later part of the Miocene. The burrowing niche was eventually filled by rodents such as gophers in North America, and other rodent groups in Eurasia, but only the moles have combined the niche of eating bugs and worms and burrowing.

CHAPTER 12

LAURASIATHERIA: CARNIVORA AND CREODONTA

Predatory Mammals

CARNIVORES, CARNIVORANS, AND CREODONTS

Predatory mammals are an essential part of the modern food web, and perform the important ecological function of keeping populations of prey species healthy and in check. They are also common as pets, from the dogs and cats that are the most popular pets of all to unusual carnivores such as ferrets. But the predatory niche has had many different occupants throughout

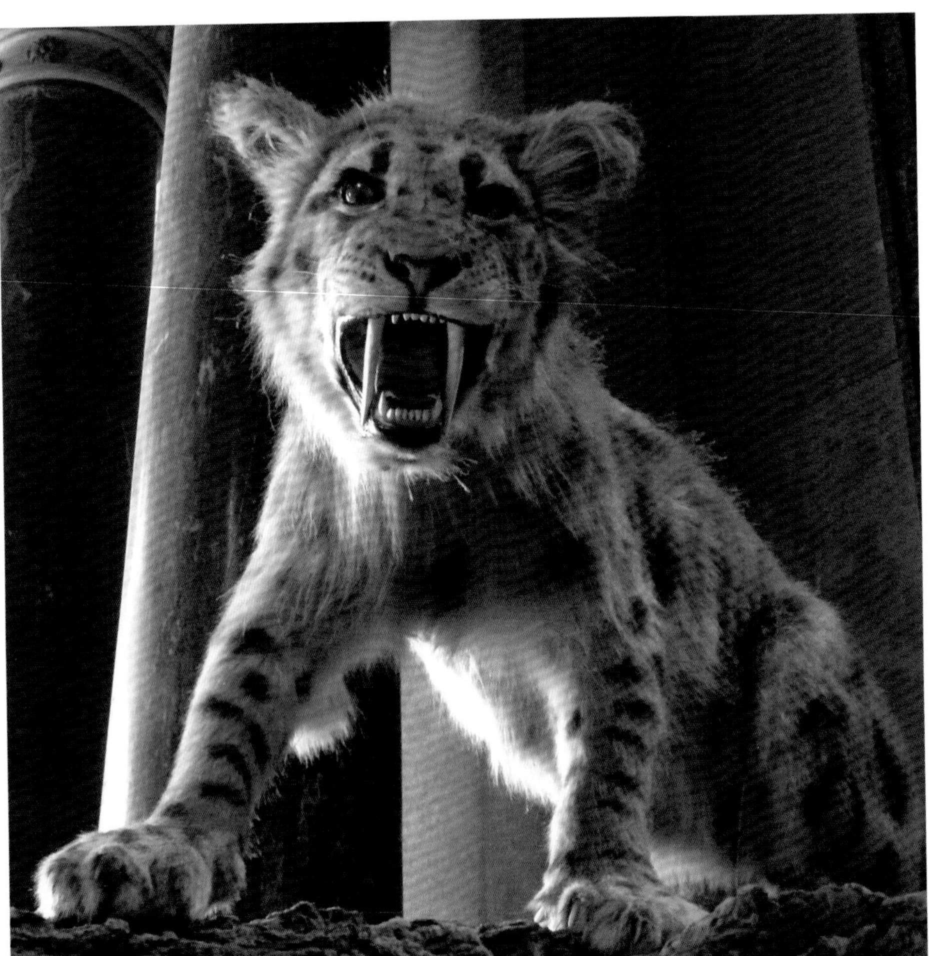

Figure 12.1. Life-like model of the Eurasian saber-toothed cat *Megantereon*.

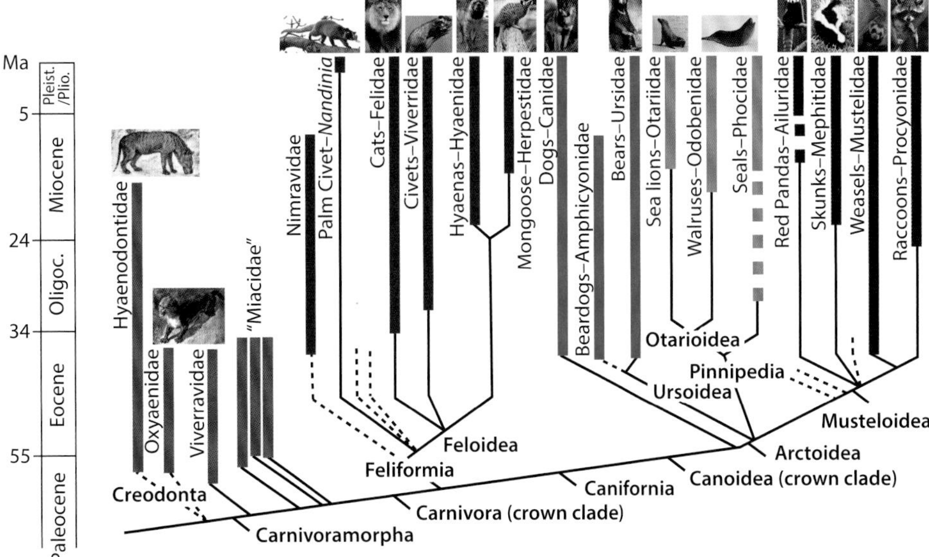

Figure 12.2. Family tree of the creodonts and carnivorans.

the Cenozoic. In the Paleocene and earliest Eocene, most of the mammalian predators were small, and the large predator niche was occupied by gigantic "terror birds" such as *Diatryma* in North America and *Gastornis* in Europe. In the Miocene and Pliocene in South America, the largest mammalian predators were only wolf-size, and another group of birds, the phorusrhacids, independently evolved into large land predators.

Among mammalian orders, several different groups have evolved to fill the role of predator. In Australia, where the only native mammals were the pouched marsupials, several groups of wolf-like, lion-like, and cat-like predators evolved from possum-like ancestors (see Chapter 3). A similar evolution occurred in South America, where marsupial predators evolved that were remarkable mimics of wolves and hyaenids (the sparassodonts, or borhyaenids) and saber-toothed cats (the thylacosmilids). In fact, the saber-tooth niche was occupied not only by marsupials, but also by creodonts and two different groups of carnivorans, including many different kinds of true cats (Figs. 1.4, 12.1). In the Northern Hemisphere archaic hoofed mammals (the mesonychids) became the largest predators of the early Cenozoic (see Chapter 13).

Through most of the Cenozoic, there have been two main groups of mammals that performed the roles of predators (Fig. 12.2). They are the extinct archaic predators of the order Creodonta and the living group of predators, the order Carnivora (cats, dogs, bears, hyenas, raccoons, weasels, seals, sea lions, and many others). In discussing these animals, we must be careful with our words. The word "carnivore" refers to any meat eater (including carnivorous plants such as the Venus flytrap), but a "carnivoran" is a member of the order Carnivora. Not all carnivorans are strictly carnivorous, either. Some are omnivores (such as bears and raccoons), and there is even one, the giant panda, that is a specialized herbivore, feeding on bamboo.

In addition to having large stabbing canine teeth, almost all predatory mammals have cheek teeth modified for slicing meat. The teeth have evolved from the simple rounded cusps in their ancestors into sets of shearing blades that occlude precisely with the opposite blade to perform a scissor-like action. In addition, all carnivorans and creodonts (but not other carnivorous mammals) develop a specialized pair of enlarged cheek teeth known as **carnassials** (Fig. 12.3). This pair of upper and lower teeth serves as the main cutting and breaking tool of the jaw. The carnassials are specially adapted for slicing tough meat and tendons and, in many carnivores, for crushing and breaking bone. A glance in the mouth of any cat or dog immediately reveals these crucial teeth. If you watch a dog chew a bone, you will see it use the side of its mouth to bring these powerful carnassial teeth into action.

Carnassials occur in both creodonts and carnivorans, but with a key difference in their position. Except for seals and sea lions (which have all their teeth modified into simple conical pegs for fish catching) and walruses (which have blunt crushing teeth), all carnivorans have carnassial shear between the last, or fourth, upper premolar, and the first lower molar (Fig. 12.3A). By contrast, the carnassial shearing pair is farther back in the jaws of creodonts (Fig. 12.3B): either between the first upper molar and the second lower molar (in oxyaenids) or the second upper molar and the third lower molar (in hyaenodonts).

The position of these crucial teeth is key to identifying which order a fossil belongs to. It may also help to explain why carnivorans became so successful and replaced creodonts. The forward position of their carnassials gave their skulls greater evolutionary flexibility in shape and feeding style, so their faces could shorten and other parts of the tooth row could be reduced without affecting the crucial carnassial teeth. Cats, for example, have almost completely lost the teeth behind their carnassials and have very short faces compared to other carnivorans. In another case, bears have modified their post-carnassial molars into teeth specialized for crushing and other functions.

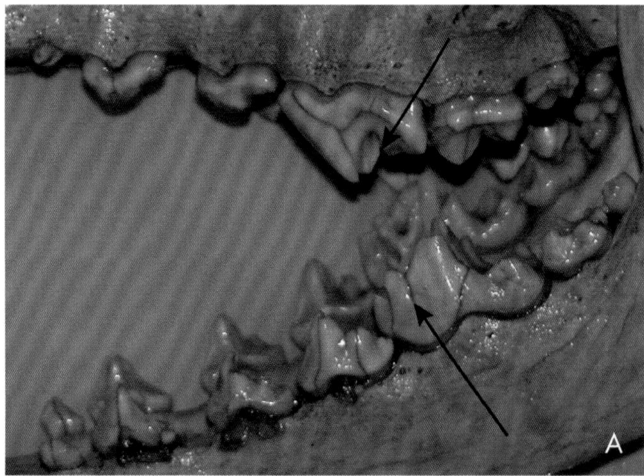

B

Figure 12.3. Teeth of carnivorous mammals. A. In carnivorous mammals a pair of cheek teeth, called the carnassials, is enlarged to form extra-strong bite force for cutting tough meat and tendons and crushing bones. In this wolf, as in all members of the Carnivora, the enlarged fourth upper premolar shears against the huge first lower molar to form a carnassial shear. On the crest of each carnassial is a V-shaped notch called an *akis*, which acts as a scissors to cut tough material. B. Side view of back part of the jaws of a creodont, *Hyaenodon*, closing in occlusion. The diamonds show the distinct akis between the V-shaped notches in each molar. Instead of a larger carnassial tooth near the front of the molar row (as in carnivorans), the larger carnassial teeth in creodonts are the last molar (second upper and third lower molars).

In the creodonts, the position of the carnassials in the back of the mouth limited the options of what evolution could do with their teeth, so they kept the same stereotyped dentition and never developed any true cat-like forms (short snout, almost no molars behind the carnassials), bear-like species, or other highly specialized groups.

ORDER CREODONTA

The creodonts were the first major group of Cenozoic mammals to adopt the predatory lifestyle. Originally mistaken for carnivorans, or considered to be ancestors of carnivorans, creodonts, more recent analyses show, were an early experiment in predator evolution. They were relatively primitive in their body form and never developed the wide range of different shapes (bear, dog, weasel, cat) seen in carnivorans.

As discussed earlier, the creodonts' carnassial teeth were positioned near the back of the mouth, in the second or third molars (Fig. 12.3B). This limited the flexibility of their teeth to perform other functions. Creodont limbs were suited only for running and jumping, and creodonts could not rotate their paws to face inward to grab prey, as cats can. Thus, creodonts had to hunt and kill using their jaws alone, another important restriction on their evolutionary flexibility.

Creodonts dominated the meat-eater niches during the Paleocene and Eocene, but by the late Oligocene had vanished from North America and Europe, though they straggled on in Asia and Africa until the late Miocene. The reason for their extinction is not known. It seems likely that the more advanced carnivorans outcompeted them in many parts of the world, while creodonts were stuck without much evolutionary flexibility in their jaws, teeth, or limbs. The last known creodonts were species of *Hyaenodon* and *Dissopsalis*, which vanished from Asia and Africa about 11 Ma.

There are two families of creodonts:

Family Oxyaenidae (Oxyaenids)

The oxyaenids were the most primitive group of carnivorous mammals, appearing in the early Cenozoic. There were four subfamilies with 13 genera and dozens of species, found in both Eurasia and North America. Oxyaenids were the first group to occupy the carnivorous niche in the Paleocene, and they persisted until the middle Eocene in both Asia and North America.

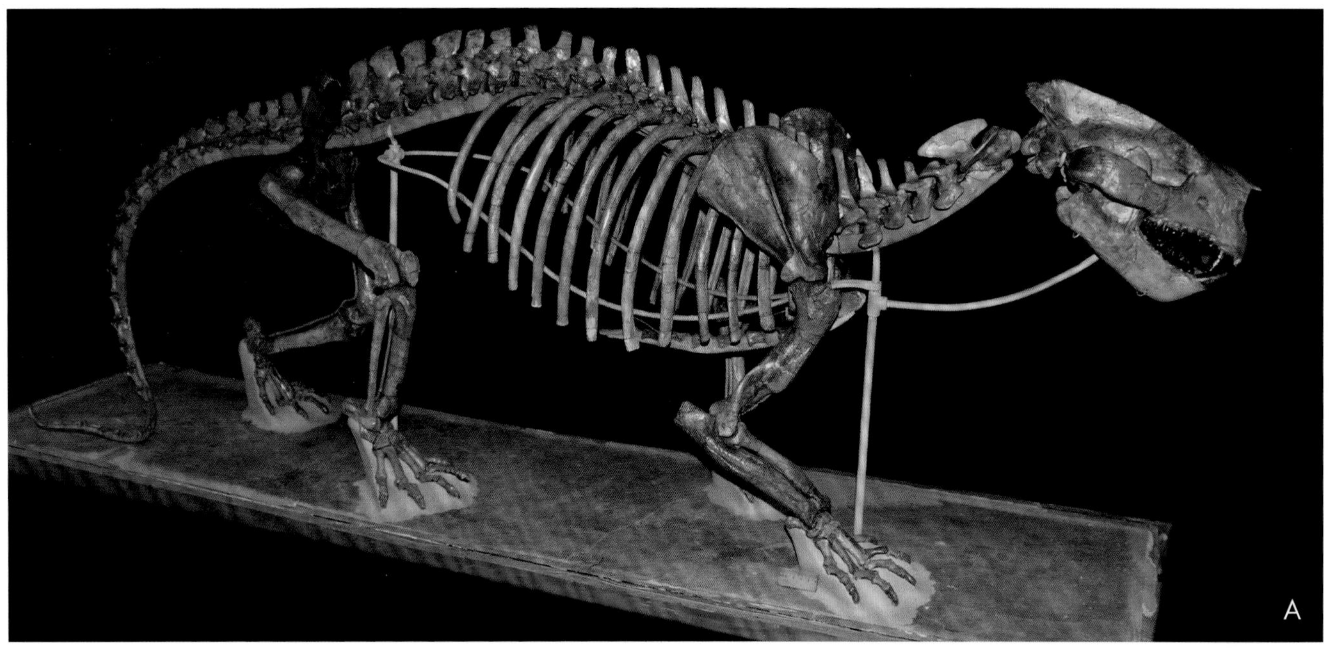

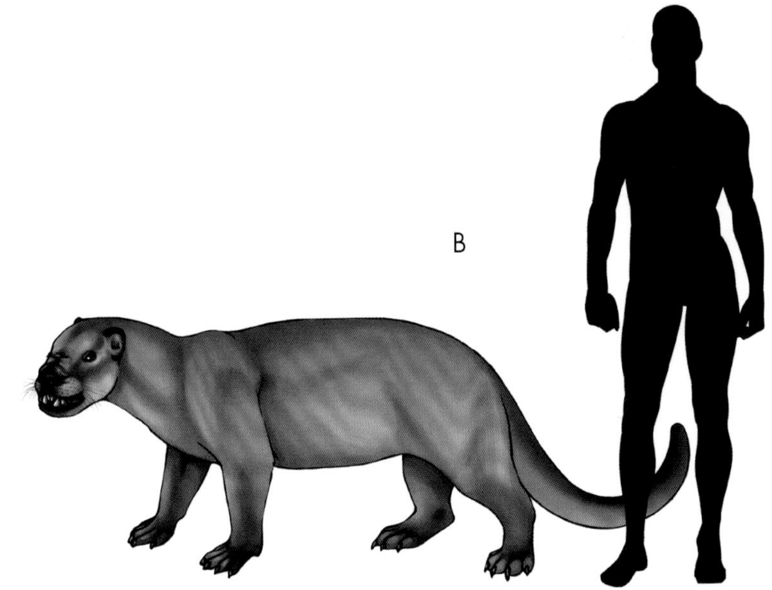

Figure 12.4. Oxyaenids. A. Skeleton of the cougar-size oxyaenid *Patriofelis*. B. Reconstruction of *Patriofelis*. (See also next page.)

Most oxyaenids had unspecialized skeletons, with low, massive, flat skulls, small brains, long tails, and generally short robust limbs, so they were ambush predators, not fast runners. The earliest oxyaenids, from the early Paleocene, such as *Tytthaena*, were small weasel-size creatures much like some of the insectivorous mammals of the Late Cretaceous. *Oxyaena*, from the late Paleocene and early Eocene, was built like a wolverine, with powerful robust limbs and a flexible body up to 1 m (3.3 ft) long. *Patriofelis*, from the middle Eocene of North America, had a big robust skull and reached up to 1.8 m (6 ft) in length and 90 kg (200 lb) in weight, making it about the size of a modern lion or panther (Fig. 12.4A, B).

The most impressive of the oxyaenids was *Sarkastodon*, from the late middle Eocene of Mongolia. It had a huge skull equipped with robust teeth that were suitable for bone cracking as well as meat eating. No postcranial bones are known, but its skull (Fig. 12.4C) was twice as long as that of *Patriofelis*, indicating that the entire animal was probably about 3 m (10 ft) long and weighed about 800 kg (1800 pounds), the size of a very big bear.

Although most oxyaenids were stereotyped in the heavy-bodied lion-like predator shape, one lineage (*Machaeroides* and *Apataelurus*) developed long saber-like canines, converging with the two lineages of saber-toothed carnivorans (nimravids and true cats) and the saber-toothed marsupial *Thylacosmilus*

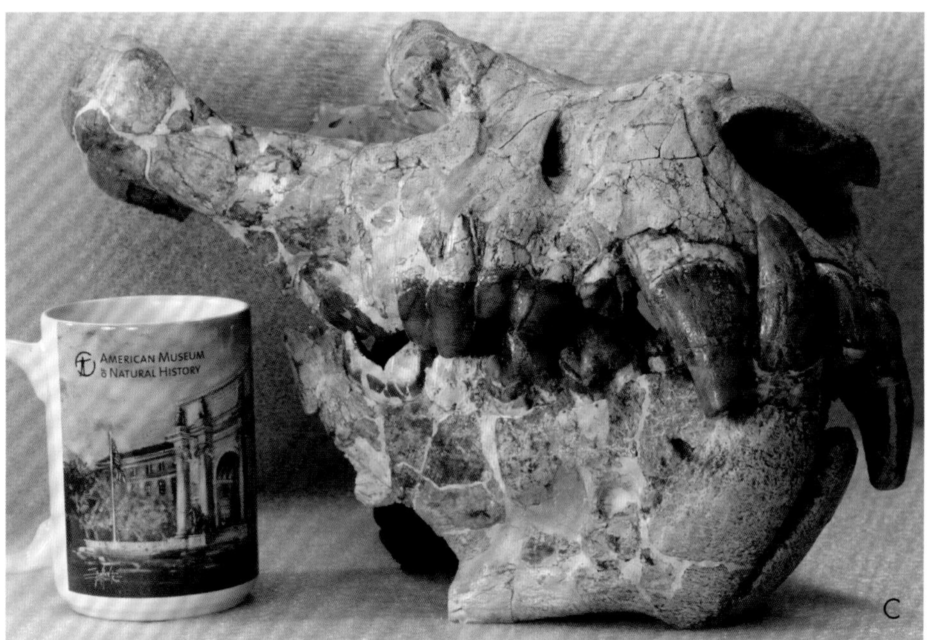

LEFT: Figure 12.4. Oxyaenids.
C. The partial skull of the huge, bear-size oxyaenid *Sarkastodon*, from the middle Eocene of Mongolia.

BELOW LEFT: Figure 12.5. The creodont *Hyaenodon* and relatives. A. Skeleton of the common large Big Badlands Oligocene species *Hyaenodon horridus*.

(Fig. 1.4). Despite the large saber-like canines, the saber-toothed oxyaenids had long snouts and never developed the short faces seen in other saber-toothed carnivores. They were also not as big as other saber-tooths. *Machaeroides* was only about the size of a bobcat, weighing about 10–14 kg (22–30 lb).

Family Hyaenodontidae (Hyaenodonts)

The other main lineage of creodonts is the hyaenodonts, whose name means "hyena tooth" in Greek. Their name is misleading, because they were creodonts and not related to hyenas, which are members of the order Carnivora. Nor was their anatomy particularly hyena-like. Instead, their teeth were adapted mostly for specialized shearing and lacked the robust bone-crushing molars found in true hyenas. Hyaenodonts had more delicate, wolf-like skeletons and slender limbs compared to hyenas and especially compared to oxyaenids.

Hyaenodonts arose in Asia in the late Paleocene and quickly spread across the northern continents in the early Eocene, when the weasel-like proviverrines were the most common predators in most size categories. Most hyaenodontids were in the 5–15 kg (11–33 lb) weight range, comparable in size to small to midsize dogs. An exception was the middle Eocene *Hemipsalodon* from the Clarno Formation of Oregon, which was cougar-size, weighing about 200 kg (440 lb).

Even bigger was the Oligocene Mongolian creodont *Hyaenodon gigas*. It was 1.4 m (5 ft) at the shoulder, about 3 m (10 ft) long, and weighed about 500 kg (1100 lb). Many different species of *Hyaenodon* evolved (Fig. 12.5), and they were highly successful predators from the middle Eocene to the late Miocene, a span of about 26 m.y. (longer than just about any other fossil mammal). Hyaenodontids were among the most common predators in the early Oligocene in North America and also occurred in Eurasia. By the late Oligocene, *Hyaenodon* had vanished from both North America and Europe, pushed out by more advanced carnivorans. But hyaenodontids persisted as important predators in Asia and Africa well into the late Miocene. The biggest of these was *Hyainolouros* (Fig. 12.5B; *Megistotherium*, also pictured, is possibly the same genus as *Hyainolouros*), a bear-size predator up to 3 m (10 ft) long and weighing about 300 kg (660 lb). It hunted large prey in Africa and Eurasia during the middle Miocene, 15–11 Ma. Although oxyaenids were extinct before 37 Ma, the hyaenodontids persisted long after them, vanishing about 11 Ma in Africa and Asia.

Figure 12.5. The creodont *Hyaenodon* and relatives. B. Reconstruction of *Hyaenodon horridus* (left foreground), *Hyaenodon gigas* (right foreground), *Megistotherium* (left background), and *Hyainolouros* (right background). The latter two may represent the same genus.

ORDER CARNIVORA

The order Carnivora is a diverse group, containing about 280 living species in 13 families and about 10 times that number of extinct species. Carnivorans have the largest range in sizes of any mammalian order, from the tiny least weasel, measuring only a few inches long and weighing as little as an ounce, to the huge polar bear, weighing up to 2000 kg (4400 lb), to the gigantic elephant seal, which can weigh 5000 kg (11,000 lb) and reach 7 m (23 ft) long. As pointed out earlier, carnivorans have a number of specialized adaptations for the predatory life, from the large canines for grabbing and stabbing prey, to the cheek teeth adapted for slicing meat. Most also have long sharp claws for fighting, slashing, and other functions, plus digestive modifications for a diet of meat and acute senses (especially smell, sight, and hearing) for detecting prey and hunting it down.

For a long time, the classification of order Carnivora was plagued by the problem of convergent evolution, since certain tooth shapes and body forms are highly influenced by a specialized diet. These shapes evolved over and over again, and were not a good indicator of evolutionary relationships. For example, the saber-tooth condition evolved independently at least four times in the mammals (Fig. 1.4): once in the marsupials, once in the creodonts, and twice independently in the Carnivora. But in recent years, paleontologists and anatomists have focused on more informative parts of the anatomy not tied to dietary specializations, such as the bones of the braincase and the middle and inner ear region. From these data, they deciphered the large-scale patterns of carnivoran evolution. Then in the late 1990s molecular biologists began to analyze all the living species and discovered a family tree among the Carnivora that matched what had been suggested by the anatomy, so the phylogeny of the order (Fig. 12.2) can be considered robust and confirmed.

These studies make clear that there are two main branches of the Carnivora: the Feliformia (cats, hyenas, mongooses, civets, and their relatives) and the Caniformia (dogs, bears, raccoons, weasels, and their relatives, plus the seals and sea lions). Both groups evolved from primitive carnivorans of the Paleocene and Eocene that are lumped into wastebasket groups: the "miacids" (the fossils closest to the Caniformia, mostly early to middle Eocene in age) and the "viverravids" (the fossils closer to the Feliformia, mostly Paleocene in age). Most "miacids" and "viverravids" were small creatures about the size and proportions of weasels or mongooses (Fig. 12.6). Like weasels, mongooses, and martens, they apparently hunted both on the ground and in the trees, and possibly burrowed as well. There were about a dozen genera in North America, where they originated; they eventually spread to Eurasia. True carnivorans were overshadowed by the much bigger creodonts during the Paleocene and most of the Eocene and did not begin to diversify in size or shape until the creodonts vanished in the late Eocene and Oligocene in North America and Eurasia.

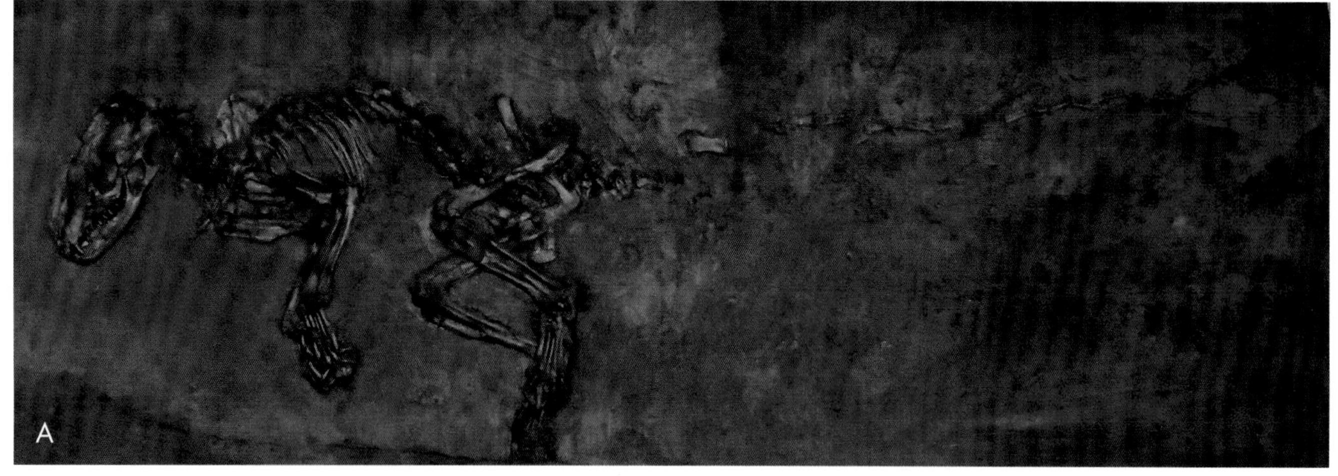

Figure 12.6. The early "miacid" carnivoran *Paroodectes*. A. Complete skeleton from the Messel lake beds of Germany. B. Reconstruction of the animal in life.

Family Nimravidae (Nimravids, or "False Cats" or "Paleofelids")

As the "miacids" began to vanish, one of the first groups of carnivorous mammals to replace them was the Nimravidae, appearing in the late middle Eocene (about 40 Ma) in both North America and Eurasia (Figs. 12.2, 12.7, 12.8). Their bodies, skulls, and teeth were extremely cat-like, so early paleontologists assigned them to the family Felidae. However, detailed studies of the inner ear, skull, and braincase anatomy show that nimravids were not true cats at all but a separate family that was the first to occupy the cat-like niche and did so before true cats (Felidae) even evolved. This means that the nimravids are among the most amazing examples of convergent evolution ever seen. Although many paleontologists think nimravids are primitive Feliformia, others point to features that ally them with Caniformia. If this conclusion is true, then they would be dog relatives that evolved to mimic cats. Since nimravids have long been extinct, there is no way to resolve the issue with molecular evidence.

Figure 12.7. Nimravids, or "false cats." A. Skeleton of the saber-toothed nimravid *Hoplophoneus*, showing its convergent evolution on the saber-toothed cat body form.

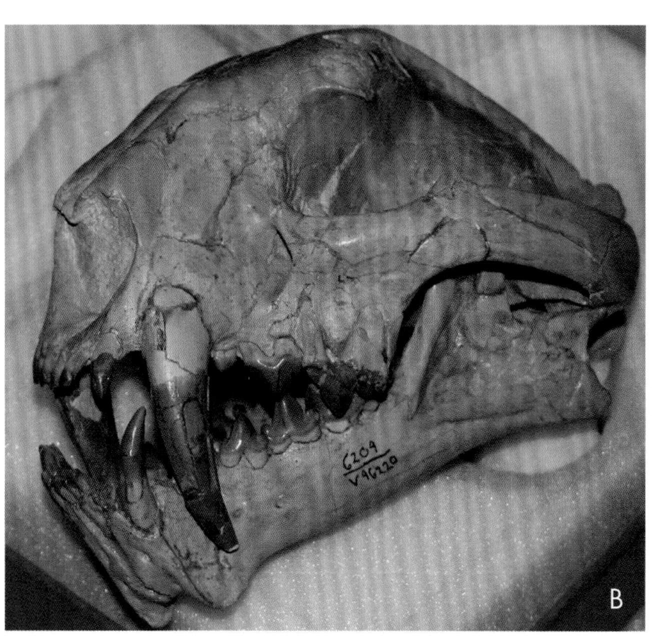

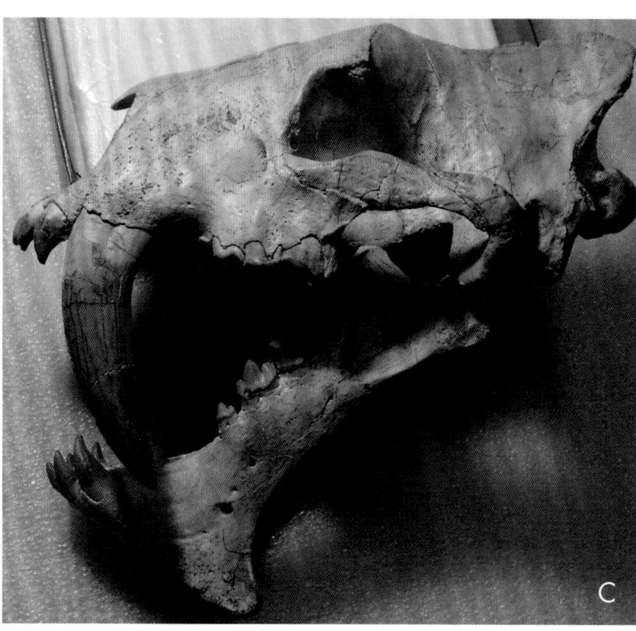

Figure 12.7. Nimravids, or "false cats." B. Skull of the "dirk-toothed" nimravid *Dinictis*, with its shorter, more conical sabers. C. Skull of *Hoplophoneus*, with its more scimitar-like narrow-bladed sabers. D. Reconstruction of *Hoplophoneus* (right) and *Dinictis* (left). (See also next page.)

D

Figure 12.7. Nimravids, or "false cats." E. The late Miocene nimravid relative *Barbourofelis*.

After their first appearance in the middle Eocene, nimravids continued to evolve through the late Eocene and Oligocene and perform the cat-like role as ambush predators, before true cats evolved. They are the most common predators in the fossil beds of the Big Badlands. Some were the size of house cats, but many were as large as leopards or even lions. A number of them (*Hoplophoneus*, *Pogonodon*, *Eusmilus*, and several other genera) developed full-length blade-like saber teeth, completely independently of the familiar saber-toothed cats, which are true felids (Figs. 1.4C, 12.7A, 12.8, C, D). Others, such as *Dinictis*, were "dirk-toothed" predators, with conical stabbing canines, like those of a lion, but longer (Fig. 12.7B, D). In North America, nimravid diversity peaked in the late Oligocene, about 28 Ma (as shown by the fossils in the John Day Formation of Oregon). Nimravids then vanished from North America around 26 Ma, resulting in a "cat gap" (total absence of cat-like predators) for 7.5 m.y. of the late Oligocene and early Miocene, until true cats arrived from Eurasia about 18.5 Ma.

Nimravids were also common in the Oligocene of Europe and Asia, where they persisted long after the North American nimravids vanished. They lasted in Eurasia until the late Miocene, finally disappearing 7 Ma. A possible relative of the Nimravidae is the Barbourofelidae (seven genera and 16 species), an extraordinary group of false saber-tooths. The barbourofelids evolved in Africa in the early Miocene and migrated from Africa to Europe three times, where they became important elements of the Eurasian middle and late Miocene carnivore fauna. One genus, *Barbourofelis* itself, escaped Eurasia and became established in North America during the middle and late Miocene (12–6 Ma). Its extraordinary skull (Fig. 12.7E) shows the full development of the saber-toothed skull, complete with flanges. Yet this saber-tooth condition evolved independently of that in the long-extinct Oligocene nimravids as well as that of the true saber-toothed cats of the family Felidae. When the barbourofelids vanished from Eurasia about 9 Ma and from North America about 6 Ma, the false saber-tooths were gone forever.

Feliformia

One major branch of the carnivorans is the feliforms, the cats, hyenas, civets, mongooses, and their relatives (Fig. 12.2). Although these animals have some features in common (retractile claws in many, and a tendency to a shortened snout and reduced teeth behind the carnassials), the Feliformia is defined by uniquely specialized features in the skull region and braincase. The validity of this group was supported in the 1990s and the years since, when molecular analyses showed that all feliforms are closely related to one another.

Family Felidae (True Cats)

The true cats (Fig. 12.8, 12.9) are familiar to us from the familiar wild cats—lions, tigers, leopards, jaguars, pumas, bobcats, lynxes—as well as our house cats. Altogether there are 41 living species of felids divided among eight genera, plus at least twice that many fossil species genera, found in every continent except Australia and Antarctica.

Figure 12.8. The evolutionary history of the nimravids, or "false cats," (branches lower left and bottom) and the felids, or true cats (branches center and upper right).

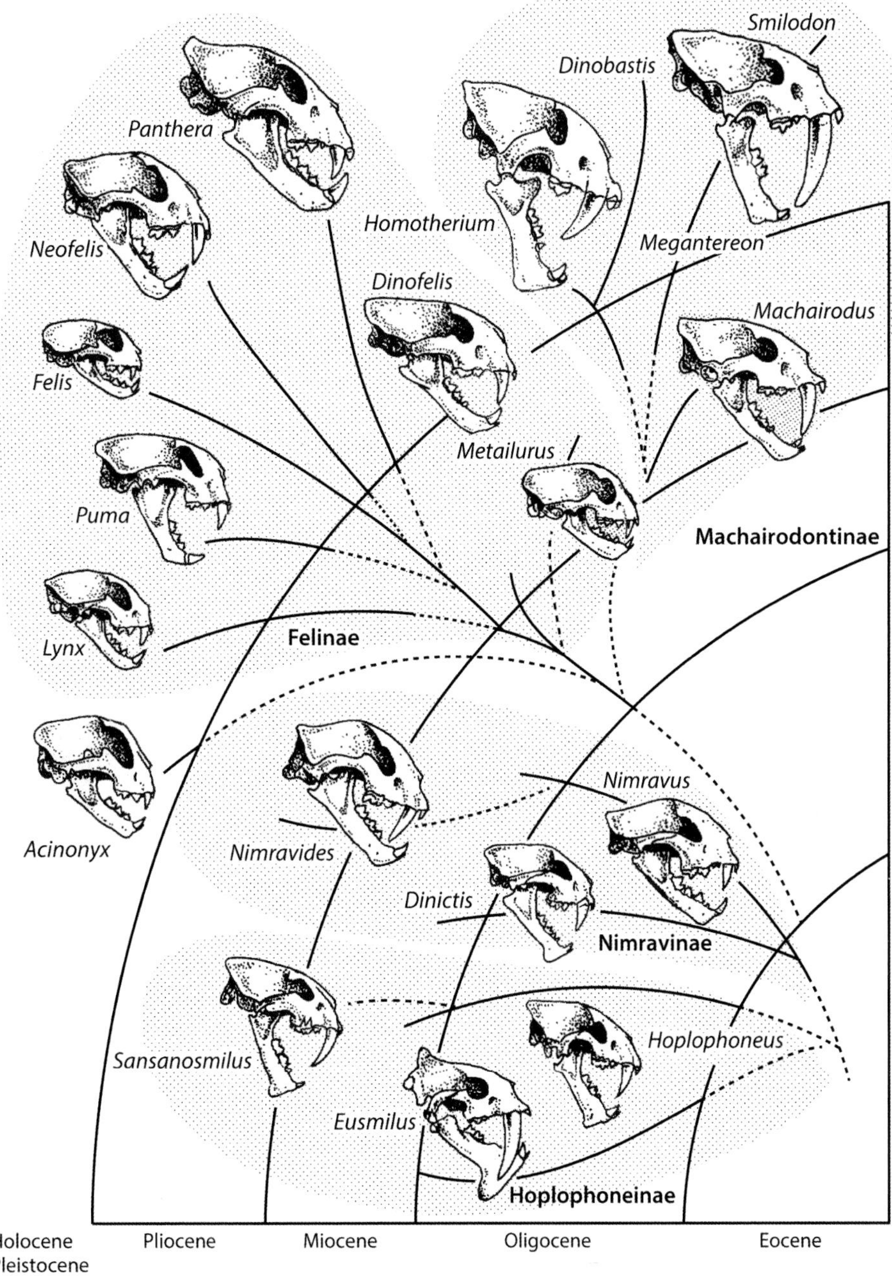

The first true fossil felid was *Proailurus* ("first cat" in Greek), which appeared in Eurasia about 25 Ma in the late Oligocene. About the size of a large house cat, it had many similarities to a mongoose or civet as well, and may have been largely a tree-climbing hunter. It evolved into *Pseudaelurus* ("false cat" in Greek, even though it is a true cat), a more advanced predator known from Eurasia from 20 Ma to about 10 Ma in the late Oligocene and lasting until the middle Miocene. When *Pseudaelurus* crossed the Bering Land Bridge to North America about 18.5 Ma, it ended the "cat gap" and began the big evolutionary radiation of cats in North America (in parallel with the radiation in Eurasia). *Pseudaelurus* was slightly larger than a house cat, but had short legs and a long trunk like a civet, suggesting that it was also a tree climber.

From *Pseudaelurus* several different lineages of cats evolved. The main lineage (subfamily Felinae) includes most of the living cats familiar to us today, from lions and tigers to cheetahs to cougars and bobcats and lynx (Fig. 12.8). A second extinct lineage, the subfamily Machairodontinae (Fig. 12.8), represents the wide spectrum of saber-toothed cats, including the familiar Ice Age fossil *Smilodon* (Fig. 12.9). At least 12 genera and 73 species of machairodonts are known besides *Smilodon*, including the

Figure 12.9. *Smilodon fatalis*, the famous saber-toothed cat from La Brea tar pits.

spectacular *Megantereon* and *Homotherium* (Fig. 12.1), which were tiger-size and even larger than *Smilodon*. Machairodonts originated in the early Miocene of Africa and then spread to the rest of the continents, including South America once the Panama land bridge formed at about 3.5 Ma (Fig. 5.4). This was the final example of the four separate occasions (Fig. 1.4) that the sabertooth lifestyle has evolved by convergent evolution, and shows how successful saber-tooths must have been.

In fact, during most of the past 54 m.y. there has usually been an occupant of the saber-toothed mammal niche on this planet, so the past 10,000 years are unusual in that there has been no living saber-tooth since *Smilodon* vanished at the end of the last Ice Age. Paleontologists have spent a lot of time speculating how saber-toothed mammals used their long and delicate saber-like canines, but clearly whatever the function, it was a very successful one if it evolved at least four times independently. Most paleontologists accept the idea that the canines were used for slashing the throats or bellies of large prey, allowing them to bleed to death or be eviscerated so the saber-tooth could feed. If saber-toothed predators were specialists in hunting the largest prey, which other carnivores could not kill, then their disappearance 10,000 years ago might be linked to the mass extinction of all the large megamammals that lived during the last Ice Age (see Chapter 18).

Family Hyaenidae (Hyenas)

Most people have a false view of hyenas as cowardly "laughing" scavengers, but this is a cartoon stereotype. In fact, spotted hyenas are successful pack hunters that kill 95% of their food themselves, can drive a lion or leopard off its prey, and can break bones and scavenge carcasses that no other predator can touch. Hyenas have a weird-sounding call that sounds vaguely like a laugh, but when they sound this call they are deadly serious and on the prowl, not laughing. Hyenas are highly social animals, living in large packs dominated by the older females. Their dominance is even expressed in their genitals, in which the vulva and clitoris are modified into a structure that looks like the male penis and is even capable of erection. When the female gives birth, this pseudo-penis is tremendously stretched to accommodate the birth of the cub and never regains its original shape afterward.

Only four species of hyenas are alive today: the spotted hyena (genus *Crocuta*), which is the largest species and has the most robust limbs and crushing jaws; the two species of the genus

Hyaena, including the striped and brown hyenas, which are smaller and more slender and are mostly scavengers; and the peculiar aardwolf (genus *Proteles*), a hyena that has its teeth reduced to simple pegs and lives entirely on ants and termites it slurps up with its long tongue. All are restricted to Africa and western Asia today, but in the geologic past there have been more than 20 genera and 70 species in three subfamilies, ranging widely over Africa and Eurasia and briefly in North America as well.

The earliest hyenas appeared in the jungles of Europe in the early Miocene, about 17 Ma, when most true cats were small tree dwellers and there was little competition from pack-hunting dogs. The earliest hyena fossil, *Provivverops*, was a slender mongoose-like creature that lived in trees and resembled its living relative, the Asian palm civet. Although this fossil is built like a civet, details of its teeth and skull anatomy show that it is really a primitive hyena.

After *Provivverops* there were several radiations of hyenas into many body types, including small civet and mongoose-like fossils, running dog-like forms, and eventually the robust bone-crackers. The first radiation of hyenas in the middle Miocene occurred across most of Eurasia, with about 30 species in the subfamily Ictitheriinae. Ictithere hyenas looked more like wolves or jackals than any modern hyena. They had slender bodies and long slim legs for long-distance endurance running (as most pack-hunting dogs do today). They were highly successful during this time as the dominant Eurasian carnivorans, outnumbering all cats and other predators in many sites.

By the late Miocene, about 7 to 5 Ma, the dog-like hyenas began to decline in diversity and number, apparently pushed out by the arrival of dogs in Eurasia from North America, where dogs originally evolved. One of these hyenas, *Chasmoporthetes*, however, evaded the competition from other dog-like forms by developing really long legs and becoming a cheetah-like sprinter (Fig. 12.10A). *Chasmoporthetes* even managed to cross the Bering Land Bridge and thrive in North America in the late Pliocene and Pleistocene. By 1.5 Ma, however, all the dog-like hyenas had vanished except for the aardwolf, which survived by specialization in an ant-termite diet.

Meanwhile, about 12 Ma, the more familiar bone-crushing hyenas began to evolve, replacing the family Percrocrutidae, the species of which were long mistaken for hyenas but are unrelated. In Africa and Eurasia, bone-crushing hyenas thrived in the bone-crushing niche, while the arrival of dogs drove out the dog-like hyenas. The bone-crushing hyenas never succeeded in North America, which already had the borophagine dogs occupying that niche. Elsewhere, however, by the late Miocene and Pliocene the bone-crushing hyenas were very successful, cleaning up the carcasses felled by large predators such as the saber-toothed cats. The most spectacular of these was the *Dinocrocuta*, a bear-size hyena that weighed about 400 kg (880 lb) and had a massive skull with immensely powerful crushing jaws (Fig. 12.10B, C). It was common in the middle and late Miocene of Africa and Eurasia. By the Pleistocene, there was *Pachycrocuta*, a mega-scavenger about the size of a lion weighing about 110 kg (240 lb). Its jaws and

Figure 12.10. Hyenas. A. Reconstruction of *Chasmoporthetes*, a long-legged running hyena from the Pliocene and Pleistocene of Eurasia and North America. (See also next page.)

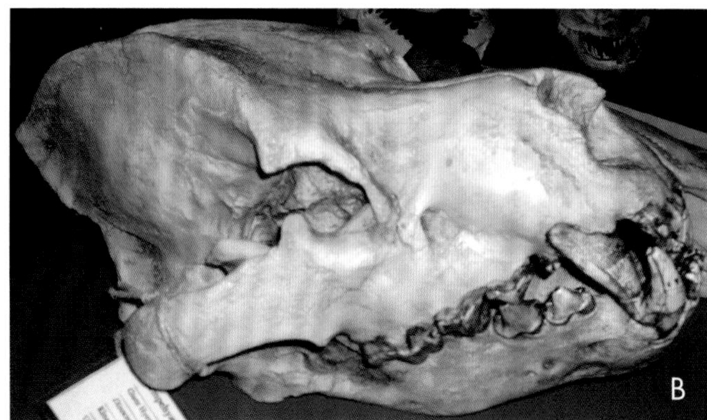

Figure 12.10. Hyenas. B. The huge skull of the bear-size hyena *Dinocrocuta*. C. Reconstruction of *Dinocrocuta*.

teeth were strong enough to break even elephant bones. In the famous "Peking man" site, Zhoukoudian, near Beijing, large numbers of *Pachycrocuta* bones showed that giant hyenas ruled the caves of *Homo erectus*. When large herbivores vanished at the end of the last Ice Age, so did *Pachycrocuta*. Meanwhile, the surviving hyenas vanished from Europe as the climate and habitat changed about 14,000 years ago. Today's smaller hyenas are adapted to living off the smaller prey that still lives in Africa and Asia today.

Family Herpestidae (Mongooses, Meerkats, and Kin)

Cats and hyenas are the most familiar of feliform carnivorans alive today, but another important family is the herpestids, or the mongooses and their relatives. Many people know of the mongoose for its reputation as an expert snake killer (described in the story "Rikki-Tikki-Tavi" from Rudyard Kipling's *The Jungle Book*). The meerkats are colonial herpestids that live in large complex burrows; they are found in many zoos but were unfamiliar to most people outside southern Africa until *The Lion King* made them famous (Fig. 12.11A). But these are only two examples of the living herpestids, which number 34 species in 14 genera spread across both Africa and Asia; almost all are known as mongooses. The small herpestids are not common as fossils but are known from Europe in the early Miocene (22 Ma) and have an extensive fossil record in Asia and Africa as well.

Figure 12.11. Herpestids and viverrids. A. The meerkat, a typical member of the mongoose family (Herpestidae). B. The genet, a typical member of the civet family (Viverridae). (See also next page.)

A related group is the Eupleridae, which includes the Malagasy civet, the cat-like fossa (of *Madagascar* movie fame), and a number of other mongoose-like creatures restricted to Madagascar. These had long been considered herpestids, but recent molecular analysis has shown that they are a separate family that evolved in isolation on Madagascar.

Family Viverridae (Civets, Genets, and Binturongs)

The civets and genets family, Viverridae, is unfamiliar to most people outside of tropical Asia or Africa (Fig. 12.11B, C). It consists of 38 species in 15 genera scattered around that region. Viverrids resemble weasels or mongooses in size and body proportions but have cat-like retractable claws and long tails—often prehensile, for grasping—and so are adapted mostly for a tree-climbing life rather than the terrestrial life of the herpestids. Many of them, especially the palm civets (genus *Nandinia*), have anatomical features that are close to those of the primitive common ancestor for all feliforms, including cats and hyenas. Molecular evidence suggests that palm civets are actually a primitive side branch of the feliforms, not closely related to any other living family.

The largest and the most unusual living viverrid is the binturong or "bear cat," a native of Southeast Asia (Fig. 12.11C). Covered by a thick coat of gray or black fur, it has the size and appearance of a large raccoon or a small bear, yet it has the claws and prehensile tail of a tree dweller. A number of schools have the "bear cat" as their mascot, and it is found in many zoos

Figure 12.11. Herpestids and viverrids. C. The binturong, or "bear cat," the largest viverrid.

around the world. Viverrids have a sparse fossil record going back to the early Miocene of Eurasia, and are known from Africa as well.

Caniformia

All living carnivorans that are not in the cat-like branch (Feliformia, including hyenas, herpestids, and viverrids) are grouped into the dog-like branch, the Caniformia (Fig. 12.2). These include not only the dogs but also the bears, raccoons, weasels and their kin, and pinnipeds (seals, sea lions, and walruses), as well as extinct groups such as the amphicyonids, or "bear dogs." Although these mammals may seem to have little in common, in fact they have certain specialized features of the braincase and bony ear region that are unique to this branch. The legitimacy of the Caniformia was confirmed in the 1990s and 2000s, when molecular studies showed that the taxa within it are all closely related to one another and only distantly related to the feliforms.

Family Canidae (Dogs, Wolves, Foxes, Coyotes, Jackals, and Their Relatives)

Everyone is familiar with "man's best friend," but very little was understood about the dog's evolution and classification until Richard Tedford, Beryl Taylor, and Xiaoming Wang completed a massive set of studies of canid fossils in the 1990s (summarized in their excellent 2008 book, *Dogs: Their Fossil Relatives and Evolutionary History*). Today there are some 13 genera and 37 species of canids, and at least 177 extinct species have been recorded so far (Fig. 12.12). However, carnivore fossils tend to be rare, so undoubtedly there were many more than this.

The relationships of all the living canids were clouded by confusion, because there have been multiple appearances of fox-like and wolf-like body forms. Recent DNA analysis demonstrates that there are three natural groupings within the canids. The first is the Vulpini, the true foxes, which includes 13 living fox species: 10 in the genus *Vulpes* (mostly Old World species) and three in the genus *Urocyon* (the American gray fox, the Channel Island fox, and the Cozumel fox). The second is the radiation of dogs in South America that began when the first canids arrived about 3 Ma, after crossing the Panama land bridge (Fig. 5.4). This group includes seven species of South American foxes, plus the maned wolf, the bush dog, and the short-eared dog. The third family is the Canini, which includes the African wild dog, the Asian dhole, and the genus *Canis* (wolves, jackals, coyotes, dingoes, and the domestic dog).

In addition, there are a number of extinct groups of dogs in the fossil record that are not members of any living group (Fig. 12.12). The oldest canid fossils known are called *Hesperocyon* ("western dog"), and come from beds in North America dating between 42.5 Ma and 31.0 Ma. The Eocene and Oligocene beds of the Big Badlands of South Dakota have produced numerous nice skulls and skeletons. *Hesperocyon* had a long body and tail, short legs, and a small skull, making it look more like a civet or coati than any dog we are familiar with. It probably captured small animals for food, as foxes and coatis do today, and may have been partially arboreal. From *Hesperocyon* evolved a great radiation of archaic dogs, the subfamily Hesperocyoninae, comprising 10 genera and 28 species. Most of these were relatively small in body size and would have looked like foxes in their external anatomy. After this great Oligocene diversification, the hesperocyonines began to decline in the early Miocene, as other more advanced dogs emerged, and finally vanished about 15 Ma.

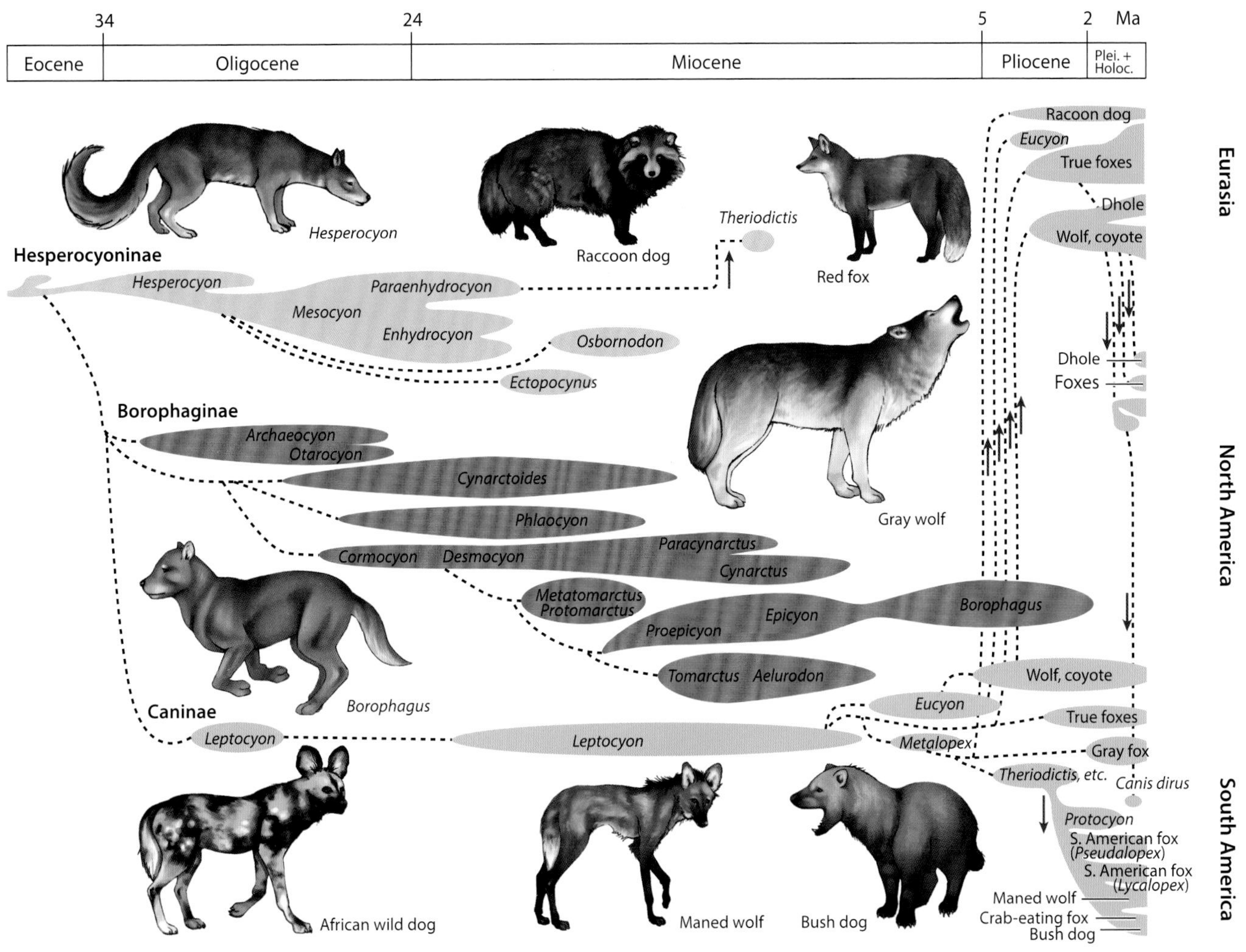

Figure 12.12. Family tree of dogs, show the primitive fox-like hesperocyonines, the bone-crushing borophagines, and the living subfamily Caninae.

The second great radiation of dogs was the bone-crushing hyena-like borophagine dogs, subfamily Borophaginae. Represented by 23 North American genera and 66 species, they first emerged in the early Oligocene about 33 Ma, when they were small fox-like creatures such as *Archaeocyon*. They began to diversify in the late Oligocene and early Miocene as the hesperocyonines declined, and by the middle and late Miocene borophagines reached the peak of their diversity. The largest forms were beasts such as *Epicyon* and *Borophagus*, reaching up to 1.5 m (5 ft) in length and weighing about as much as a brown bear (Fig. 12.13). These borophagines had huge, powerfully built skulls and jaws, with a domed forehead and large areas for strong jaw muscles. They were capable of tremendous bite force, and had large crushing teeth much like those of other bone crackers such as hyenas. Borophagines were by far the most numerous and diverse carnivorans of the Miocene in North America. They dominated the predator niches that had few cat-like forms (during the "cat gap" discussed earlier) and very few other large mammal competitors. The last borophagine died during the early Pleistocene, possibly as a result of the loss of the large mammals like mammoths, horses, and camels on which it depended for food.

The final group of dogs is the subfamily Caninae, to which all modern species belong (Fig 12.12). They first appeared along with borophagines in North America in the early Oligocene, about 33 Ma, but only one genus is known from the Oligocene, *Leptocyon*, which survived until about 10 Ma (late Miocene), at which time the Caninae were beginning to replace the declining borophagines. At the beginning of the Pliocene (about 5 Ma), the dogs (*Canis*), Old World foxes (*Vulpes*), and other genera had escaped from North America and spread quickly across Eurasia. Then they diversified in Africa as well (African wild dogs, jackals). Eventually the Old World was populated by a variety of native dogs, foxes, and jackals, as the canids became established.

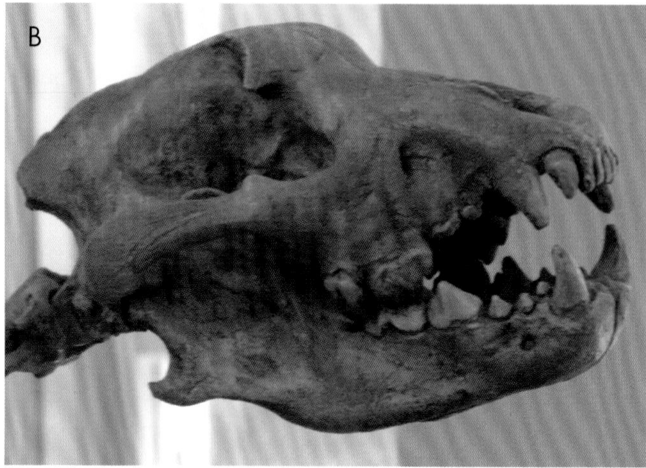

Figure 12.13. Borophagine dogs. A. The huge bone-crushing borophagine dog *Epicyon*, which reached the size of a bear and had massive jaws and blunt teeth for crushing bone. B. Close-up of the skull of *Epicyon*, showing the robust build, steep forehead, massive jaws, and huge crushing teeth for breaking bone.

The invasion of canids across the Panama land bridge about 3.5 Ma led to the radiation of South American foxes, maned wolves, and their relatives. Finally, when humans came to Australia about 40,000 years ago, they brought dingoes with them, so canids finally invaded that continent as well and thus are established on every continent except Antarctica.

Family Amphicyonidae (Bear Dogs)

An important extinct family of carnivorans is the Amphicyonidae, the "bear dogs." This name is misleading: they were neither bears nor dogs but a separate family with a completely independent history, the members of which only happened to look a bit dog-like and a bit bear-like. The amphicyonids' relationships to other carnivorans have long been controversial, although recent work suggests that they are part of the dog branch (Canoidea) rather than the bear-raccoon-weasel-seal branch (Ursoidea) of the Caniformia (Fig. 12.2). Even though they vanished only 1.8 Ma, during the early Ice Ages, the amphicyonids originated back in the late middle Eocene (44 Ma) and had a long successful run of nearly 42 m.y. that gave rise to at least 37 genera and almost 100 species.

The earliest amphicyonid was *Daphoenus lambei*, a small fox-size creature weighing about 5 kg (11 lb) from the late middle Eocene of North America. It closely resembles the earliest dogs, except for subtle differences in the teeth. In the lower Oligocene beds of the Big Badlands of South Dakota we have many complete skeletons of the species *Daphoenus dodgei*, which was about the size of a coyote and quite dog-like in appearance. From the early Miocene there are many good skeletons of the wolf-like *Daphoenodon* from places like Agate Springs fossil beds in Nebraska. Some of these were found with their pups in underground burrows, showing how they raised their young.

In the middle Miocene there were bear-size amphicyonids such as *Ischyrocyon* and the especially huge *Amphicyon*, which may have weighed almost 600 kg (1320 lb) and was one of the largest predatory mammals ever known in North America (Fig. 12.14). As they got larger, amphicyonids switched from the more slender dog-like build and mode of walking on the tips of the toes (**digitigrade**) to a fully robust bear-like build and walking on the palms of the feet (**plantigrade**).

Amphicyonids were apparently very mobile predators as well. In the late Eocene (35 Ma) *Cynodictis* spread from its North American homeland to Eurasia. About 19 Ma the Eurasian *Ysengrinia* and *Cynelos* spread back to North America, followed by *Amphicyon* at 18 Ma, *Pliocyon* at 17 Ma, *Pseudocyon* at 16 Ma, and *Ischyrocyon* at 14 Ma, where they lived alongside New World natives such as the daphoenines. By 8 Ma amphicyonids had vanished from North America and Europe, apparently outcompeted by borophagine dogs and by true bears (Ursidae). Meanwhile, amphicyonids made it to Africa in the late Miocene, from whence they vanished about 4.5 Ma.

Family Ursidae (Bears)

Bears are a familiar part of our culture, from the Goldilocks story to teddy bears, the Berenstain Bears, Baloo the bear from *The Jungle Book*, Winnie the Pooh, and also as national symbols and sports mascots. But these creatures are also the largest land carnivorans on every continent on which they live. Polar bears and Alaskan brown bears are the largest land predators alive today, and some extinct bears are the largest mammalian terrestrial predators ever known.

Bears are also unusual carnivorans in that they are omnivores, eating large amounts of fruits, including berries, and other vegetation in addition to meat, carrion, and fish. Consequently, the exclusively meat-slicing teeth of some carnivorans are modified in bears into more crushing and grinding molars, especially in the back of the jaw (Fig. 1.9). Bears are also stereotyped for the solitary large-bodied predator role, having immensely strong jaws and robust skeletons adapted mostly for

Figure 12.14. Skeleton of the giant amphicyonid, or "bear dog," *Amphicyon ingens*, from the middle Miocene of North America and Eurasia, with paleontologist Ashley Fragomeni Hall for scale.

slow walking and ambush predation rather than rapid running and pack hunting.

Only five genera and eight species of bears are alive today: the sun bear *Helarctos*, the sloth bear *Melursus*, the spectacled bear *Tremarctos*, the genus *Ursus*, with four species (black bear, brown or grizzly bear, polar bear, and Asiatic bear, and numerous subspecies), and the giant panda *Ailuropoda*. The panda has been variously linked to bears and raccoons, but recent DNA research shows that it was an early side branch of ursid evolution that split from the rest about 19 Ma.

However, the bear family has a long history of at least 13 genera and dozens of species going back to the late Eocene (38 Ma). The earliest bears (*Parictis*), from the late Eocene of North America, were small and looked much like raccoons, with a similar diet of omnivory. By the early Oligocene, there were similar bears (*Amphicynodon*) in Eurasia, emigrants from North America. The first bear subfamily to evolve from these primitive amphicynodontines was the hemicyonines, with the dog-like Eurasian *Cephalogale* appearing in the early Oligocene and later genera such as *Phoberocyon* (20 Ma) and *Plithocyon* (15 Ma) migrating from Eurasia back to North America. About 20–18 Ma there was also an odd-looking aquatic bear, *Kolponomos*, known from the late Oligocene–early Miocene rocks of Oregon and Washington. It had a peculiar downturned snout and teeth adapted for crushing mollusks.

Meanwhile the main lineage, subfamily Ursinae, began with *Ursavus*, which originated in Asia in the early Miocene and then spread to North America about 20 Ma. Different species of this genus ranged from cat-size to wolf-size, but they already had the distinctive crushing omnivorous teeth found in all living bears. One branch, the tremarctines, became the short-faced bears of North and South America and the spectacled bears, which spread to South America across the Panama land bridge about 3.5 Ma (Fig. 5.4). The short-faced bears (*Arctodus*), larger than any living bear, are the largest land carnivorans ever known (Fig. 12.15B); they stood 3.7 m (12 ft) tall on their hind legs, had a 4.3 m (14 ft) arm span, and weighed almost 1000 kg (2200 lb).

The remaining Ursinae experienced a rapid evolutionary radiation in the late Miocene and Pliocene, resulting in the lineage of the sloth bear. During the Plio-Pleistocene, the rest of the species of *Ursus* evolved. The most impressive of these was the cave bear, *Ursus spelaeus*, fossils of which have been found in numerous European Ice Age cave deposits; it probably fought our early ancestors often. Cave bears had a distinctive domed forehead (Fig 12.15A), and weighed up to 500 kg (1100 lb), although their

Figure 12.15. Extinct bears. A. Skeleton of the huge cave bear *Ursus spelaeus*, which towered above the humans that fought it in the Ice Ages. B. Skeleton of the gigantic short-faced bear *Arctodus simus*, which weighed over a ton and was larger than any living bear.

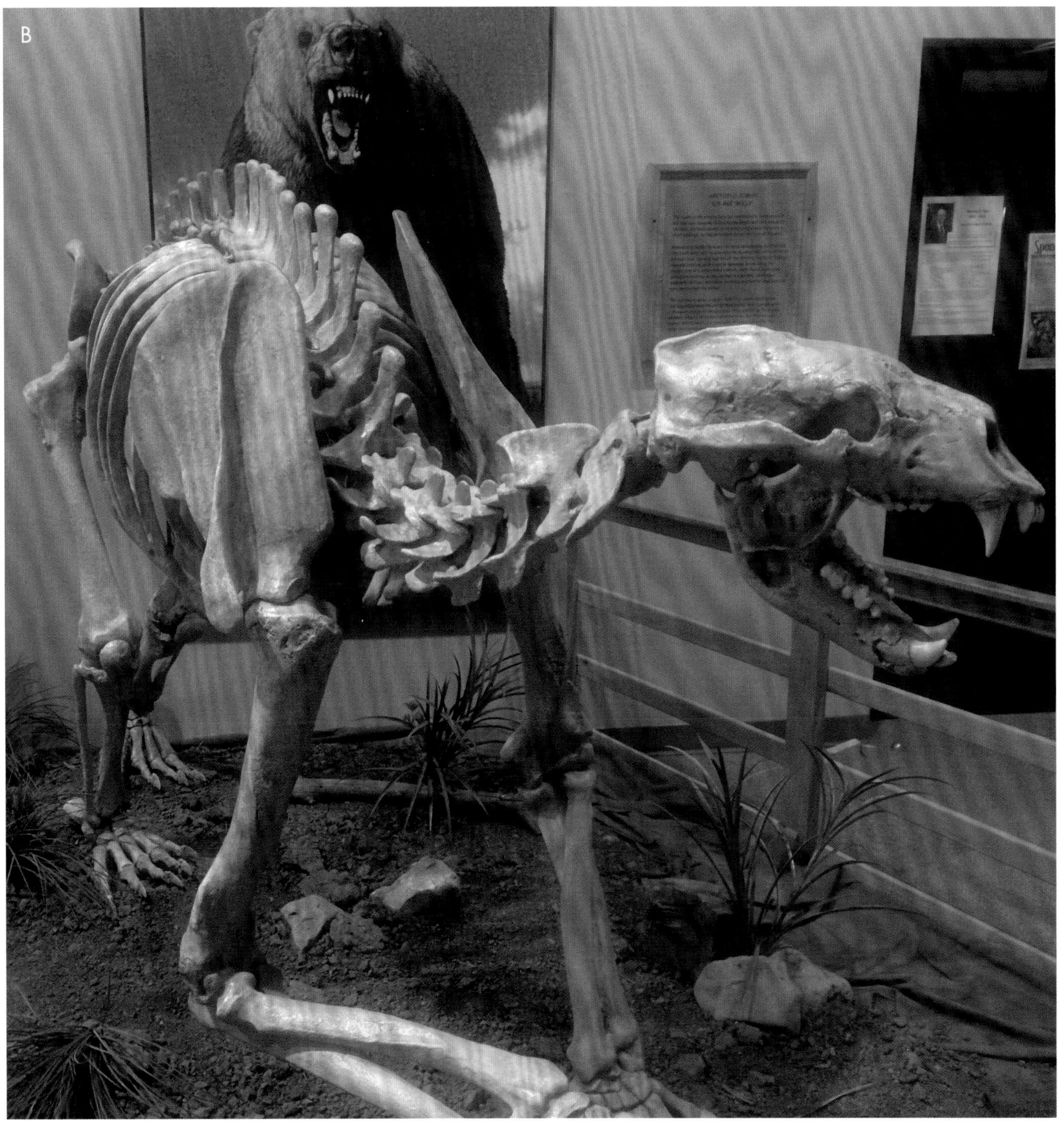

size varied, becoming larger during glacial advances and smaller during interglacial episodes. Although cave bears vanished in the late Pleistocene, their relatives are still found on all the northern continents today.

Pinnipedia (Seals, Sea Lions, and Walruses)

The seals, sea lions, and walruses (formerly placed in their own order, "Pinnipedia") are marine carnivorans that had land ancestors among the bear family and eventually made the transition to a marine lifestyle. From those bear-like origins, their bodies have been modified into a streamlined form for fast swimming, and their hands and feet modified into flippers for propulsion in water (the word "pinniped" means "fin foot"). Pinnipeds also developed adaptations for deep diving without developing the bends (decompression sickness), and have layers of blubber to protect themselves from the freezing water in which they often live. Most live on fish, squid, and other marine prey, although walruses rake up mollusks with their tusks, suck them into their mouths, and crush them with their peg-like teeth. Even though pinnipeds are marine for most of their lives, they all come to land to breed, to rest, and also to escape marine predators.

Most of the families are quite familiar to us from movies, television, and marine parks. There are 33 living species in 22 genera, spread among three living families. The true seals, which have no external ears and are unable to walk with their hind flippers, make up the family Phocidae. The sea lions, which have external ears and hind flippers that can rotate forward for walking, are in the family Otariidae. The third family is the Odobenidae, the walruses, which have no external ears but can rotate their hind flippers forward.

Additionally there are more than 50 extinct species of pinnipeds known from over a dozen extinct genera. For many years there was controversy as to whether pinnipeds were descended from a single common ancestor, or whether seals came from the weasel branch and sea lions came from the bear branch of the Carnivora. But recent anatomical studies, complemented by molecular data, have confirmed the idea that they all descended from a primitive bear relative. The earliest known fossil pinniped is *Pujila darwini*, an otter-like creature from the lower Miocene lake beds of the Canadian Arctic. The next transitional fossil is *Enaliarctos*, from the early Miocene (24–22 Ma) of California and Oregon (Fig. 12.16A). *Enaliarctos* superficially resembles a seal, with its flippers and streamlined body, as well as its large eyes, whiskers, and ears suitable for hearing underwater. However, its hands and feet were not yet fully modified into the classic flipper, and its teeth and braincase were still quite bear-like. *Enaliarctos* apparently swam with both front and hind flippers, a transitional stage between the hind-limb propulsion of true seals and the forelimb propulsion of sea lions.

After *Enaliarctos* the three pinniped families diverged. The earliest sea lions split in the middle Miocene, about 16 Ma; most of their fossil record places them in the North Pacific before they eventually spread to the Southern Hemisphere oceans. Walruses first appeared about 18 Ma, with fossils such as *Proneotherium* of North America and *Prototaria* of Japan, which looked like sea lions with large canines. The evolution of walruses is well documented by fossils, including the four-tusked *Gomphotaria* of the middle Miocene (Fig. 12.16B). Then came the Pliocene *Valenictis*, which had intermediate-length upper tusks and a mouth that had not yet developed the peg-like teeth and suction-pump mechanism of modern walruses (Fig. 12.16C, D).

The true seals can also be traced back to the early Miocene with the extinct family Desmatophocidae, which appears in transitional fossils between modern seals and *Enaliarctos*. Desmatophocids still swam with both front and hind flippers and did not yet have the skull and tooth specializations seen in modern seals. The modern phocids first evolved in the Atlantic

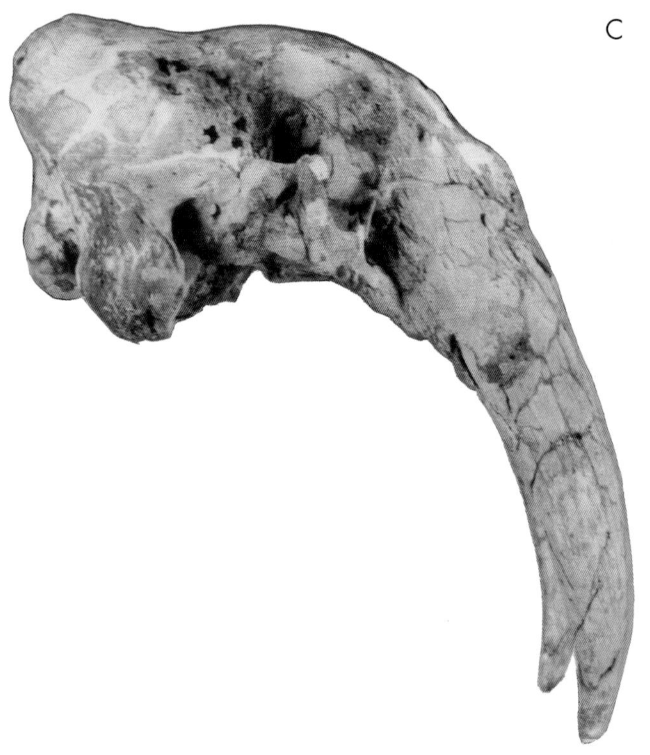

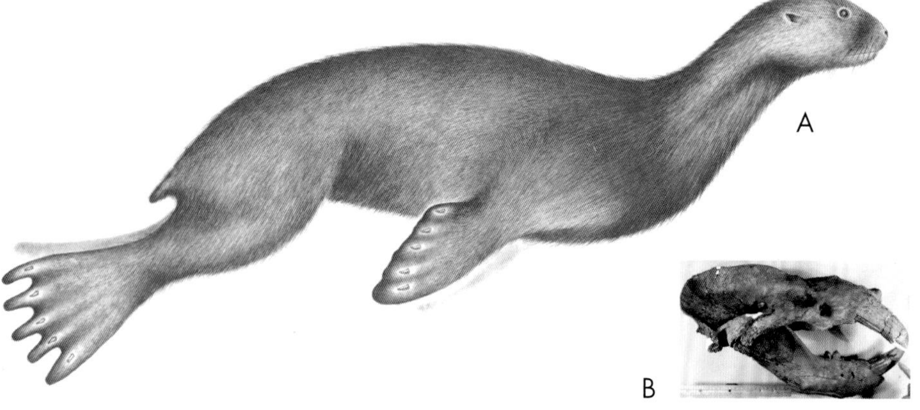

Figure 12.16. Fossil pinnipeds
A. Reconstruction of the ancestral pinniped *Enaliarctos mealsi*, from the early Miocene of California. B. Skull of the Miocene four-tusked walrus *Gomphotaria*. C. Skull of the primitive two-tusked walrus *Valenictis*, from the Pliocene of San Diego. D. Skull of the modern walrus, *Odobenus rosmarus*, with the extremely long down-turned tusks used for raking up mollusks.

Basin, before spreading worldwide around the world's oceans by the middle Miocene, and even into the Mediterranean Sea, Black Sea, Caspian Sea, and the Arctic Ocean.

Superfamily Musteloidea

Detailed anatomical studies of the middle ear bones and braincase, and also recent molecular data, support the idea that the weasel family (Mustelidae), the skunks (Mephitidae), the lesser or red "panda" (Ailuridae), and the raccoon family (Procyonidae) are closely related to one another; all placed in a group called Musteloidea (Fig. 12.2).

Family Ailuridae (Red or Lesser "Panda")

For almost two centuries the relationships of the red "panda" (*Ailurus fulgens*; Fig. 12.17) were controversial; most zoologists placed it either in the raccoon family or in the bears with the giant panda. Recent molecular analyses, however, show that red "pandas" are a separate branch within the Musteloidea (Fig. 12.2), not related to bears or giant pandas at all, so *Ailurus* should not be called a "panda."

In the geologic past there were at least eight other genera and more than a dozen species of ailurids, going back to Eurasian forms such as *Amphictis* and *Viretius* from the early Miocene, and to *Simocyon* from the middle Miocene to the Pliocene of Eurasia, Africa, and North America, which reached the size of a cougar.

Family Procyonidae (Raccoons, Coatis, Ringtails, Kinkajous, and Their Kin)

The procyonids are familiar to all of us not only for the clever raccoons (*Procyon*), but also the long-nosed coatis (*Nasua*), the monkey-like kinkajou (*Potos*) of the jungles of Latin America, and the rarely seen cacomistle and ringtail (*Bassariscus*) of the American deserts (a genus known from the middle Miocene and unchanged since). Most procyonids are small-bodied terrestrial and arboreal omnivores, and their molars are modified for grinding and crushing in a manner similar to that seen in bear teeth. Kinkajous were the first group to split off from the procyonid common ancestor about 22 Ma, and they are the most unusual. Kinkajous are completely arboreal, with monkey-like prehensile tails, and have teeth adapted to a diet of fruit and flowers (and occasional eggs and small vertebrates).

Procyonids have a sparse fossil record of jaws and teeth, with the earliest being the primitive ringtail *Probassariscus* from the middle Miocene of North America. It was an immigrant, descended from the early Miocene Eurasian fossil *Broiliana*. From the middle Miocene onward, procyonids remained a New World group, giving rise to multiple genera in the late Miocene to the modern genera of raccoons and coatis by the Pliocene.

Families Mustelidae (Weasels and Their Kin)

The Mustelidae, or weasel family, is the largest group in all the Carnivora, with 57 species in 22 genera alive today, and many more known from fossils. Its members include not only familiar weasel-like forms such as ferrets, minks, polecats, martens, fishers, and stoats, but also unusual forms such as the big, ferocious wolverine, the digging badgers, the unrelated honey-badgers, and the otters, which make up a separate subfamily of seven genera and 12 species. All are predators on small animals, mainly birds, rodents, and fish. Some are adapted for catching prey on the ground and in burrows (weasels) or in trees (martens and fishers) or for digging up rodent prey (badgers), while others take many kinds of prey by chase (wolverines) or hunt fish and invertebrates in the water (otters).

Mustelids first appeared in the late Eocene of North America, as shown by the well-known Big Badlands fossil *Palaeogale*, and

Figure 12.17. The red "panda," *Ailurus fulgens*.

then migrated to Eurasia in the late Oligocene, evidenced by the fossil *Plesictis*. During the Miocene there was a tremendous radiation of fossil mustelids in both Eurasia and North America, resulting in a wide variety of sizes and shapes of weasel-like, wolverine-like, badger-like, and otter-like creatures. One of the best known is *Sthenictis*, found in both the Miocene of Eurasia and North America (Fig. 12.18A). Some were quite remarkable, such as the American *Megalictis*, a huge, bear-size mustelid built like a wolverine, which lived in the late Oligocene during the "cat gap," when there were no cat-like predators in North America. There was also a long-legged wolf-size mustelid known as *Ekorus* from the late Miocene of Africa, which weighed almost 44 kg (97 lb). One of the most impressive was the giant "bear otter," *Enhydriodon dikikae*, which lived side by side with our earliest ancestors in the Pliocene of Ethiopia (Fig. 12.18B). It reached 2 m (6.5 ft) in length and may have weighed as much as 180 kg (400 lb). Mustelids migrated back and forth between North America and Eurasia frequently during the Miocene, making them excellent time markers because each immigration event is another time plane.

Family Mephitidae (Skunks)

The skunks (family Mephitidae) were long placed within the Mustelidae, but recent work has shown that they are a separate family, closely related to procyonids plus mustelids rather than a group within the mustelids. Skunks are famous for defensively spraying a powerful smelly oil from their anal scent glands. However, such scents are common to nearly all mustelids, for marking territory and occasionally driving away predators; skunks have just taken this ability to an extreme.

Today the mephitids include 12 species in four genera: the two species of striped skunks (*Mephitis*) and the four spotted skunks (*Spilogale*), both primarily North American; the four species of hog-nosed skunks (*Conepatus*), found in both North and South America; and the two stink badgers of Indonesia (*Mydaus*). Their fossil record goes back to the early Miocene in Eurasia, and during the middle Miocene they migrated to North America and became diverse there as well.

Figure 12.18. Family Mustelidae, weasels and their relatives. A. Complete skeleton of *Sthenictis*, known from the Miocene of North America and Eurasia. B. Reconstruction of the gigantic "bear otter" from East Africa, *Enhydriodon dikikae*, including a detail of its snout bones in the upper right.

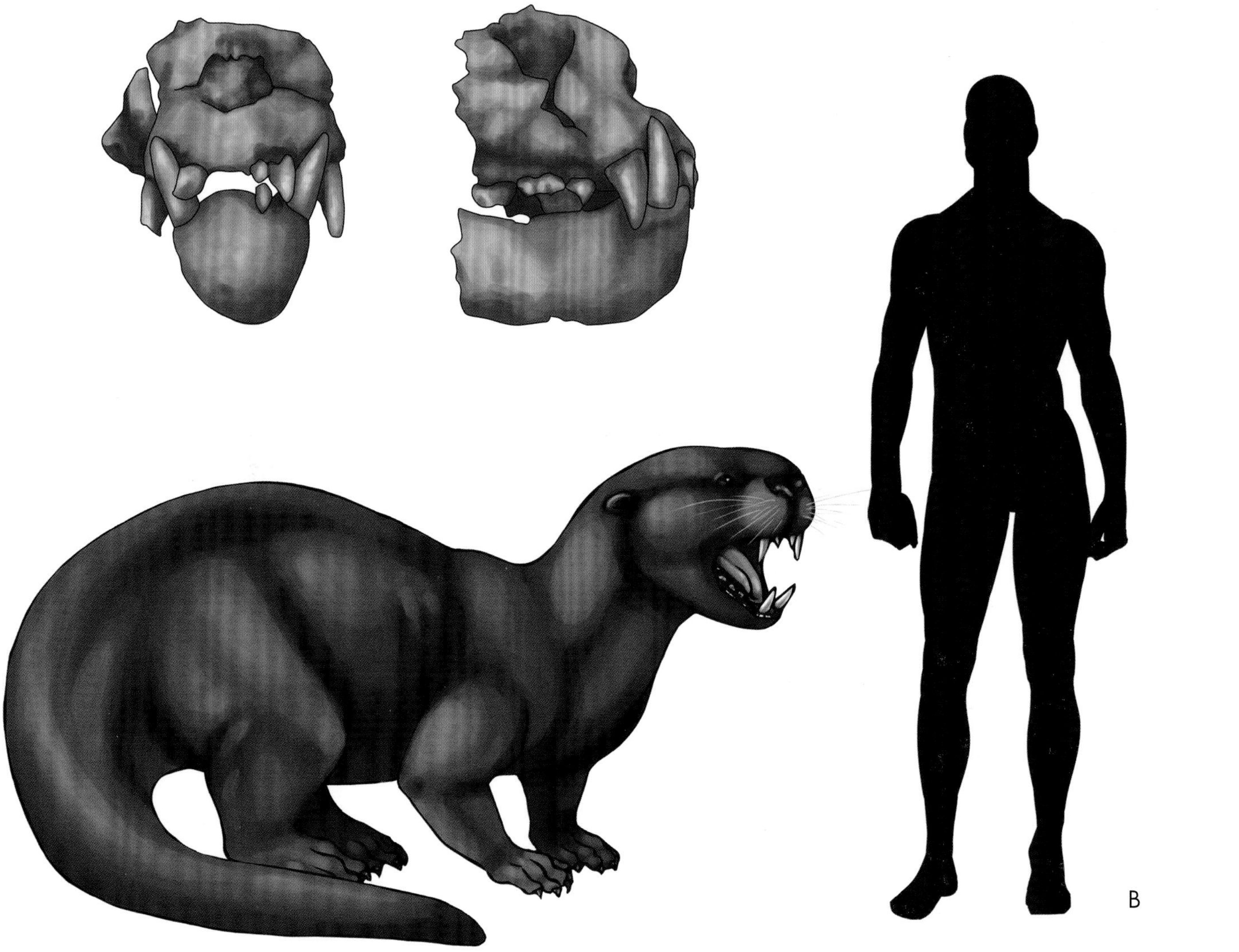

CHAPTER 13

LAURASIATHERIA: UNGULATA

Hoofed Mammals and Their Relatives

The ungulates, or hoofed mammals, are the dominant group of large land mammals on the planet. This has been true since the end of the Age of Dinosaurs. The living hoofed mammals include two large orders, the Artiodactyla or "even-toed" hoofed mammals (pigs, hippos, camels, pronghorns, giraffes, deer, cattle, sheep, whales and dolphins, and their extinct relatives), and the Perissodactyla or "odd-toed" hoofed mammals (rhinos, horses, tapirs, and their extinct relatives). Most scientists agree that these two orders are closely related and place them in a larger group, the Ungulata.

Some other mammals also have hoof-like structures on their feet, such as elephants, hyraxes, and their relatives. As recently as the late 1990s, the scientific consensus placed them with the Ungulata as well. However, molecular evidence puts elephants,

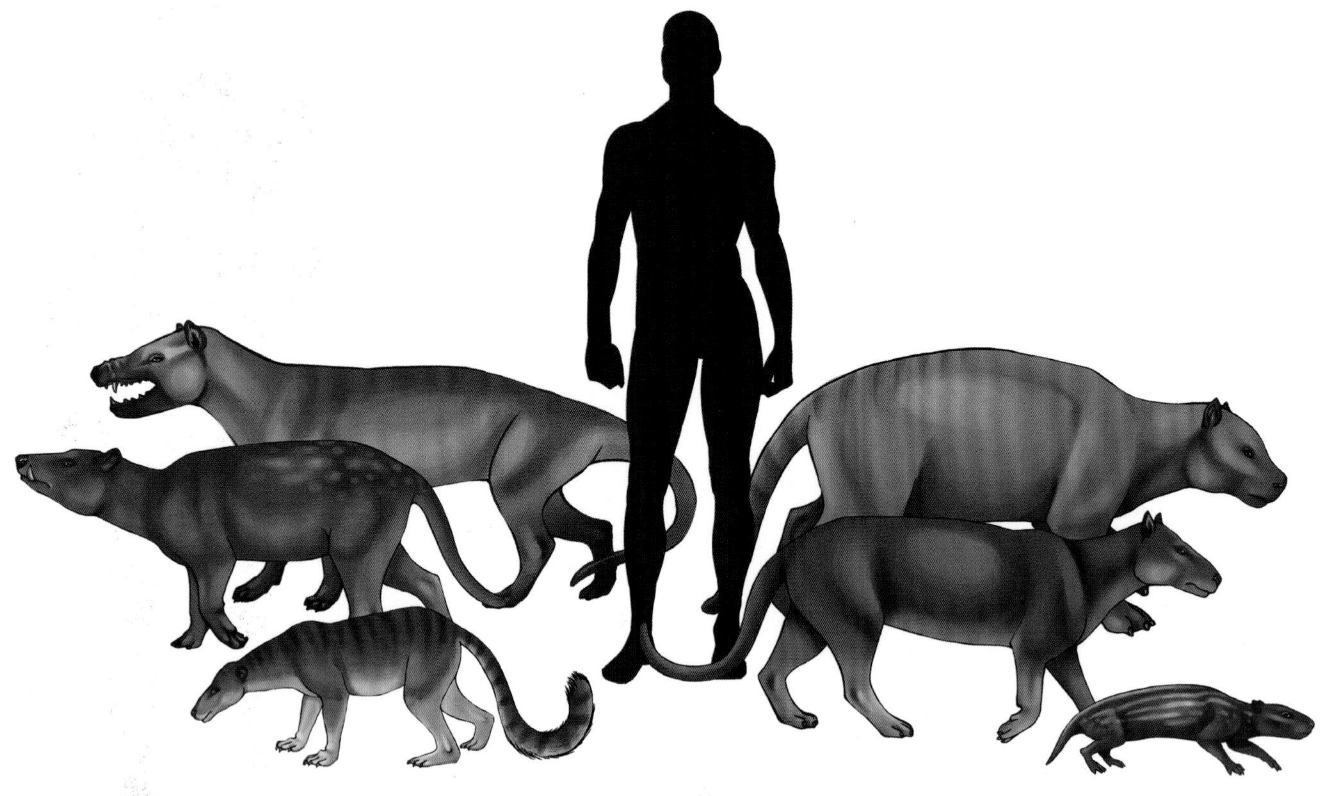

Figure 13.1. A composite of archaic ungulates. On the right (front to back) are the dachshund-shaped *Hyopsodus*, the perissodactyl relative *Phenacodus*, and (in back) the periptychid *Ectoconus*. On the left (front to back) are the primitive arctocyonid *Chriacus* (shaped like a coati), the predatory hoofed mammal *Arctocyon*, and the large, wolf-size *Mesonyx*.

hyraxes, and many other hoofed groups with the other African mammals, the Afrotheria. At present, most mammalian specialists agree with this idea, so the only true ungulates remaining are the perissodactyls and artiodactyls and their extinct kin.

"CONDYLARTHS"

The earliest ungulates (such as the zhelestids of the Late Cretaceous of Asia and *Protungulatum* of the latest Cretaceous and earliest Paleocene of North America) are identified not only by the broad tips of their finger and toe bones (once covered by some sort of hoof made of keratin) but by other features as well. These creatures tended to have relatively low-crowned blunt cheek teeth, with squared crowns in top view, suited for an omnivorous diet with an emphasis on vegetation. By contrast, most mammals of the Late Cretaceous ate a more insectivorous diet (based on their teeth) or were seed and fruit eaters (multituberculates). Not much of the skeleton of the earliest ungulates is known, but their ankle bones tended to be shorter and more robust than those of other contemporaneous mammals. During the latest Cretaceous and the early Paleocene, archaic ungulates like *Protungulatum* were the largest mammals as well, with the most robust skeletons.

As we discussed earlier, in the early days of taxonomy, it was customary to create taxonomic groups (wastebasket groups) for assemblages of animals that were not in fact closely related. For example, the hoofed mammals were typically classified into two living orders, the Perissodactyla and the Artiodactyla, both of which are clearly natural groups with distinctive evolutionary specializations that demonstrate their common ancestry. However, the many other fossil ungulates that were not clearly artiodactyls or perissodactyls were tossed into the taxonomic wastebasket group "Condylarthra." "Condylarths" had no unique features of their own, but were distantly related to the two living orders of ungulates. The definition of "condylarth" was effectively "any extinct hoofed mammal that is not a member of a living order."

In 1988 Earl Manning, Martin Fischer, and I argued that there was no longer any point in using an outdated unnatural wastebasket group like "Condylarthra." Since then most scientists have dropped this meaningless taxon in favor of the less misleading "archaic ungulates." However, once in a while the obsolete term "condylarth" pops up, usually in research from paleontologists who have not kept up to date. In addition, Manning, Fischer, and I showed that once we eliminated the obscuring effects of tracing the ancestry of different ungulate groups to some vague "condylarth" ancestor, we were able to decipher the pattern of how ungulates are interrelated. Thus the term "Condylarthra" turned out to be not merely obsolete but actually to hamper thinking and prevent us from making progress in research.

Today we recognize five natural groups of archaic ungulates (Fig. 13.1) that were once tossed into the "condylarth" wastebasket but now are sorted out as close relatives of different living orders. They are the following.

Family Arctocyonidae (Arctocyonids)

The earliest and most primitive group of archaic ungulates is the arctocyonids (Greek for "bear dogs"). The early Paleocene European *Arctocyon*, as the name implies, looked vaguely like a small bear or dog, because it had large canine fangs and sharper cheek teeth than is typical for most hoofed mammals (Figs. 13.1, 13.2A). However, its molars are still low-crowned and adapted for a bearlike omnivorous diet of plants and meat. Arctocyonids apparently had hooves rather than claws on its hands and feet. Only a few of the arctocyonids, such as *Chriacus*, a late Paleocene to early Eocene fossil from western North America, are known from more than teeth and jaws. *Chriacus* had relatively primitive

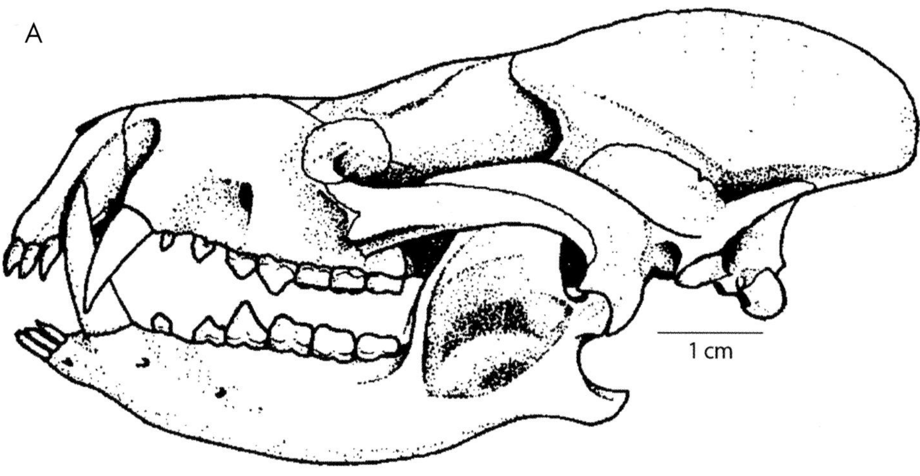

Figure 13.2. Arctocyonids. A. The skull of *Arctocyon ferox*, with its large canine tusks and dog-like appearance. (See also next page.)

Figure 13.2. Arctocyonids. B. Reconstruction of *Chriacus*, an arctocyonid built much like the living coati.

skeletons with robust limbs that could be used to climb trees and to walk but to not run as fast as a modern deer or horse. *Chriacus* also had a long bony tail, a feature lost in nearly all later ungulates. Reconstructions of *Chriacus* (Fig. 13.2B) make it appear much like the living coatis, a genus of raccoon relatives from the US Southwest to the tropical jungles of Central and South America that eats a widely varying diet of meat, insects, seeds, and plants, and a generalized skeleton that can dig, climb trees, and walk across the landscape. Some paleontologists have argued that *Chriacus* makes a good ancestor for the artiodactyls, although the evidence is not conclusive. If this were accepted, then arctocyonids could be clustered with the artiodactyls, removing them from the assorted archaic ungulates. More recent analyses indicate only that artocyonids are distant relatives of the entire ungulate radiation and not close to any living order.

Families Periptychidae and Hyopsodontidae (Periptychids and Hyopsodonts)

The periptychids were one of the first archaic ungulate groups to diversify after the extinction of the dinosaurs. The family includes about 14 genera and more than two dozen species, classified into three subfamilies. Their distinctive fossils are among the most common and diverse in the lower Paleocene beds of Wyoming and New Mexico (Fig. 1.2). Most of them were quite small and are known only from teeth and a few jaws. Despite this limitation, their teeth are very distinctive and easy to recognize. Most periptychids have lots of wrinkles and crenulations on the surface of the enamel of their molars, which may have increased the grinding surface of the teeth as the tips wore down.

Unlike most periptychid species, which are represented only by teeth, the genera *Ectoconus* (Fig. 13.1) and *Periptychus* are known from nearly complete skeletons. They were sheep-size animals with a primitive robust skeleton, small braincase, and long sturdy tail. *Ectoconus* and *Periptychus* were not particularly well adapted to running but may have been good diggers. After dominating the early and middle Paleocene in North America, periptychids vanished completely by the late Paleocene, leaving no descendants or even close relatives among living mammals.

Another diverse family, often clustered with the periptychids, is the Hyopsodontidae. This family had six genera and about a dozen species, most from the late Paleocene, except *Haplomylus*, which lasted into the early Eocene, and *Hyopsodus*, which managed to survive all the way to the late Eocene (36 Ma), a duration of almost 20 m.y. This is by far the longest time range of any archaic ungulate and one of the longest in all the mammals. It was also the last of the archaic ungulates to survive, at least 15 m.y. after most of the rest, which had vanished by the end of the early Eocene.

Most hyopsodonts are known only from their distinctive cheek teeth, which are low-crowned and quadrangular, with four distinct cusps; in many genera, the cusps show a tendency to connect with cross-crests. The genus *Hyopsodus* is known from complete skeletons. It had a long slinky body and tail, and relatively short limbs, somewhat resembling a dachshund or a weasel in shape (Fig. 13.1). The great abundance of its fossils suggests *Hyopsodus* was successful in a niche not occupied by other early mammals, possibly slinking through the low underbrush and avoiding predators. Detailed studies of the skull and ear region suggest they had relatively large brains and a good sense of smell.

The relationship of hyopsodonts and periptychids to other mammals is still controversial. Some recent studies based on newly discovered specimens and anatomical features suggest both are distant relatives of the perissodactyl branch, although this is not strongly supported yet.

Family Phenacodontidae (Phenacodonts)

The best known of all the archaic ungulate families is the Phenacodontidae. Only four genera occurred in North America, but they also spread to Europe. Phenacodonts are among the most common fossils of the middle Paleocene (*Tetraclaenodon*) and late Paleocene (*Ectocion*), and even in lower Eocene beds, where *Phenacodus* (Figs. 13.1, 16.7) lived alongside a diverse array of primitive horses, tapirs, rhinos, and the running artiodactyls such as *Diacodexis*. In fact, *Phenacodus* survived all the way to the late middle Eocene, about 46 Ma, much later than any other archaic ungulate except *Hyopsodus*.

Phenacodonts are very distinctive from other archaic ungulates. Their teeth have simple rounded cusps that are beginning to connect with cross-crests and distinctive combinations of ridges. They look like very primitive versions of the first horses, which has led many to suggest that the perissodactyls arose from the phenacodonts. Closer relatives of perissodactyls, such as the Paleocene Asian fossil *Radinskya*, have now been found, but phenacodonts could be more distant relatives.

Phenacodus, a sheep-size mammal, is represented by a number of complete skeletons. There are partial skeletons from other phenacodonts, including *Tetraclaenodon*, which was about the size of a dog. Phenacodont skeletons were very generalized and primitive, or maybe slightly adapted for running, with a long robust body and limbs and a long tail, and no obvious specializations of the body for digging or climbing. Their hands and feet are also unspecialized, with five well-developed fingers and toes, although the thumb and big toe are small and begin to vanish in some species (almost all advanced ungulates have lost their thumb and big toe altogether). Based on these features, it appears that walking and low-speed running were the phenacodonts' main mode of locomotion. Enough complete skulls and jaws are known to suggest that there was some difference between males (larger canine tusks) and females. Some skulls have deep notches in their nasal opening, suggesting the presence of a prehensile lip or short proboscis.

Family Mesonychidae (Mesonychids)

Very different from the relatively small unspecialized archaic ungulates we have discussed is a group known as the mesonychids (Figs. 13.1, 13.3). Most were relatively large, dog-size and even bear-size predators, with skulls bearing large canine fangs, and a skull shape that is very dog-like or bear-like. Their fossils used to be lumped in with other primitive carnivorous mammals, including the creodonts. Both the mesonychids, such as *Dissacus*, and the creodonts were the dominant predatory mammals of the middle to late Paleocene in North America and Asia, and mesonychids were the largest predators of all during that time. They remained so through much of the Eocene, when true carnivorans were small weasel-size creatures.

Mesonychids first appear during the early Paleocene in China, with a fossil known as *Yangtanglestes*. By the late Eocene, there were several genera in Asia, including *Dissacus*, *Sinonyx* (Fig. 13.3), and *Jiangxia*, where there was no competition from creodonts or carnivorans. *Dissacus*, a jackal-size predator, immigrated to North America and Europe. One of its descendants in North America was the bear-size *Ankalagon*, the largest predatory animal of the middle Paleocene in North America. During the early Eocene, another Eurasian immigrant, *Pachyaena*, developed a body much like that of a hyena. This shape is the basis for their name, which means "heavy hyena" in Greek, although they were not as long-limbed as modern hyenas. Some species were larger than bears. From the late early Eocene through the middle Eocene, more bear-size mesonychids roamed Asia and North America alongside *Pachyaena*, including *Synoplotherium*, *Harpagolestes*, and *Mesonyx* (Fig. 13.1). All of these genera were extinct in North America by the end of the middle Eocene (37 Ma). The last of the mesonychid group was *Mongolestes*, from the upper Eocene Ulan Gochu beds of Mongolia (often incorrectly listed as "early Oligocene" in outdated literature), which died out before the end of the Eocene (34 Ma).

Given their dog-like or bear-like builds and huge fangs, it was natural to assume that mesonychids were related to other carnivorous mammals. However, complete skeletons show they had hooves, not claws, so mesonychids are another branch of archaic ungulates, more adapted for a predatory or scavenging mode of life than most archaic hoofed mammals, which were omnivores. Furthermore, mesonychids had cheek teeth with huge conical cusps (Fig. 13.3), and lacked the shearing blade-like teeth that fully carnivorous mammals have (Fig. 12.3), evidence that they were not efficient meat eaters. Their teeth were better suited for crushing prey and gulping it down whole, or for catching fish,

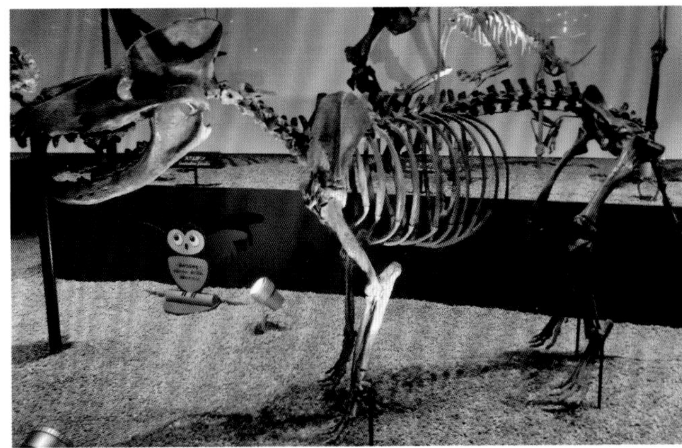

Figure 13.3. The bear-size Chinese mesonychid *Sinonyx*.

without cutting it up into small pieces—or breaking bones and crushing carcasses as scavengers.

These big hoofed mammals that were full-fledged scavengers and predators—"wolves with hooves"—have long been puzzling. Since the 1960s a number of researchers have pointed to similarities in the skull shape and the triangular blade-like cheek teeth of the mesonychids and the early archaeocete whales, and argued that whales were descended from mesonychids. In the early 2000s it became clear from both fossils and molecular evidence that whales were descended from the artiodactyls and were most closely related to hippos among the living artiodactyls. Thus the mesonychids were at best only distantly related to the artiodactyls (including whales), and the distinctive triangular teeth in both whales and mesonychids must have evolved by convergence.

Then there is the puzzling fossil known as *Andrewsarchus* (Fig. 13.4). Found in 1923 in Mongolia in the upper middle Eocene Irdin Manha beds, it is known from only a single complete skull of immense size: almost a yard long (83 cm/33 in) and nearly two feet wide (56 cm/24 in)—more than twice the length of the largest living bear skull! Only a few other tooth fragments are known, and there is no lower jaw or skeleton yet. Its teeth are huge: giant canine fangs and enormous blunt conical cheek teeth that are suited for crushing but not much slicing. It was the largest mammalian land predator that ever appeared, but it is still a puzzle. If *Andrewsarchus* had a mesonychid-like body, it would have been about 3.4 m (11 ft) from snout to rump, and about 1.8 m (6 ft) tall at the shoulder. On the other hand, if it were more like the huge amphibious whale ancestors such as *Ambulocetus*, its body would have been very different in shape. Some scientists think it is a bizarre entelodont (Chapter 14), with a massive body shaped like that of a pig and long legs. Scientists must be skeptical of most reconstructions of the beast, because they make assumptions about its body and its size that are not based on any fossil evidence.

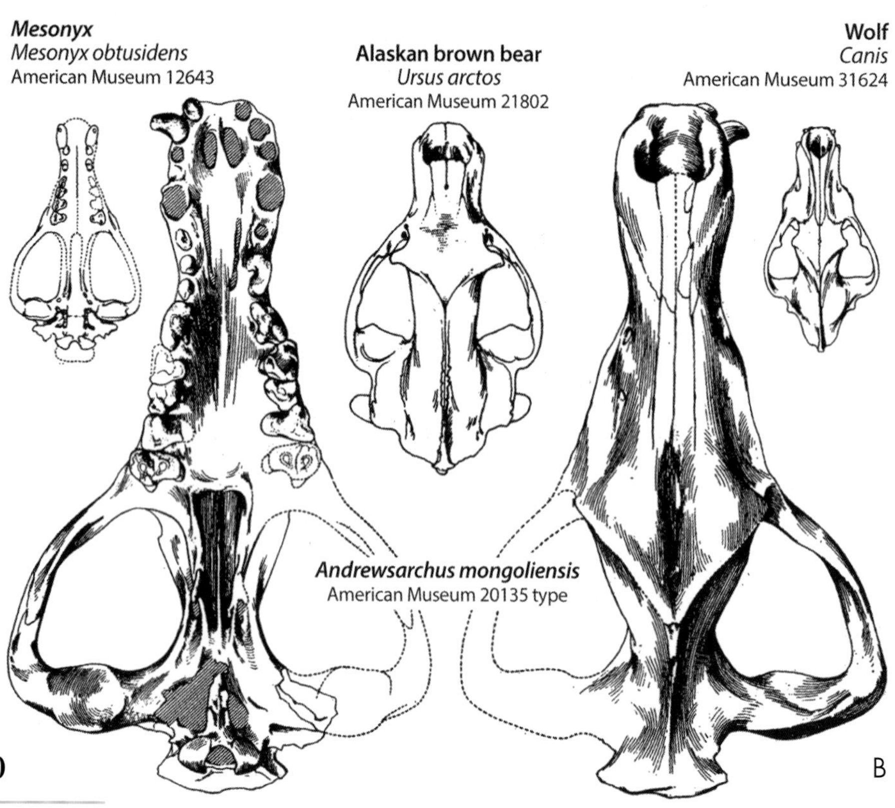

Figure 13.4. The giant mesonychid *Andrewsarchus*. A. The enormous skull of *Andrewsarchus*, much larger than the skull of any living land carnivore and bigger than most Eocene whale skulls. B. Henry Fairfield Osborn's original illustration of the *Andrewsarchus* skull, comparing it to that of *Mesonyx* (upper left) and to the largest known skulls of bears and dogs.

CHAPTER 14

LAURASIATHERIA: ARTIODACTYLA

"Even-Toed" Hoofed Mammals: Pigs, Hippos, Whales, Camels, Ruminants, and Their Extinct Relatives

The order Artiodactyla, or the "even-toed" hoofed mammals, is the most diverse and abundant group of large mammals on the planet today (Fig. 14.1). They are called "even-toed" because the axis of their hands and feet (Fig. 14.2) runs between their middle, or third, digit (equivalent to your middle finger and middle toe) and their fourth digit (equivalent to your ring finger or fourth toe). The hand and foot are symmetrical along this axis between the third and fourth digits, and they usually have four toes (two on each side of the axis, as in pigs and hippos) or only two toes (just digits three and four). They are "cloven-hoofed" in the biblical sense. This foot symmetry and pattern of two or four toes is unique to the Mammalia, and comes with other related features of the limb. For example, the main hinge bone of the ankle (the astragalus) has a pulley-like surface at each end, making it very flexible in a front-to-back plane for fast running. Many of the living artiodactyls (especially antelopes, pronghorns, deer,

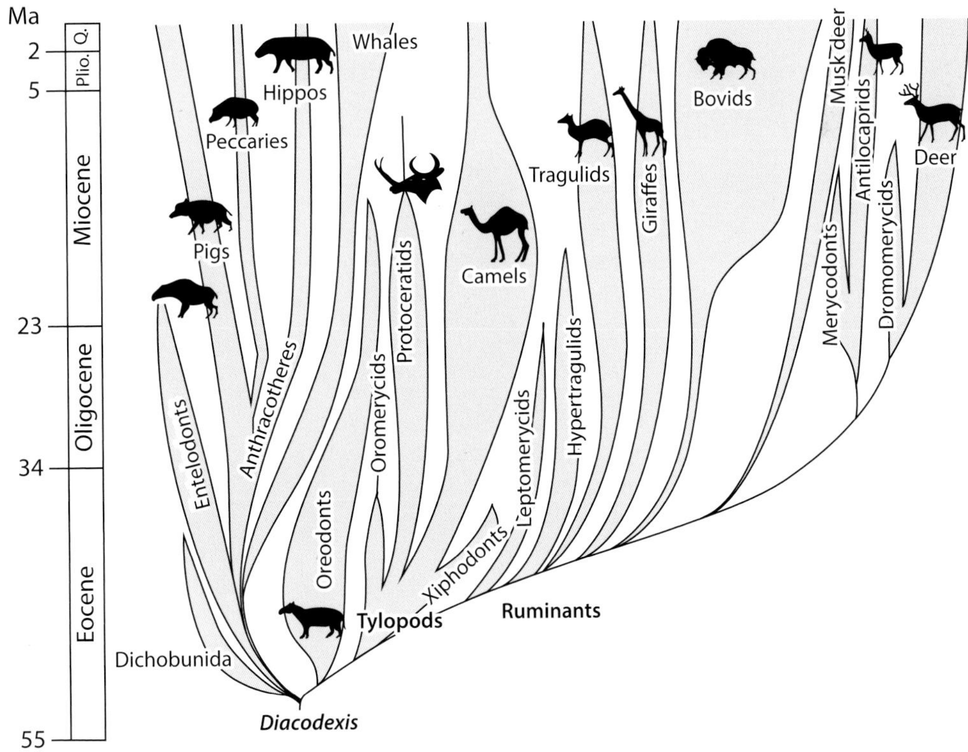

Figure 14.1. Family tree of artiodactyls. "PLIO" = Pliocene; "Q" = Quaternary.

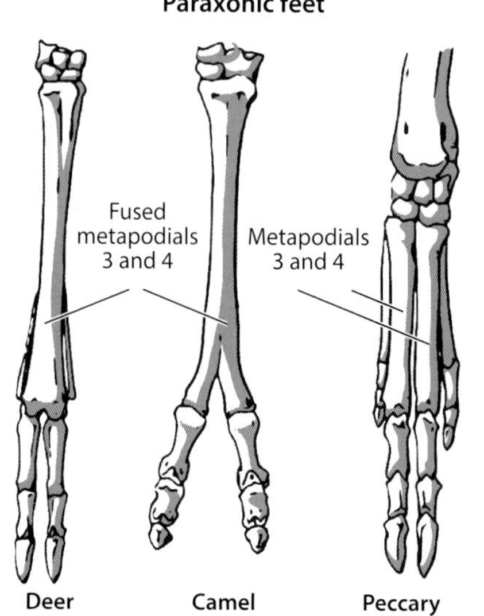

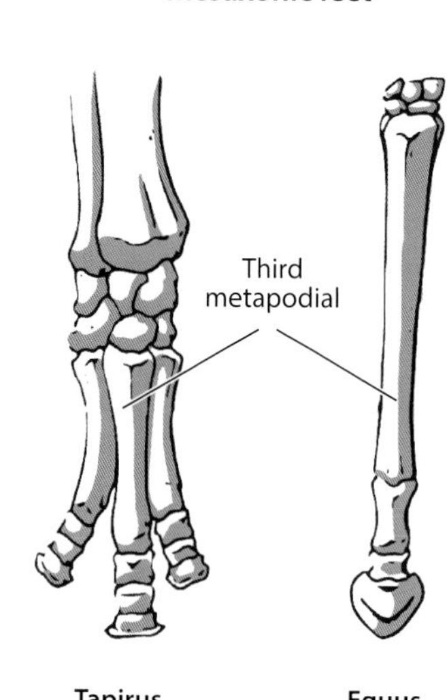

Figure 14.2. Foot symmetry of artiodactyls and perissodactyls. Artiodactyls (left) have even-toed or "paraxonic" fore and hind feet, with either two toes or four toes per foot. The axis of the foot goes between the third and fourth toe. The ankle bones (metapodials) are sometimes fused together, forming what is called a cannon bone. Perissodactyl, or "mesaxonic" feet (right), have an odd number of toes, either one (in horses) or three (in tapirs, rhinos, and most extinct forms). The axis of the foot goes through the middle of the third toe.

and giraffes) have become specialized for speed; they have not only lost their side toes (so they are completely two-toed), they have also elongated the bones of their hands and feet (carpals and tarsals) into long limb segments that give them better running and jumping ability. In many artiodactyls, these elongate pairs of wrist or ankle bones (metacarpals and metatarsals, or metapodials) have been fused into a single bone, known as a cannon bone (Fig. 14.2).

The living Artiodactyla fauna includes more than 280 species of land mammals grouped into 10 families. These include pigs (19 species), peccaries (four species), hippos (two species), and camels (six species). The rest of the artiodactyls are all ruminants: chevrotains (six species), musk deer (seven species), pronghorns (one species), giraffids (two species), true deer (90 species), and the huge family of bovids (more than 140 species of cattle, sheep, goats, antelope, and their kin). In addition, there are at least 25 other families of extinct terrestrial artiodactyls known from the fossil record, adding many hundred additional extinct species to the already diverse living group.

Today these animals are among the largest mammals on the landscape on every continent except Australia and Antarctica. In North America, for example, the biggest wild land animals are the bison, pronghorns, and deer (including moose and elk), while in the African savanna there is a huge diversity of giraffes, hippos, pigs, cattle, and antelopes. Nearly all the large mammals we raise domestically for food are artiodactyls: cattle for beef and milk, sheep for mutton and wool, goats for meat and milk, pigs for pork and bacon, and camels, usually raised for transportation but sometimes food and milk. Domesticated cattle, sheep, goats, and pigs have been introduced to nearly every continent except Antarctica, and they are now more abundant than any other kind of large mammal. In many places, the large native wild animals have been replaced with domesticated herds. Sadly, this means that many wild artiodactyls are endangered or threatened by the pressures imposed by domesticated artiodactyls, such as cattle, pigs, and goats, which destroy their habitat and eat their food sources.

In addition to the terrestrial artiodactyls, the whales and dolphins and their relatives are descended from one branch of the artiodactyls, the extinct relatives of the hippopotamus known as anthracotheres. The similarity in the molecule structures of whales and hippos was first noticed in the 1980s and 1990s, and the connection was conclusive when the DNA from whales and hippos (as well as many other mammals) was compared.

In 2002 two groups of paleontologists independently found that the fossils of the earliest whales, which were still four-legged land animals, also have the characteristic "double-pulley" ankle bone, or astragalus, that is diagnostic of artiodactyls. Consequently the order Cetacea, which once stood isolated and unrelated to other animals, is now a subgroup of order Artiodactyla, specifically the hippo branch. Some scientists think the entire order should be renamed "Cetartiodactyla" (Cetacea plus Artiodactyla), but that implies that the two groups are equal relatives that evolved independently, not that one is a subgroup of the other. Since all recent research clearly establishes that Cetacea is a subgroup within Artiodactyla, it is pointless and unnecessary to rename Artiodactyla to indicate additional groups placed within it. Creating the group "Cetartiodactyla" is comparable to renaming the Dinosauria the "Avedinosauria" to show that birds are a group within dinosaurs (a taxonomic name no one supports).

ARTIODACTYL ORIGINS

The earliest fossils that are clearly artiodactyl are known from the earliest Eocene, when they show up simultaneously in Eurasia and North America. Lumped into a wastebasket group known as the "dichobunids," they were tiny creatures about the size of a large rabbit or a small duiker antelope, about 50 cm (20 in) long, with long delicate legs and a slender body (Fig. 14.3). They still had five fingers and four toes, although the side digits are highly reduced. Their hind legs are very long, which suggests they were capable of leaping like rabbits. In contrast to these skeletal features, their skulls and teeth already have many of the specializations that are unique to artiodactyls.

The origin of these earliest Eocene "dichobunids" is still not clear. There are primitive groups of archaic hoofed mammals in the Paleocene, such as the arctocyonid known as *Chriacus* (Fig. 13.2B), that have many skeletal similarities to the most primitive artiodactyls, although they do not have the specialized teeth. Most molecular evidence suggests that Artiodactyla and Perissodactyla are more closely related to each other than to any other group of mammals, a pairing that defines the group known as "ungulates," or hoofed mammals.

It has been suggested that artiodactyls evolved in isolation in some place like the island continent of India during the Paleocene, then suddenly spread all across Eurasia and then to North America in the earliest Eocene, as India collided with Asia after pulling away from Gondwana during the Cretaceous. So far, however, we have no Paleocene fossils from India to support this, and the earliest Eocene fossils from southern Asia do not seem to show this either. In 2007 a fossil known as *Ganungulatum* that bears strong similarities to primitive artiodactyls was reported from the late Paleocene of Mongolia. This suggests that artiodactyls arose in eastern Asia before India docked with Asia.

Following their spread all around Eurasia and North America in the early Eocene, "dichobunids" soon evolved into many different families in the middle Eocene. Europe was flooded and then became subdivided into a number of small islands that were separated from Asia by the Obik Seaway, which cut Eurasia in half along the line of the Ural Mountains. On these isolated islands, a number of uniquely European artiodactyl families evolved during the middle and late Eocene, only to vanish in the early Oligocene when a wave of Asian immigrants invaded Europe and apparently outcompeted the natives. This great extinction event and faunal replacement is known as *La Grande Coupure* (the "great cutoff" or "great break").

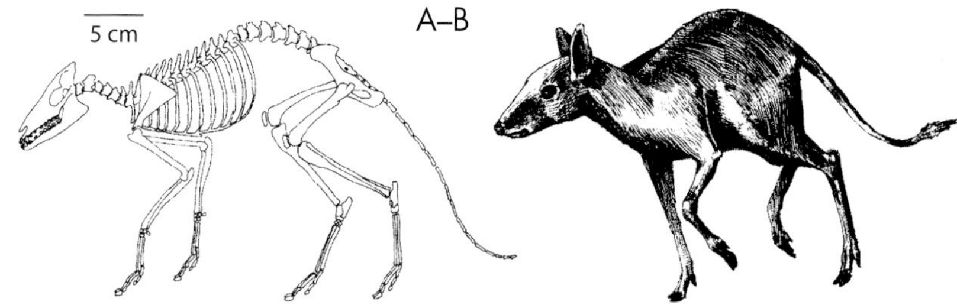

Figure 14.3. Primitive artiodactyls. A–B. The primitive dichobunid *Diacodexis* from the early Eocene. C. The primitive middle Eocene dichobunid *Messelobunodon*, from the lake shales of Messel, Germany.

Figure 14.4. Reconstruction of *Anoplotherium*, an odd group of artiodactyls from the isolated Eocene islands of Europe. This model was built by Waterhouse Hawkins for the Crystal Palace of the Great Exposition of 1851, and is currently on display in Sydenham Park near London.

As a result, European Eocene artiodactyls evolved in isolation from the rest of the world, and evolution experimented with many body forms. The xiphodontids were much like the earliest camels, and at one time paleontologists thought the two might actually be close relatives. Other European Eocene artiodactyls, namely the choeropotamids and cebochoerids, somewhat resembled the earliest relatives of pigs. Anoplotheres (Fig. 14.4) were sheep-size but had long limbs and tails and a camel-like snout. The cainotheres, mixtotheres, and amphimerycids retained the more deer-like limbs of the primitive "dichobunid" ancestors. Some looked nothing like any mammals from any other continent at any time in geologic history. In addition to artiodactyls, there were unique endemics among many different Eocene mammal groups; for example, Europe did not have horses or tapirs but rather the horse relatives known as palaeotheres (Fig. 15.3) and the tapir relatives known as lophiodonts.

SUOID ARTIODACTYLS

The pig-like groups of living artiodactyls include two families: the true pigs (family Suidae) and the peccaries or javelinas (family Tayassuidae). Along with these two surviving families, the fossil record yields many other creatures that are pig-like in their teeth and jaws and skulls but may not actually be closely related to pigs. These include a middle to late Eocene family, the helohyids, whose members had low rounded cusps on their cheek teeth, as do many other pig-like creatures. The best known of these, *Achaenodon*, has a skull that vaguely resembles a pygmy hippo, though it differs in every important detail.

Family Entelodontidae ("Killer Pigs")

Another spectacular group of pig-like creatures was the entelodonts (Fig. 14.5). Thanks to the television series *Walking with Beasts* and other documentaries they have become media stars and acquired such nicknames such as "hell pigs," "killer pigs," and "terminator pigs." Like many pigs and their relatives, entelodonts had large heads, with blunt rounded cusps on their cheek teeth and big canine tusks, along with a chunky body supported by four robust limbs with four fingers and toes. However pig-like these features were, there are no clear unique evolutionary specializations that unite entelodonts with the pigs, peccaries, hippos, or other suoid artiodactyls. Most of the features that place entelodonts close to other suoids on recent phylogenies are primitive features of the teeth and skeleton, or features that could have evolved by convergent evolution, so the true biological affinities of entelodonts are still unclear. The latest analyses place them closer to the hippo-whale branch than to the pigs.

Entelodonts first appear in the late middle Eocene of China about 40 Ma, with a small sheep-size creature known as *Eoentelodon*. Shortly thereafter, a similar-size creature appeared not only in China and Mongolia but also in North America. Known as *Brachyhyops* ("short-faced pig"), it is known from a well-preserved robust skull from the early late Eocene of Wyoming;

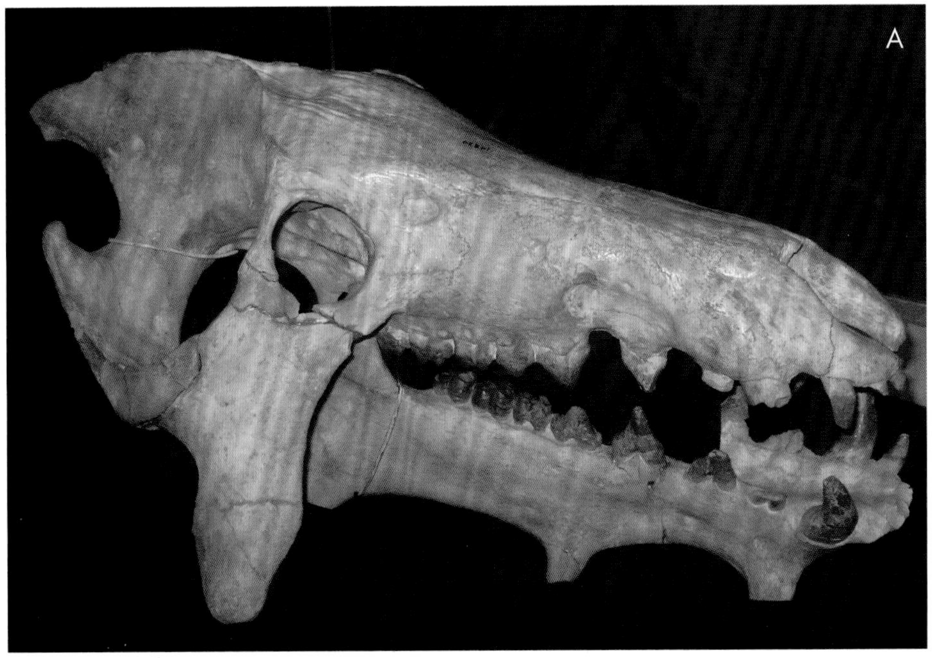

Figure 14.5. The bizarre pig-like entelodonts. A. Skull of the Oligocene entelodont *Archaeotherium*, showing the huge blunt teeth, the knobs and flanges on the lower jaw and cheekbones, and the large tusks. B. Reconstruction of the rhino-size entelodont *Daeodon*, from the lower Miocene Agate Springs beds in Nebraska.

B

other specimens have been found in Utah, Texas, and Saskatchewan. By the latest Eocene and early Oligocene, entelodonts had evolved into the famous Big Badlands fossil mammal known as *Archaeotherium* (Fig. 14.5A) and its Eurasian equivalent, *Entelodon*. About 1.2 m (4 ft) tall at its high humped shoulder, and weighing about 270 kg (600 lb), it was the size of a very large domestic hog. However, *Archaeotherium* had longer, more slender legs than a pig. Its huge bony skull had broad flanges around the eyes and cheekbones, and odd bony knobs along the bottom of the lower jaw. These strange bony structures on entelodont skulls are analogous to similar structures found on the broad flaring cheekbones of living pigs such as warthogs and giant forest hogs. These skull features help boars show their dominance and maturity, and protect their eyes and faces during combat, especially with other boars in their sounder, or herd. Many *Archaeotherium* skulls have bite marks and other battle scars on their faces that are evidence of such combat between adult males. The teeth of *Archaeotherium* and most entelodonts show extreme wear on their blunt cusps, and even on the tips of their canines, showing that they had a diet of very hard, abrasive objects. Most of the evidence suggests that they were omnivores and scavengers, and even bone-breakers, eating almost anything they could, including some vegetation and gritty roots and tubers. A site in Wyoming with multiple fossil skeletons of early Oligocene camels and other specimens offers evidence the animals were slaughtered by *Archaeotherium* and suggests this predator may have even stored carcasses in a meat cache.

By the late Oligocene, the descendants of *Archaeotherium* in North America got even larger and more bizarre-looking. Many had even more bony knobs and widely flaring flanges on their cheekbones and around their eyes and additional bumps on their lower jaws. The culmination of this trend was the rhino-size entelodont *Daeodon* (formerly called *Dinohyus*) from the lower Miocene (18–19 m.y. old) Agate Springs fossil beds in Nebraska (Fig. 14.5B). These monsters were almost 2.1 m (7 ft) tall at the shoulder and weighed up to 431 kg (930 lb). Their skull alone is almost 90 cm (3 ft) long. Closely related to *Daeodon* was a huge creature known as *Paraentelodon* from the late Oligocene of Asia (including Kazakhstan, Georgia, and China), which was even larger that the largest specimens of *Daeodon*. Once these beasts vanished, about 17 Ma, they were extinct worldwide, and nothing replaced them. The reasons for their extinction are unknown, although some paleontologists have pointed out that they died out about the same time that mastodonts first escaped Africa and invaded Eurasia. At the same time large bear dogs and other huge African predators that were unknown in Asia or North America during the Oligocene escaped as well, and probably wiped out many of the groups that had survived most of the Oligocene.

Family Suidae (True Pigs)

Everyone is familiar with domesticated pigs, which have been part of human culture for thousands of years. They are not only a source of pork and bacon, some pigs are also used to sniff out valuable fungi such as truffles. Today there are 16 species of wild pigs (family Suidae), including the warthog, giant forest hog, red river hog, the babirusa, with its upper and lower canine tusks curled up above its head, and a variety of other wild hogs (Fig. 14.6). The wild suids have always been restricted to the Old World and never lived in the New World until Europeans introduced domesticated pigs, starting with the voyages of Columbus. The ecological niche for pigs in the New World was instead

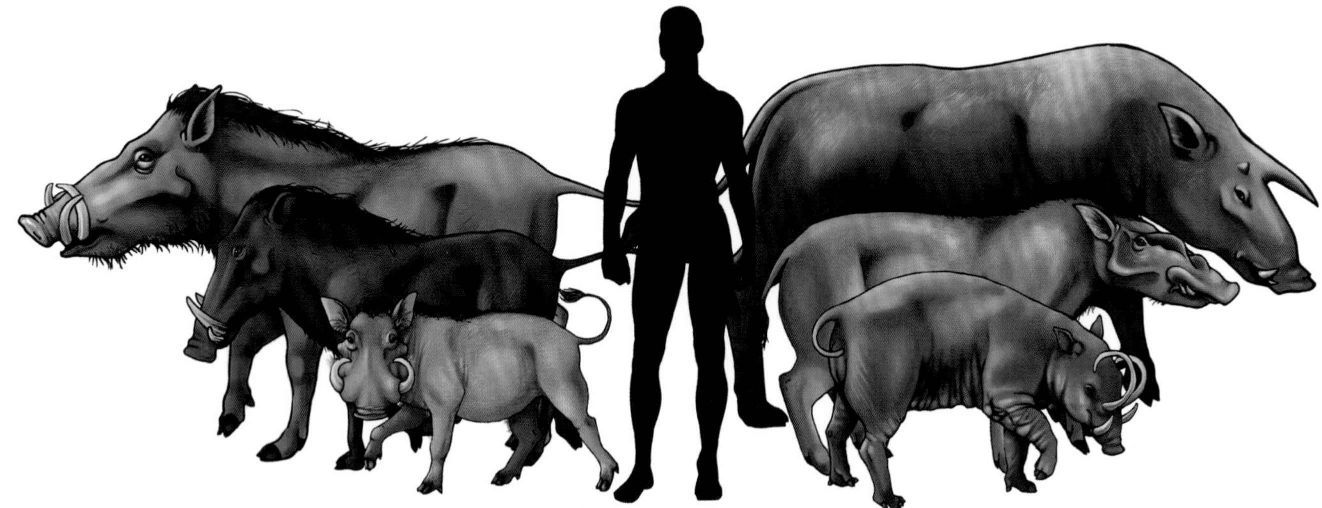

Figure 14.6. A spectrum of living and extinct pigs from the family Suidae. In the left foreground is the living warthog *Phacochoerus*, with the extinct warthogs *Metridiochoerus* (the smaller one) and *Notochoerus* (the bigger one) behind it. On the right is the living "deer pig," the babirusa, with the extinct multi-horned *Nyanzachoerus* behind it, and the horned *Kubanochoerus*, or "unicorn pig" in the background.

occupied by the javelinas, or peccaries (family Tayassuidae), which are not closely related to the true pigs (Suidae).

Pigs and peccaries can be traced back to a number of primitive fossils from the middle Eocene of China and Thailand. Contrary to statements that have been made by people unfamiliar with these fossils, these primitive "palaeochoerids" (including *Eocenchoerus*, *Egatochoerus*, *Siamochoerus*, and *Odoichoerus*) cannot be clearly assigned to either the Suidae or the Tayassuidae. More of these primitive forms, such as *Palaeochoerus*, *Propalaeochoerus*, and *Yunnanochoerus*, are known from the Oligocene of Asia and Europe, but true pigs do not really appear and begin to diversify in Eurasia until the early Miocene. At that time, they diverge into numerous subfamilies of the Suidae. These include the listriodont pigs, in which the simple blunt, rounded cusps of the cheek teeth are modified into cross-crests for more efficient leaf eating. These pigs appear to have become specialized herbivores and begun to abandon the diverse omnivorous diet of most other pigs. Another subfamily is the tetraconodont pigs, which have the last two premolar teeth hugely expanded, adapted for crushing bones or nuts, and have evolved four well-developed cusps on their molars (hence their name). The sanithere pigs have developed cusps and wrinkles on their molars that approach the W-shaped pattern seen on ruminant molars, and the wear on their teeth suggests that they adopted a grazing mode of life.

Another group, the kubanochoerines, were very large and long-legged, and the only pigs to develop horns. *Kubanochoerus* (Fig. 14.6) had a large blunt horn that protruded forward from between its eyes and may have served as horns and antlers do in ruminants: for status in males and for sparring with other boars. The largest species of *Kubanochoerus* was 1.2 m (3.9 ft) tall and weighed almost 500 kg (1100 lb), so it was cow-size.

By the late Miocene, pigs were evolving rapidly not only in Eurasia but also in Africa. There are so many different species in different lineages that extinct pigs are among the best fossils for biostratigraphic dating of fossiliferous beds in the Old World. In the late Miocene and Pliocene lived *Nyanzachoerus*, which had an array of large knobs and ridges on its face (Fig. 14.6). The giant warthog *Metridiochoerus* (Fig. 14.6), from the Pliocene and Pleistocene of Africa, had two pairs of tusks flaring out from the side of its jaws and incredibly complex convoluted grinding surfaces on its molars. It was over 1.5 m (5 ft) tall and weighed about 150 kg (330 lb). In its jaws it had a single elongate molar on each side, rather than multiple molars. Another, even larger African pig was *Notochoerus*, which has its four enlarged upper and lower tusks pointed forward and sideways (Fig. 14.6). Most of these lineages of pigs from the Miocene are now extinct, but wild pigs such as the warthog, giant forest hog, red river hog, and babirusa are the remnants of the more than 40 genera of suids that lived in the past.

Family Tayassuidae (Peccaries, or Javelinas)

While true pigs (family Suidae) evolved exclusively in the Old World, peccaries, as we have said, occupied the ecological niche for pigs in the New World. They were probably descended from the same middle Eocene Asian suoids (*Eocenchoerus, Egatochoerus, Siamochoerus*, and *Odoichoerus*) that also gave rise to true pigs. However they originated, peccaries were established in North America by 37 Ma and have evolved independently ever since.

Peccaries look superficially like pigs, but they have many important anatomical differences. The easiest to recognize is in their canine tusks (Fig. 14.7). No matter how bizarre their skulls become, peccary tusks always point straight up or down, while those of most pigs flare out to the side of the snout. Peccary skulls are narrower, shorter, and deeper than pig skulls, and the bottom of the peccary skull forms a flat plane, rather than the curve of the bottom of a pig skull from the tooth row to the braincase.

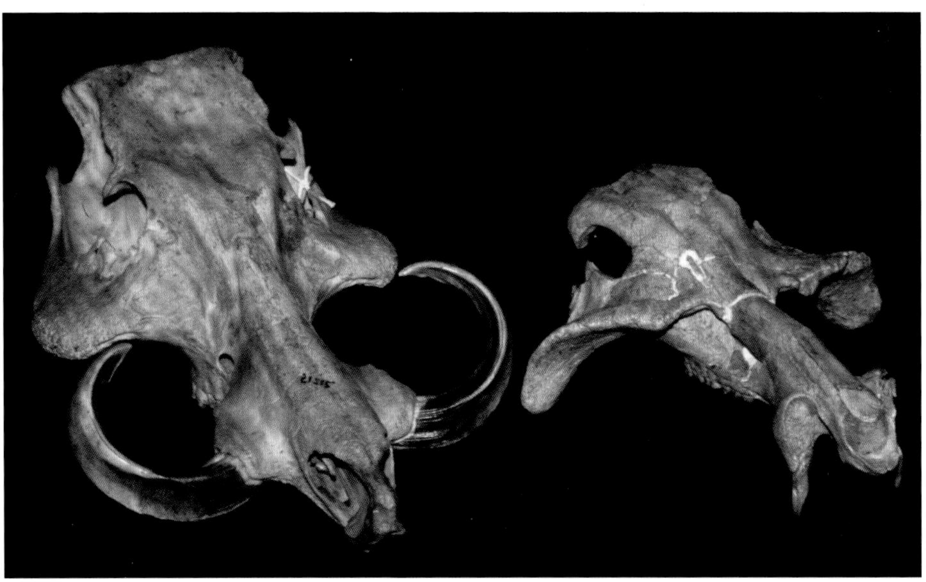

Figure 14.7. Comparison of a peccary skull (*Skinnerhyus*, on the right) and pig skull (a warthog, on the left). The most obvious difference is that pig tusks flare out widely to the sides of both the upper and lower jaws, while peccary tusks lie in the vertical plane and occlude precisely with one another. Peccaries have a smooth straight forehead, while pigs have a convex skull top.

Today there are only three (or possibly four) species of living peccaries (Fig. 14.8): the collared peccary (*Pecari tajacu*), the white-lipped peccary (*Tayassu pecari*), and the Chacoan peccary (*Catagonus wagneri*), which was known only from fossils before being discovered living in the dense brush of the Gran Chaco in Paraguay in the early 1960s. A possible fourth species, the giant peccary (*Pecari maximumus*) has been reported from the Amazonian jungles, although whether it is truly a distinct species is still disputed.

Peccaries first appeared in North America 37 Ma, with *Perchoerus minor*, from the Big Badlands, a very primitive peccary about the size of a house cat. Peccaries are also known from the middle Oligocene (32–30 Ma), from the many more skulls of the biggest *Perchoerus* species, *P. probus*, and from the late Oligocene, when a slightly more advanced peccary, *Thinohyus*, lived. During the early Miocene, there was a great evolutionary radiation of peccaries, including the advanced subfamily of hesperhyine peccaries, and the dwarfed peccaries placed in an invalid wastebasket genus, "*Cynorca*." In the late Miocene there was a huge radiation of different genera that had many bizarre flanges and ridges on their cheekbones (Fig. 14.8). As with the huge cheekbones on warthogs, these features probably served to show the dominance of adult males and also protected the eyes during fights among boars. At the end of the Miocene, most of these extreme peccaries vanished, leaving a lineage of flat-headed peccaries (*Platygonus*), which was more herbivorous than omnivorous and showing the beginnings of cross-crests on its molars, and the large, primitive long-nosed peccary *Mylohyus*, which had more generalized omnivorous teeth (Fig. 14.8). Both of these lineages lasted until the end of the last Ice Age in North America, then vanished, along with all the other large mammals, about 10,000 years ago (see Chapter 18).

Some North American peccaries managed to cross Central America and reach the Amazon basin by 10 Ma. From that point on, South America had a radiation of its own peccaries, resulting in the ancestral groups for most of the Central and South American peccaries alive today.

Family Anthracotheriidae (Anthracotheres)

Another group of pig-like creatures is the anthracotheres (Fig. 14.9), whose name means "coal beasts" (the first specimens of *Anthracotherium* came from a French coal mine). Anthracotheres had relatively primitive pig-like skeletons, yet unlike other suoid artiodactyls (which usually have teeth with blunt rounded cusps), they had upper cheek teeth with a strange mix of sharp W-shaped crests and small rounded cusps. Their teeth were suited to eating soft vegetation and possibly roots of aquatic plants. Most anthracotheres were the size of a large pig or a small hippo, but they had

Figure 14.8. A spectrum of living and extinct peccaries. In front (left to right) are the three living species, the Chacoan peccary (*Catagonus wagneri*), the white-lipped peccary (*Tayassu pecari*), and the collared peccary (*Pecari tajacu*). In the right middle is the ancestral Eocene–Oligocene peccary *Perchoerus*. Behind it is the more advanced Miocene peccary *Prosthennops*. On the left in the middle is the flat-headed peccary *Platygonus*, and behind it the larger long-nosed peccary *Mylohyus*. Behind the human is the *Skinnerhyus*, with its broadly flaring cheekbones.

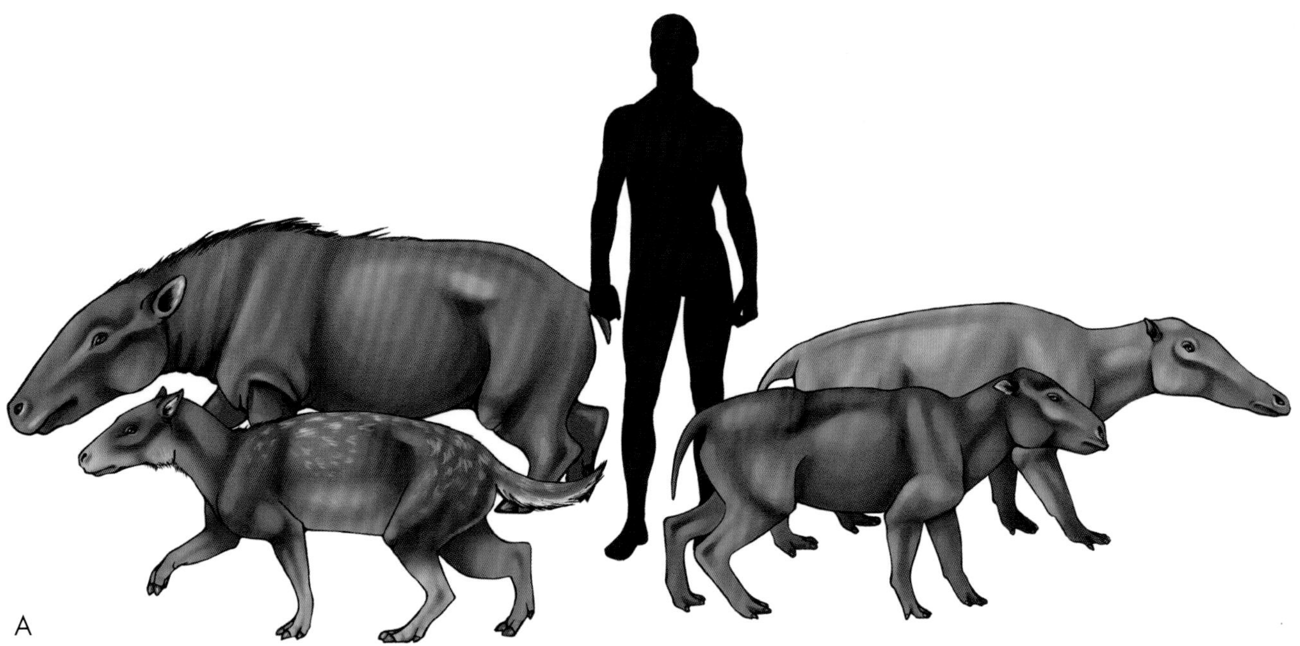

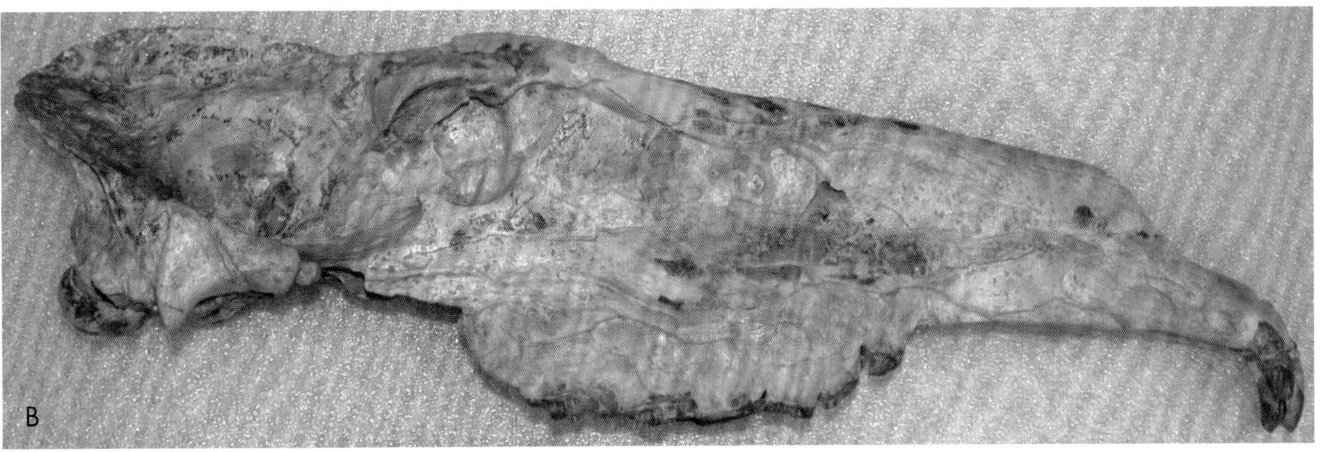

Figure 14.9. Anthracotheres. A. Reconstruction of a group of different kinds of anthraocotheres. On the left is the long-snouted *Aepinacodon* and the much larger Eurasian *Anthracotherium*. In the right foreground is the primitive form *Elomeryx*, and behind it the much larger *Bothriodon*. B. The bizarre skull of *Aepinacodon*, which had a long tubular snout and eyes elevated on the top of its head like periscopes.

a long, narrow head and snout that made them unlike either, or indeed unlike any other mammal that ever lived. They are usually thought to have been semiaquatic animals, like the pygmy hippo, and their fossils are usually found in ancient river and pond deposits as well as the coal swamps that gave them their name.

Anthracotheres first appeared in the late middle Eocene, about 40 Ma, with small genera such as *Siamotherium* and *Anthracohyus*, found in Burma and Thailand. Shortly thereafter, *Anthracotherium* appeared across Eurasia, and *Heptacodon* fossils appear in the middle Eocene beds of North America. From the late Eocene onward, anthracotheres were a common element in Eurasian faunas, migrating all over the world from their Eurasian base. There were occasional immigration events to North America in the late Eocene (*Bothriodon*; *Aepinacodon*, Fig. 14.9B), and again in the late Oligocene (*Arretotherium*). Anthracotheres were among the first Eurasian groups to spread to Africa in the late Eocene, when it was still the domain of natives such as proboscideans and hyraxes. They went through another big evolutionary radiation in Eurasia in the early Miocene, but by the late Miocene only one genus, *Merycopotamus*, remained. When it vanished at the end of the Miocene, this entire family became extinct. The true reason for the extinction may never be known, but some paleontologists think that anthracotheres were such specialized aquatic creatures that they could not survive the global cooling and drying events of the Pliocene. Or they may have been outcompeted by their descendants, the hippos.

WHIPPOMORPHA

Recent evidence shows that not every relative of the anthracothere lineage is extinct. Instead, the group evolved into hippopotamuses and into whales. The molecular evidence clusters whales and hippos entirely within a group that has been called Whippomorpha.

Family Hippopotamidae (Hippopotamuses)

Everyone is familiar with hippos from zoos and nature documentaries, but there is more to their story than is generally known. They are not just huge pigs specialized for aquatic lifestyles. Their name comes from the Greek words *hippos* (horse) and *potamos* (river), and they were called "river horses" by the Greeks and Romans. However, they have no connection to horses at all, but instead are artiodactyls and usually placed close to other pig-like forms. The latest combinations of molecular data and anatomical studies of fossils show that hippos are descended from anthracotheres and that their closest relatives are the whales, which are descended from a different kind of anthracothere relative, the raoellids (Fig. 14.10).

Despite the common myths, hippos do not eat water plants. They live and sleep in large lakes and rivers in the daytime, but at night they roam the banks of the waterside to graze on grasses that grow on dry land. The pygmy hippo, *Choeropsis liberiensis*, has a much narrower, smaller head and snout than the common hippo *Hippopotamus amphibius*. Pygmy hippos spend far more time on land and live mostly in the dense undergrowth of the Congo Basin of Africa.

The earliest known hippopotamus fossils appear in Africa in the early Miocene, with *Morotochoerus* and *Kulutherium*, and in the middle Miocene there are *Kenyapotamus*, *Archaeopotamus*, and their relatives (Fig. 14.10). As several scientists have noted, these early hippo skulls are extremely similar to the last of the anthracotheres, such as *Merycopotamus*. The kenyapotamines underwent a Miocene evolution and in the late Miocene were replaced by the modern subfamily Hippopotaminae. Most of these groups evolved in Africa, although one extinct genus, *Hexaprotodon*, occurred in southern Asia as well. Most of these fossil hippos were small, primitive forms, more like the living pygmy hippo than the huge aquatic hippos.

During the Ice Ages, hippos spread widely across Europe (including England), before vanishing at the end of the Pleistocene from everywhere except in their African homeland. In the Pleistocene, dwarfed hippopotamuses inhabited many islands, including Madagascar, Cyprus, Malta, and Crete (Fig. 14.11). Hippos on islands were forced to live on a much smaller resource base than mainland hippos, and so their body sizes were reduced, which was also a good strategy because they were not under threat from mainland predators.

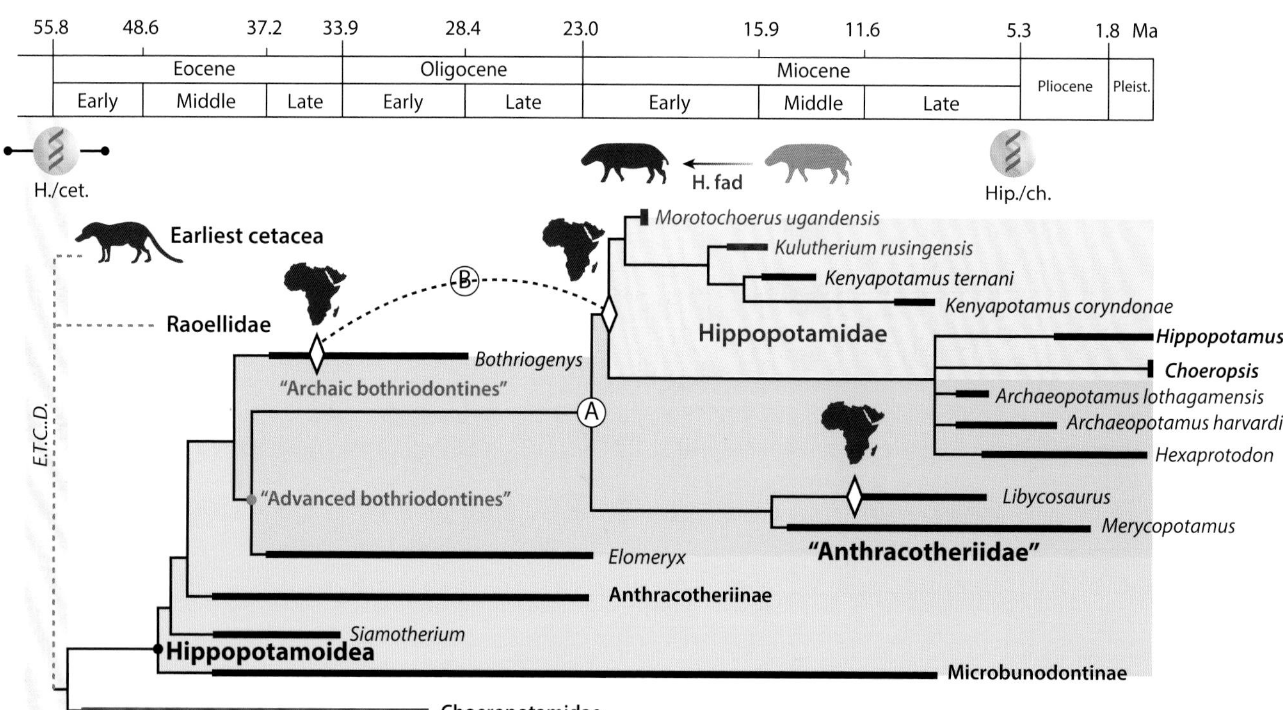

Figure 14.10. Evolution of the hippopotamids (blue area) from the anthracotheres (green and yellow areas). The branching event that separated raoellids and whales from anthracotheres is found in the left side of the diagram.

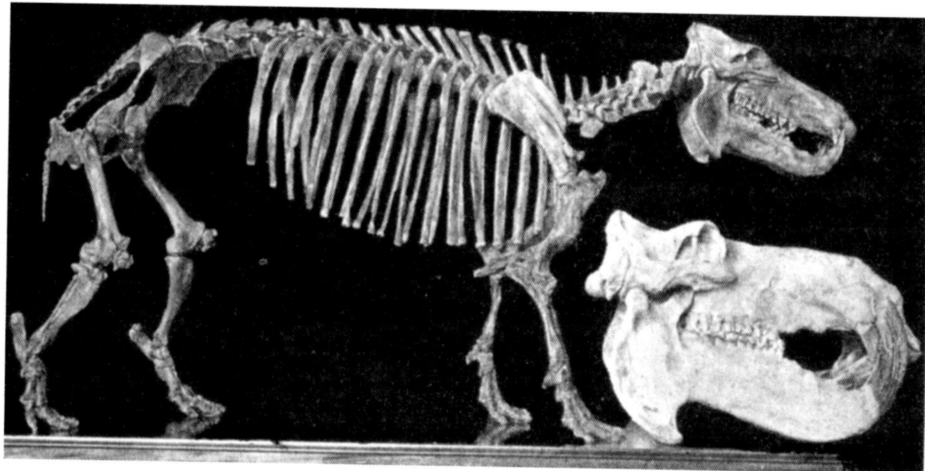

Figure 14.11. The skeleton of the dwarfed Madagascar Ice Age *Hippopotamus lemerlei*, with the skull of the living *Hippopotamus amphibius* for comparison.

Cetacea (Whales and Dolphins)

The idea that whales are mammals (and not fish) was first recognized by the Swedish biologist Linnaeus in 1758. A century later Charles Darwin speculated on how whales might have evolved from some large bear-like land predator that waded out to catch fish. But not until the 1980s and 1990s were the first really good fossils found that conclusively showed how whales originated from the artiodactyls. By the late 1990s, molecular data pointed to the whales being closest to hippopotamuses among living mammals, and in the early 2000s the connection between whales and the anthracothere-hippo lineage was established by a variety of fossils.

The oldest transitional whale is a fossil very close to the early anthracothere relatives known as *Indohyus*, a creature from the early Eocene of Kashmir. Even though it was barely larger than a rabbit and had long hind legs for leaping and the body of a chevrotain, it had distinctive anatomical features that make it the link between whales and other artiodactyls. Its ear region shows many features that are found only in the whales. It also has limbs of very dense bone (just as in the early whales, hippos and many other aquatic groups), which provides ballast to help it wade or dive under water without floating out of control. Chemical analysis of the bones showed that *Indohyus* was aquatic, but the chemistry of its teeth proves it ate land plants.

Just slightly younger than *Indohyus* was *Pakicetus*, from lower middle Eocene rocks in Pakistan (Fig. 14.12). Although most of its skeleton was wolf-like, with four long limbs for walking, it also had a skull that resembled that of the archaeocete whales, including the large serrated triangular teeth. Its brain was small and primitive, with no special features in the ear for hearing underwater and detecting faint sonar echoes, but it did have dense ear bones and other features suggesting some underwater hearing. *Pakicetus* was found in river sediments dating to about 50 Ma. These sediments indicate that it was a mostly terrestrial animal that spent a lot of time in the water. Indeed, its long legs with long hands and feet are adapted mostly for both running and jumping, but also for swimming. However, its limb bones, unusually dense to provide ballast in the water, suggest that it was primarily a wader, not a swimmer.

The next step was the discovery of *Ambulocetus natans*, whose name means "walking swimming whale." Its description is based on a nearly complete skeleton and additional fossils, all recovered from a nearshore marine deposit in Pakistan from about 47 Ma, which show that is truly halfway between a whale and a land mammal (Fig. 14.12). It was about 3 m (10 ft) long, the size of a large sea lion. In front it bore a long toothy snout like that of other primitive whales, with the same distinctive triangular blade-like teeth. Its ears were still not very specialized, nor were they suited for echolocation, but it probably used them instead for hearing vibrations through the ground or water. Behind the head it had long, strong limbs with really long fingers and toes (probably for webbed hands and feet). Thus, it is a four-legged whale that could both walk and swim, hence its name.

Studies of its spine showed that *Ambulocetus* could undulate its back up and down, as an otter does, as well as paddle with its feet, but it did not swim with a rigid torso like a seal or penguin. This kind of up-and-down spinal motion is very similar to that used in some whales, although most whales have a rigid torso and use only their tails for propulsion.

However, *Ambulocetus* was not a fast swimmer. Its crocodile-like proportions support the idea that it was an ambush predator. *Ambulocetus* probably lurked motionless under the water until prey came close, then caught its food with a rapid lunge. The location of the specimens in the nearshore marine rocks of the Upper Kuldana Formation suggests that these animals lived on the margins of lakes and rivers as well as the ocean shore. Chemical analysis of their teeth further proves that they lived mostly in fresh water.

A few years after the discovery of *Ambulocetus*, another nearly complete whale skeleton known as *Dalanistes* was found. Like *Ambulocetus*, it had fully functional forelimbs and hind limbs, but had even longer fingers and toes to support webbing on its feet. Its snout was much longer and even more whale like, as was its robust tail. It is followed in the fossil record by *Rodhocetus*, and

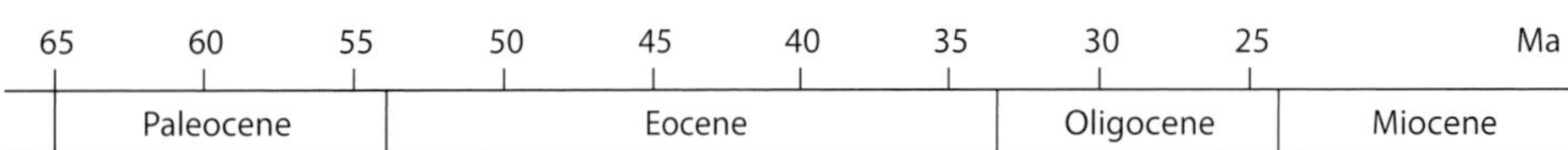

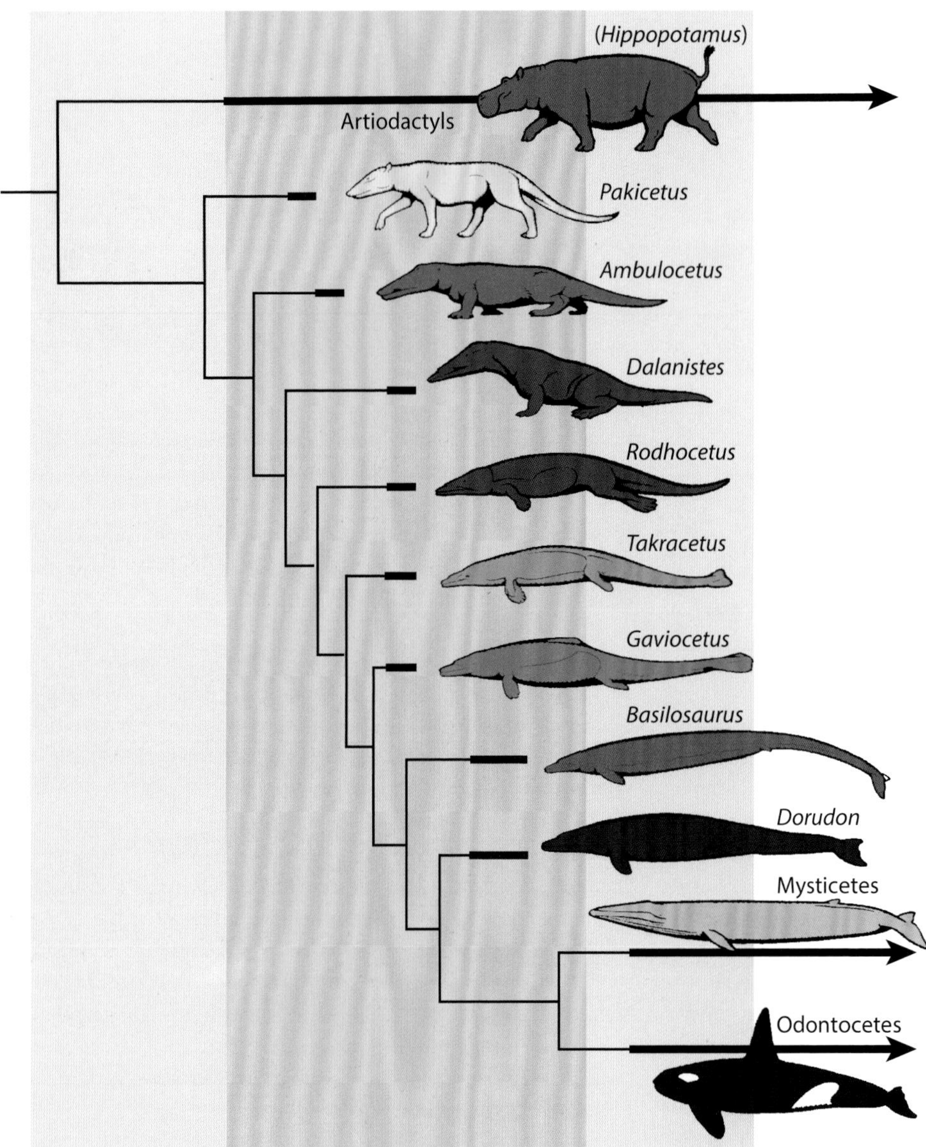

Figure 14.12. Evolution of whales from more primitive terrestrial forms and their anthracothere-hippo relatives to the fully aquatic whales without hind limbs.

since the discovery of that species numerous other transitional whales, such as *Takracetus* and *Gaviocetus*, have been found that have more and more reduced feet, and hands that had developed into whale-like flippers (Fig. 14.12). Their bodies are also more dolphin-like and show further development of tail propulsion (as in living whales).

This trend culminated with the huge basilosaurid archaeocete whales of the middle Eocene (Fig. 14.13), which were about 24 m (80 ft) long and weighed about 5400 kg (12,000 lb). They resemble some modern whales in having a long, pointed snout with triangular teeth for catching fish, but they are much more primitive than any living whale. For one thing, they do not have a blowhole near the top of the head (as all modern whales do) but have nostrils on the tip of the snout (as most other mammals do). Archaeocete ears are also very primitive, with no specialized ear bones like those found in modern whales that are adapted for echolocation in water.

Archaeocete hands and arms were modified into paddles, but no hind legs were found on the incomplete specimens dug up in the United States. Then in 1990 complete articulated archaeocete whale skeletons with their hind limbs still in place were found in Egypt. Only about the size of a human arm, on a whale over 24 m (80 ft) long, these hind limbs were no longer functional but mere vestiges from the time when whales still walked on four legs (although they still anchor the muscles to the pelvic floor). If you see a skeleton of a modern whale on

Figure 14.13. Skeleton of the Eocene archaeocete whale *Basilosaurus*.

display in a museum, look in the hip region just below the backbone and behind the ribcage. If the specimen is complete and mounted correctly, you will see the tiny nonfunctional remnants of hip bones and thigh bones, buried deep in their bodies, doing very little except proving that whales were descended from four-legged land animals.

By the late Eocene the roots of the modern whale families were evolving as the ancient archaeocetes declined. Modern whales have a variety of unique evolutionary specializations archaeocetes do not share. For example, the nasal opening (and the bones around it) has shifted from the tip of the long snout to the top of the head to form a blowhole, pushing back and deforming all the bones of the top of the skull behind the nose. The two groups of modern whales have both completely lost any external hind limbs (unlike archaeocetes), and both show many other anatomical specializations.

The first major group of living whales is the toothed whales, or odontocetes (Fig. 14.14), which today include the huge predatory sperm whales (of *Moby Dick* fame), the beaked whales, dolphins, porpoises, orcas, narwhals, belugas, and many others. Toothed whales use their conical teeth to catch active swimming prey, mostly fish, squid, seals, and other fast-moving aquatic lifeforms. Odontocetes have a number of unique specializations, including a single blowhole on their head (mysticete whales have a pair), an asymmetric head, specialized hearing and sound production to improve their echolocation abilities, and a fatty organ over the snout called the **melon**, which is used like a sound lens to direct beams of sound outward. According to some classifications, there are at least 10 different families of odontocetes, containing 46 genera and 70 species, as well as many more fossil families, genera, and species.

The oldest well-known fossil odontocete was *Squalodon*, the "shark-toothed whale," from the late Oligocene to middle Miocene (33–14 Ma). It was a medium-size whale, about 3 m (10 ft) long, distributed from Eurasia to North America to Argentina to New Zealand. Even though it still had the primitive serrated triangular blade-like teeth of archaeocetes, it had the nasal openings almost completely on the top of the head and a much shorter neck than any Eocene whale.

By the middle Miocene, most of the odontocete families were established, although with much more primitive representatives than their modern-day descendants. These include the sperm whales (family Physeteridae), which can be traced to fossils such as *Ferecetotherium* from the Oligocene of Azerbaijan and to a dozen other genera by the middle Miocene. The most impressive of these were *Acrophyseter* of the middle Miocene (Fig. 14.15A)

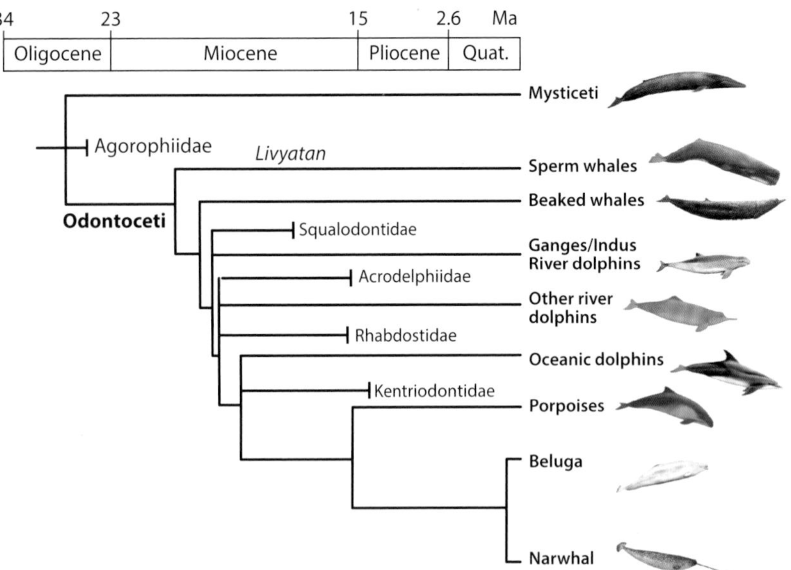

Figure 14.14. Evolution of the toothed whales, or odontocetes.

and the late Miocene sperm whale *Livyatan melvillei* (Fig. 14.15B, C). Unlike modern sperm whales, which have a toothless upper jaw and teeth only in their narrow lower jaw, these ancient sperm whales had wicked-looking curved conical teeth in both jaws. This suggests they were active predators of smaller whales and fishes, not a specialized feeder on giant squid as the modern sperm whale is. *Livyatan* was truly gigantic (18 m/60 ft long), the largest mammalian predator ever to swim the oceans. Its genus name is a homonym of the biblical sea monster Leviathan, and the species name refers to Herman Melville, the author of *Moby Dick*.

There are nine additional families of odontocetes, most of which have an extensive fossil record (Fig. 14.14). These include four families of river dolphins, plus the Phocoenidae, or true porpoises; the Ziphiidae, or beaked whales; the Delphinidae, or typical dolphins, plus orcas and related forms; the Monodontidae, or narwhals and belugas; and the Kogiidae, or the pygmy sperm whales. Most of these groups have fossil representatives by at least as early as the middle Miocene.

The other main group of whales is the baleen whales, or mysticetes. They are much less diverse than the odontocetes, with about 15 species in six genera in four families (Fig. 14.16). Rather than catch large swimming prey like fish or squid by rapid swimming and echolocation (as odontocetes do), baleen whales are filter feeders (Fig. 14.17), and they do not use echolocation. They gulp a huge volume of water containing large numbers of tiny plankton (especially the tiny crustaceans known as krill) into their mouths, then use their huge tongues and throat muscles to squeeze the water out of their mouth cavity. As they do this, they force the expelled water through a large filtering device known as baleen (made of fibers similar in composition to hair or fingernails) hanging from their upper jaw, which traps the food while letting the water through. As a consequence of this feeding adaptation, they have no use for teeth, and their jaws are huge arches of toothless bone built to support the baleen on the upper jaw (Fig. 14.17), and the jaw and throat muscles that surround it. They lose their embryonic teeth before birth.

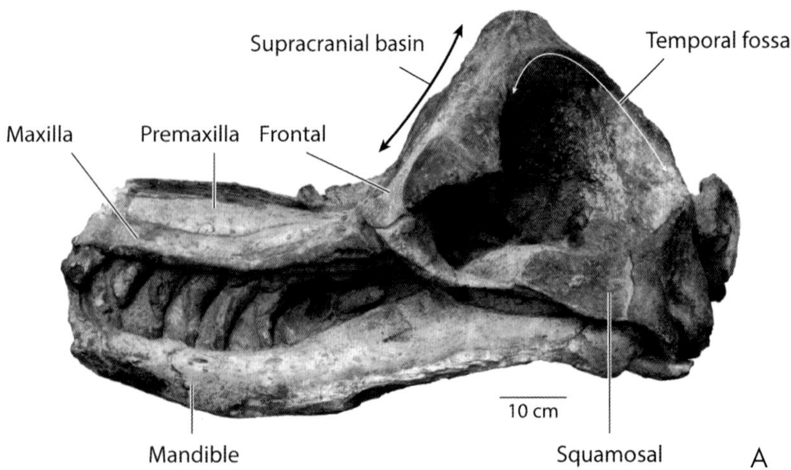

Figure 14.15. Fossil sperm whale relatives. A. Skull of the middle Miocene archaic sperm whale *Acrophyseter*.

Figure 14.15. Fossil sperm whale relatives. B. Reconstruction of the skull of the gigantic extinct sperm whale *Livyatan melvillei*, showing the huge curved teeth, which are very different from the simple conical teeth found only on the lower jaw of the living sperm whale. C. Reconstruction of *Livyatan* preying on a smaller whale.

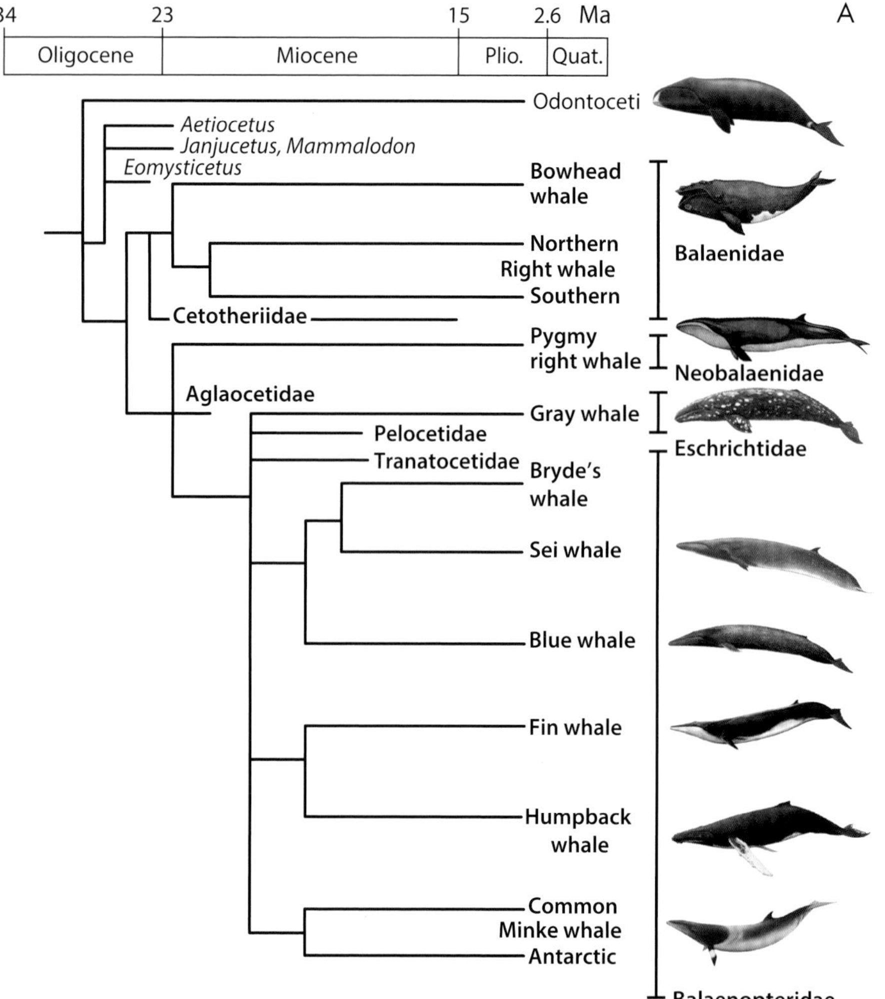

Figure 14.16. Evolution of the baleen whales, or mysticetes. A. Simplified family tree, showing the major families. B. Detailed family tree, showing most of the fossil species as well as their living relatives.

The earliest known mysticete, *Llanocetus*, from the late Eocene of Antarctica, has both its ancestral triangular archaeocete teeth and baleen in the mouth at the same time. By the late Oligocene, there were even more advanced toothed mysticetes with both baleen and teeth, such as *Janjucetus* and *Mammalodon* from Australia (Fig. 14.18A). In the early Oligocene, an archaic family of toothed mysticetes, the aetiocetids, appeared in Australia, but soon evolved to become common whales in the North Pacific region (Japan and North America) during the late Oligocene to Pleistocene (although the pygmy right whale may be a surviving descendant). Meanwhile, the first fossils of toothless mysticetes (*Eomysticetus*) appeared in the late Oligocene of South Carolina.

By the middle Miocene all four living families of mysticetes were well established, and numerous fossils are known from every ocean and continent (Fig. 14.16). These are the Balaenidae (right whales and bowhead whales); the Neobalaenidae (the pygmy right whales, possibly relict cetotheres); the Eschrichtiidae (gray whales); and lastly, the mighty rorquals of family Balaenopteridae, which includes the humpback whale, fin whale, minke whale, sei whale, and the immense blue whale (Fig. 14.18B). At 30 m (100 ft) in length and 190 metric tons (210 short tons) in weight, the blue whale is the largest animal alive—and is larger than any animal that has ever existed, including the biggest dinosaurs and the most monstrous of all the marine reptiles.

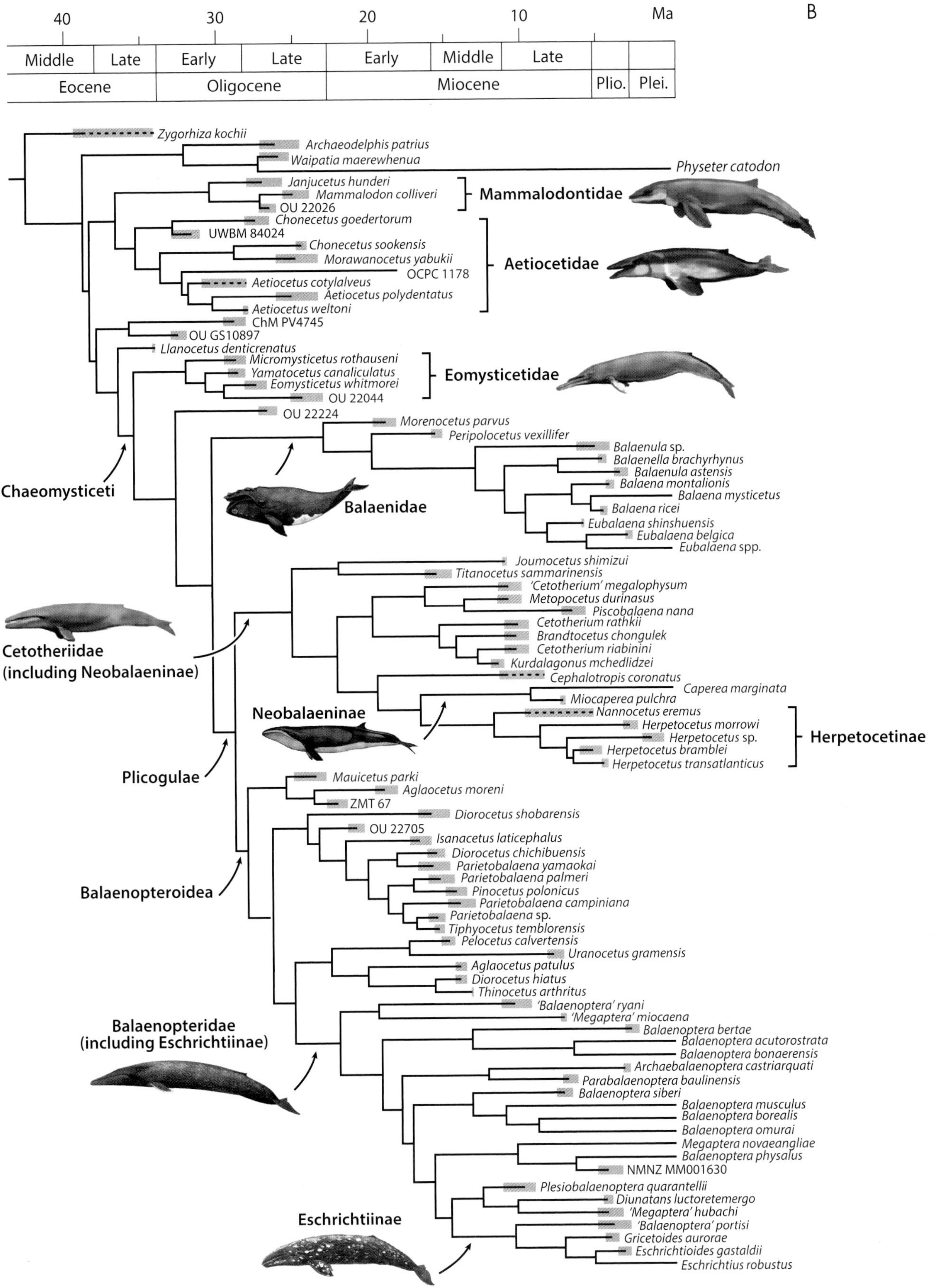

B
40
30
20
10
Ma
Middle
Late
Early
Late
Early
Middle
Late
Eocene
Oligocene
Miocene
Plio.
Plei.
Zygorhiza kochii
Archaeodelphis patrius
Waipatia maerewhenua
Physeter catodon
Janjucetus hunderi
Mammalodon colliveri
OU 22026
Mammalodontidae
Chonecetus goedertorum
UWBM 84024
Chonecetus sookensis
Morawanocetus yabukii
OCPC 1178
Aetiocetus cotylalveus
Aetiocetus polydentatus
Aetiocetus weltoni
Aetiocetidae
ChM PV4745
OU GS10897
Llanocetus denticrenatus
Micromysticetus rothauseni
Yamatocetus canaliculatus
Eomysticetus whitmorei
OU 22044
Eomysticetidae
OU 22224
Morenocetus parvus
Peripolocetus vexillifer
Balaenula sp.
Balaenella brachyrhynus
Balaenula astensis
Balaena montalionis
Balaena mysticetus
Balaena ricei
Eubalaena shinshuensis
Eubalaena belgica
Eubalaena spp.
Chaeomysticeti
Balaenidae
Joumocetus shimizui
Titanocetus sammarinensis
'Cetotherium' megalophysum
Metopocetus durinasus
Piscobalaena nana
Cetotherium rathkii
Brandtocetus chongulek
Cetotherium riabinini
Kurdalagonus mchedlidzei
Cephalotropis coronatus
Caperea marginata
Miocaperea pulchra
Nannocetus eremus
Herpetocetus morrowi
Herpetocetus sp.
Herpetocetus bramblei
Herpetocetus transatlanticus
Herpetocetinae
Cetotheriidae (including Neobalaeninae)
Neobalaeninae
Plicogulae
Mauicetus parki
Aglaocetus moreni
ZMT 67
Diorocetus shobarensis
OU 22705
Isanacetus laticephalus
Diorocetus chichibuensis
Parietobalaena yamaokai
Parietobalaena palmeri
Pinocetus polonicus
Parietobalaena campiniana
Parietobalaena sp.
Tiphyocetus temblorensis
Pelocetus calvertensis
Uranocetus gramensis
Aglaocetus patulus
Diorocetus hiatus
Thinocetus arthritus
Balaenopteroidea
'Balaenoptera' ryani
'Megaptera' miocaena
Balaenoptera bertae
Balaenoptera acutorostrata
Balaenoptera bonaerensis
Archaebalaenoptera castriarquati
Parabalaenoptera baulinensis
Balaenoptera siberi
Balaenoptera musculus
Balaenoptera borealis
Balaenoptera omurai
Megaptera novaeangliae
Balaenoptera physalus
NMNZ MM001630
Plesiobalaenoptera quarantellii
Diunatans luctoretemergo
'Megaptera' hubachi
'Balaenoptera' portisi
Gricetoides aurorae
Eschrichtioides gastaldii
Eschrichtius robustus
Balaenopteridae (including Eschrichtiinae)
Eschrichtiinae

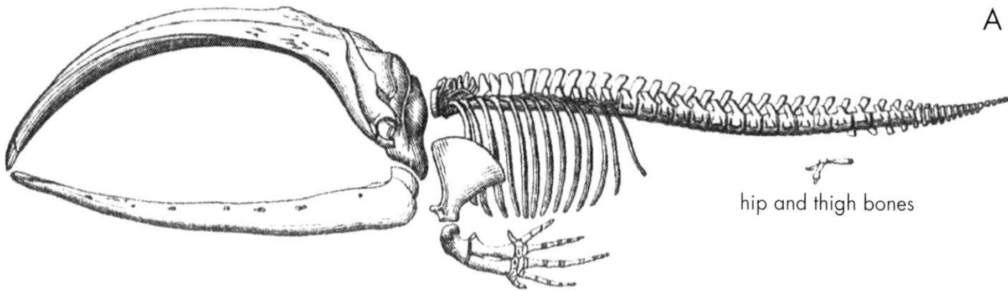

Figure 14.17. Baleen whale anatomy. A. Baleen whale skeletons have toothless jaws that form long arcs of bone. Hanging from these toothless jawbones is a screen of baleen that helps filter food out of the seawater. The whale takes a gulp of seawater full of prey and then forces the water back out through the baleen filter using its throat muscles and huge tongue. The vestigial hip and thigh bones are visible. B. Photograph of a fin whale mouth feeding, showing the baleen hanging from the upper jaw.

Figure 14.18. Baleen whales. A. Skull of the archaic baleen whale *Janjucetus*, which still had archaeocete-like teeth but also had the beginnings of the features of baleen whales, including a screen of baleen on their upper jaw.

Figure 14.18. Baleen whales. B. The blue whale is the largest animal that ever lived, bigger even than any dinosaur.

TYLOPODS

Another major branch of the artiodactyls is the Tylopoda ("swollen foot" in Greek, in reference to the way their ankle and toe bones spread at their tips). The only living group of tylopods is the camels, llamas, and their relatives, but a number of fossil groups seem to be related to them. Tylopods have traditionally been closely allied to the Ruminantia (discussed later), since living camels have a ruminating stomach (although not the completely four-chambered stomach of true ruminants). The tylopods have many additional similarities with the ruminants, especially in the details of their long limbs with fused hand and foot bones and their similar teeth. However, most of the recent molecular analyses of artiodactyl relationships put the tylopod branch as separate and distinct, not only from the suoids and whippomorphs but even from the ruminants. In this arrangement, tylopods are the first group to split off from the basal artiodactyl tree, while suoids and ruminants form a separate branch of closer relatives. Whatever the eventual resolution of this controversy, tylopods have always been considered a distinct suborder within the order Artiodactyla.

Family Camelidae (Camels, Llamas, and Their Relatives)

Everyone is familiar with stereotypes about camels as "ships of the desert" that can store water in their humps and walk for days without drinking. Most of these ideas are myths. The humps store excess body fat to help them during periods of starvation, but they are not water reservoirs. More important, four of the six living species of the Camelidae—the llamas, alpacas, vicuñas, and guanacos, all from South America—do not have humps. Only the living dromedary camels native to Africa and Arabia and the Bactrian camel from central Asia have humps. Likewise there is no evidence from the skeletons of most fossil camels to suggest that they had humps. The proper mental image of a "typical camelid" is not the humped Old World camel but the humpless llamas, guanacos, or vicuñas of the New World instead.

Despite our association of camels with African and Asian deserts, or with South America's llamas and their relatives, camels appeared first in North America, about 40 Ma, and were restricted to this continent through most of their history

(Fig. 14.19). They did not escape to the Old World until about 6.5 Ma and did not cross into South America through the Panama land bridge until about 3.5 Ma (Fig. 5.4). The earliest known camel is the tiny rabbit-size creature known as *Poebrodon* from the late middle Eocene of Utah and California. Even though it looks like the primitive artiodactyls known as diacodexeids (Fig. 14.3) in many features, *Poebrodon* shows several of the distinctive features of the skull and ear region (especially a large bony bubble-like chamber around the ear region) found in the camel family but not in any other early artiodactyl. In addition camels had very high-crowned teeth at the beginning of their evolution, and even as other artiodactyls also developed such teeth, camels often had the highest crowns of any group. This might suggest that they were already adapted to eating very abrasive gritty vegetation. But living camels are opportunistic feeders and will eat almost any green vegetation, not just gritty grasses.

During the late Eocene and early Oligocene there were primitive camels such as *Poebrotherium*, best known from the Big Badlands of the Great Plains, which was about the size and proportions of a small gazelle. *Poebrotherium* was one of the longest-limbed, fastest-running animals of its time—long before horses or other mammals developed long slender limbs.

By the late Oligocene, camels had radiated into a variety of distinct groups that further diversified during the Miocene (Fig. 14.19B). One group, the stenomylines, became even smaller and more delicate and gazelle-like than more primitive late Oligocene camels such as *Paratylopus*. The most unusual feature of stenomylines, however, is their large, extremely high-crowned molars, with roots that go to the very bottom of their jaws. Another group of early Miocene camels, the floridatragulines, developed very long slender snouts that almost made them look like a cross between a camel and a crocodile. They lived only in the subtropical regions of Florida, coastal Texas, and Central America. A third lineage, the oxydactylines, began to get larger by developing extremely long slender legs and a longer and longer neck. One genus, *Aepycamelus*, was the giraffe equivalent in North America, since true giraffes never lived here (Fig. 14.19C). With its long neck, *Aepycamelus* towered 5.5 m (18 ft) above the ground, and it weighed as much as 700 kg (1600 lb).

By the middle Miocene camels were extremely diverse all over North America, with at least nine genera and 17 species, found especially in fossil beds of the Great Plains and Rocky Mountains. Along with the gazelle-like stenomylines and giraffe-like aepycamelines, there were the relatively short-legged camels with slightly lower-crowned teeth known as miolabines (*Miolabis*, *Nothotylopus*) and another group with slightly longer legs and longer snouts known as protolabines (*Protolabis*, *Tanymykter*, *Michenia*). In the Pliocene and Pleistocene, some camels got truly gigantic.

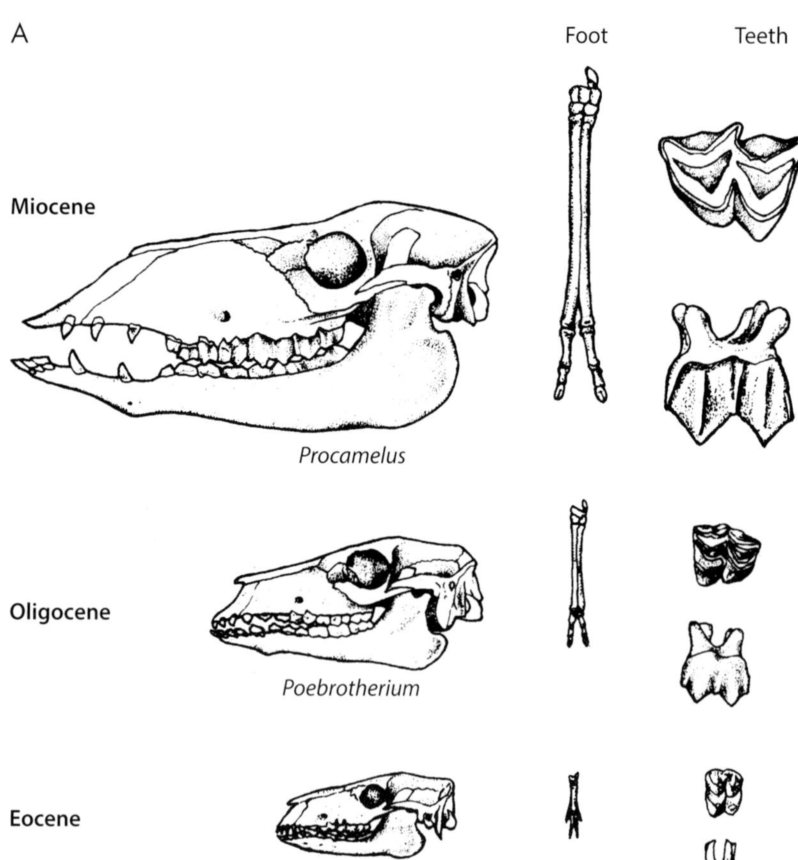

Figure 14.19. Camel evolution. A. Simplified view of camel evolution from the early 20th century, showing the increase in size, skull and snout length, limb length and fusion, and higher crowned teeth. (*Protylopus* is not a camel, but an oromerycid similar to the most primitive camels.)

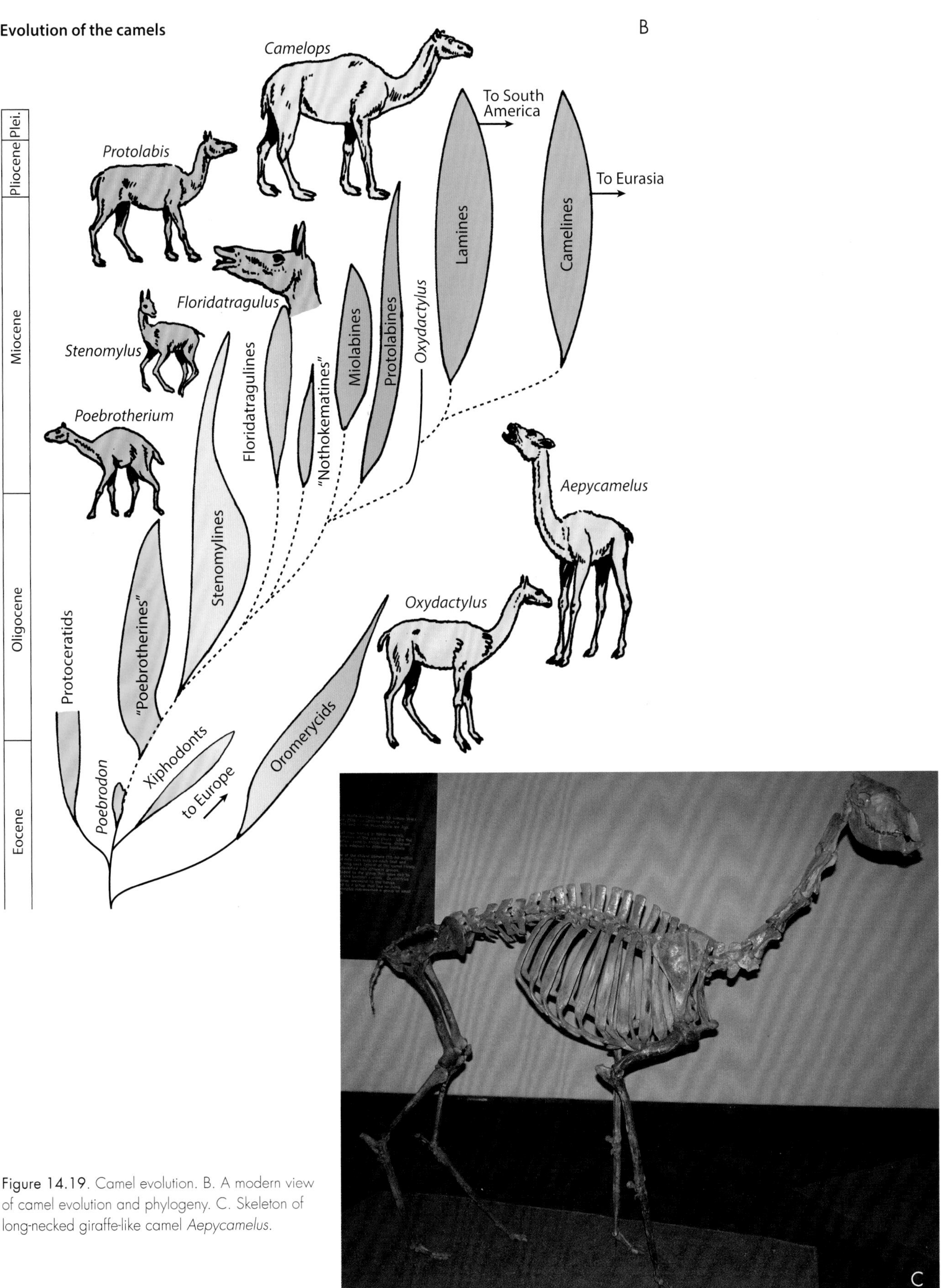

Figure 14.19. Camel evolution. B. A modern view of camel evolution and phylogeny. C. Skeleton of long-necked giraffe-like camel *Aepycamelus*.

Titanotylopus and *Gigantocamelus*, as their names suggest, were huge, measuring almost 3 m (10 ft) to the top of the head and weighing up to 2500 kg (5500 lb). Just as the modern African savanna supports a wide variety of antelopes, wildebeests, zebras, giraffes, and other beasts, the Miocene savannas of North America supported a variety of camels performing the roles of giraffes, antelopes, and gazelles, along with American horses in the zebra roles, rhinos performing in the role of hippos, peccaries instead of pigs, and gomphotheres instead of elephants.

This great diversity of camels was decimated during the extinction at the end of the Miocene, apparently due to a global climate change resulting in colder and drier habitats that also wiped out many of the groups that dominated the Miocene plains in North America. These extinct groups included American rhinoceroses, protoceratids, palaeomerycids, musk deer, oreodonts, most pronghorns and horses, as well as a group of rodents with horns. One surviving camel group, the llamines, became important in the Pliocene and Pleistocene of the Americas, and spread across the Panama land bridge to evolve into llamas, alpacas, vicuñas, and guanacos (Fig. 5.4). The other group, the camelines, was restricted to North America except when some of them managed to escape to Eurasia about 6.5 Ma. That lineage evolved into the modern dromedary camels in Africa and Arabia, and the Bactrian camels of central Asia. Both llamines and camelines then vanished from their native homeland of North America at the end of the last Ice Age (along with the rest of the megamammals), leaving descendants only in South America and in the Old World.

Family Oromerycidae (Oromerycids)

In the middle and late Eocene there was a small family (only five genera) of primitive relatives of camels called oromerycids. For a long time, oromerycids were mistakenly considered camels, and older diagrams in popular books show one of them, *Protylopus*, as a very primitive camel (Fig. 14.19A). But in recent years, much more complete fossils have been found, and more detailed studies conducted, and today it is clear that oromerycids were a separate North American family closely related to the earliest camels. They have distinctive features of their teeth and skulls that show they are not camels, yet they are so similar that the mistake is easy to make. Only a few oromerycids are known from more than teeth, and only two (*Eotylopus* and *Montanatylopus*) are known from partial skeletons. Oromerycids parallel true camels in many ways, including developing more high-crowned teeth (though never as high-crowned as in any true camel), but their diminutive skeletons are still very primitive, with short limbs and no fusion of the leg and toe bones or reduction of side toes seen in ever the earliest camels. Oromerycids vanished during the Eocene-Oligocene extinction event, along with many other groups that could not survive the transition from the late Eocene forests to the drier scrublands of the early Oligocene.

Family Protoceratidae (Protoceratids)

Another distinctive North American group is the family known as the Protoceratidae (Fig. 14.20). Some scientists regard the protoceratids as tylopods and thus related to camels (based on features of the skull and skeleton), while others place them with ruminants (based on their teeth and ear bones). The earliest protoceratids of the middle and late Eocene looked much like primitive camels, although some apparently had a proboscis similar to that of a tapir hanging down from their upper lip. By the upper Oligocene, the beds of the Big Badlands yield striking creatures, named *Protoceras*, that had short blunt knoblike horns on the snout, over the eyes, and on top of the head. In the Miocene protoceratids began to diversify in developing different odd combinations of horns. *Syndyoceras* had a V-shaped horn over its nose and an inward-pointing curved horn over each eye (Fig. 14.20B). Its late Miocene descendant *Synthetoceras* had a slingshot-shaped horn on its nose and curved horns over each eye that pointed up and inward. Even more bizarre was a very small protoceratid, *Paratoceras*, which had curved horns over its eyes and a Y-shaped horn on top of its head that looks like a propeller on a beanie. The very last protoceratid, *Kyptoceras*, had a pair of short horns pointing up from its snout, and a pair of long horns pointing up and then forward from the back of its head.

Protoceratids are always rare fossils, and most are found in rocks that suggest they lived in dense brush habitats, especially around water. Their teeth always remain low-crowned, so they probably ate leaves and other soft browse. Their broad snout resembles that of a modern moose, also consistent with a diet of soft swampy vegetation. Their limbs remained short, with little fusion of the foot bones, so they were good for walking in marshes and not specialized for running like the limbs of the artiodactyls found in the dry plains. Protoceratids died out at the end of the Miocene, when most of the mammals of the American savanna also vanished, as the landscape became more grass-covered and the climate grew drier and colder.

Family Merycoidodontidae (Oreodonts)

By far the most commonly fossilized mammals in the upper Eocene–middle Oligocene beds of the Big Badlands, and through most of the Oligocene and Miocene of North America, were the oreodonts (Fig. 14.21). Despite their numbers and the great diversity of sizes and shapes that they demonstrate, evident in the huge number of fossils, they vanished near the end of the Miocene, along with some of the other mammals of the North American savanna. But their skulls and teeth are so common in the lower Oligocene beds of the Big Badlands that the old name for these rocks was the "turtle-oreodon beds," after the two most abundant fossils. Their classification and names were a mess for many years due to incompetent scientists misinterpreting crushed specimens and mistaking normal variability in a population for valid species distinctions, but today 25 genera and about 60 species are recognized and grouped in 12 distinctive subfamilies. These groupings tell us they were very diverse and

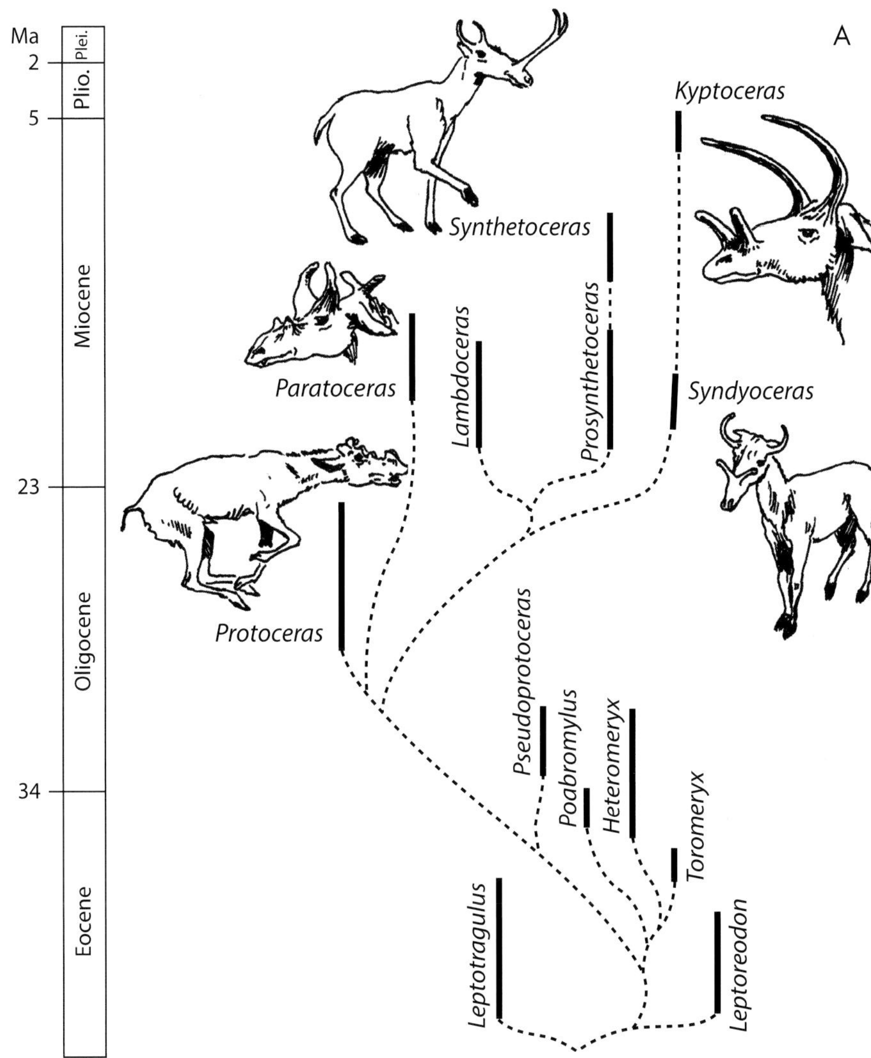

Figure 14.20. Protoceratids. A. Evolution of the protoceratids. B. Skeleton and reconstruction of the early Miocene protoceratid *Syndyoceras* with the V-shaped horns on its nose and head.

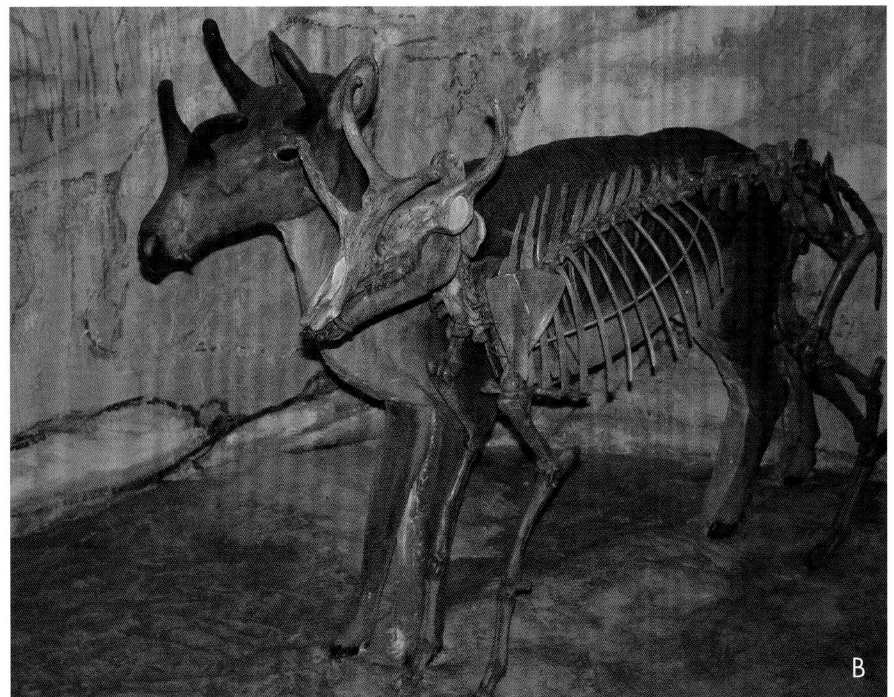

ecologically variable through a history that lasted about 33 m.y., from 40 Ma to about 7 Ma.

Oreodonts were typically sheep-size to pig-size, but they were not closely related to sheep or pigs. Most analyses place them closer to camels, although in older research they were considered suoids. Oreodonts did not resemble any living mammal, so comparing them with sheep or pigs or camels is not very informative or accurate.

During the late Eocene, oreodonts resembled many of the other primitive artiodactyls of the time, with unspecialized skulls and skeletons and short unfused limbs suitable for a variety of gaits. However, their teeth already had the crescent-shaped grinding crests seen on camels and ruminants; yet, most oreodonts retained low-crowned teeth throughout their long history and never developed into specialized grazers or fast runners. They were probably generalized feeders on almost any kind of vegetation in any habitat, which might explain why they are so abundantly fossilized. One side branch of oreodonts, the agriochoeres, such as *Protoreodon* and *Agriochoerus* (Fig. 14.21), had not only the primitive short legs and skull features but also a long tail and claws that suggest they may have been adapted to climbing in trees.

By the late Oligocene and early Miocene, the more primitive generalized oreodonts (such as the common Big Badlands oreodont *Merycoidodon*, Fig. 14.21, the fossils of which are sold in rock shops around the world) were replaced by a number of more specialized lineages. One group of tiny oreodonts, the leptauchenines, had relatively large heads, with the eyes and ears set high on the top of the skull, and openings around the nasal region suggesting some sort of specialized nasal apparatus (Fig. 14.21). Their skulls were also very deep, with higher-crowned teeth than seen in almost any other oreodont group. Another lineage, the merycochoerines, had long pig-like bodies, with shorter legs and a barrel-shaped trunk, and broad flaring cheekbones and snouts. Their build, similar to that of a small hippo or tapir, suggests they may have been more aquatic than the more generalized types. At the end of oreodont evolution there were highly specialized oreodonts, including *Brachycrus* (Fig. 14.21), which has highly retracted nasal bones, providing a large area for the attachment of a muscular proboscis, like that seen in modern tapirs. The last of the oreodonts, *Ustatochoerus*, from the middle Miocene, also had a short proboscis. The disappearance of oreodonts is a mystery, since they were the most abundantly fossilized mammals during the Eocene, Oligocene, and early Miocene, and then declined rapidly in the middle Miocene. Perhaps the brushy and swampy habitats favored by later oreodonts had disappeared; or perhaps they were outcompeted or were easy prey for the many new predators that invaded North America from Asia throughout the Miocene. Whatever the reason, they became extinct and we do not know why—and we still do not even know who their closest relatives were.

Figure 14.21. A diverse array of oreodonts. In left foreground is the tiny rabbit-size leptauchenine known as *Sespia*. In the right foreground is the small form *Miniochoerus*. Behind it is the long-legged *Merychyus*, which had high-crowned grazing teeth, and behind that is *Agriochoerus*, with the primitive long tail and claws instead of hooves. In the left foreground is the tapir-like *Brachycrus*, and behind it is *Leptauchenia*, which had the eyes and ears high on its head and extremely high-crowned teeth. In the back is the common Big Badlands oreodont *Merycoidodon*.

RUMINANTIA

Most living and fossil artiodactyls belong to the huge group known as ruminants, which are characterized by their distinctive, four-chambered stomach. They get their name from the step in their digestive process in which they "ruminate," or chew partially digested food, known as their cud (Fig. 14.22). When ruminants crop a mouthful of vegetation, they swallow this food and it enters the first two stomach chambers, the rumen and reticulum. Here it ferments in a vat of bacteria that help break down the indigestible cellulose that makes up most of the cell walls of plants. Once the ruminant has an opportunity, it regurgitates this material back to the mouth cavity and chews it some more ("chew its cud"), which breaks it down further. The cud then returns to the rumen and reticulum for further fermentation, before moving on to the third stomach chamber, the omasum, where much of the water and inorganic minerals are extracted from the food mass. Finally, it ends up in the abomasum (equivalent to the regular stomach in nonruminating animals), where stomach acids further break down the food. By the time it passes on to the small intestine, it has been completely broken down, so a very high percentage of the food eaten is digested. This foregut fermenting system, so called because the fermentation and breakdown of cellulose happens at the beginning of the process, is a very efficient process that gets maximum nutrition out of every bite of plant material. This means ruminants can survive on relatively small amounts of vegetation when times are hard.

By contrast, most other herbivorous mammals are hindgut fermenters (Fig. 14.22). When a horse or rhino or elephant eats a large volume of plant matter, the cellulose is not broken down until it has already passed through the stomach and small intestine, where it meets a dead-end chamber known as the caecum before it passes to the large intestine. The caecum starts the process of bacterial fermentation, but it is relatively late in the digestion process, so most of the cellulose passes through the gut undigested. That is why horse droppings are full of undigested plant fibers (as are elephant and rhino feces). Hindgut fermenters must compensate for this less efficient digestive process by eating larger volumes of plant material and passing it through their guts quickly to get the same amount of nutrition as a ruminant gets out of a smaller amount of food. Rabbits are another example of a hindgut fermenter, but they have an interesting compensation strategy—they eat their own fecal pellets, so the second time the food goes through their digestive tract, its cellulose is already partly broken down from its first trip through the gut.

Ruminants all have similar digestive tracts, but there are a number of other features they have in common as well. Almost all of them have their foot and toe bones (metacarpals and metatarsals) fused into a single element called a metapodial, or cannon bone (Fig. 14.2). This is the third and longest segment of their legs, and it contributes to their running speed. Most ruminants have highly reduced or completely lost side toes, so only two toes are important in their limbs.

The ruminants show a wide variety of cranial appendages, or "headgear," as well. Some, such as musk deer and chevrotains, have no horns or antlers, but the males have large canine tusks instead. Others, including members of the family Bovidae, grow permanent horns made of a bony core surrounded by a sheath of keratin (the substance that makes up our hair and fingernails). Giraffes have short horns (**ossicones**) made of a bony core covered by keratin and skin. Pronghorns have a bony core in their horns, covered by a keratinous sheath that is shed and regrown each year. Antlers occur only in the true deer (family Cervidae) and only in

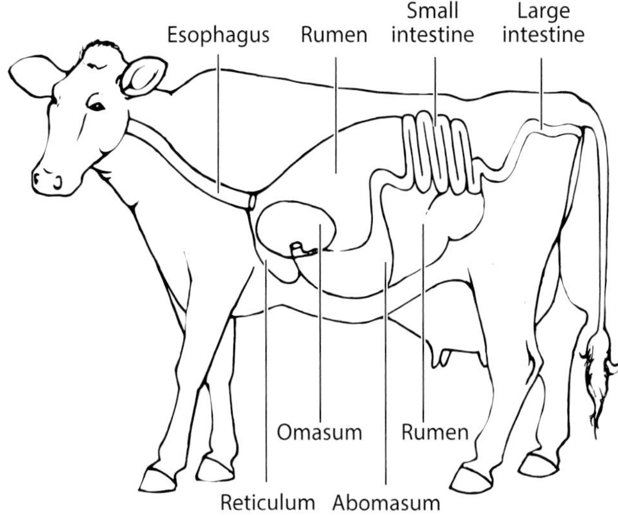

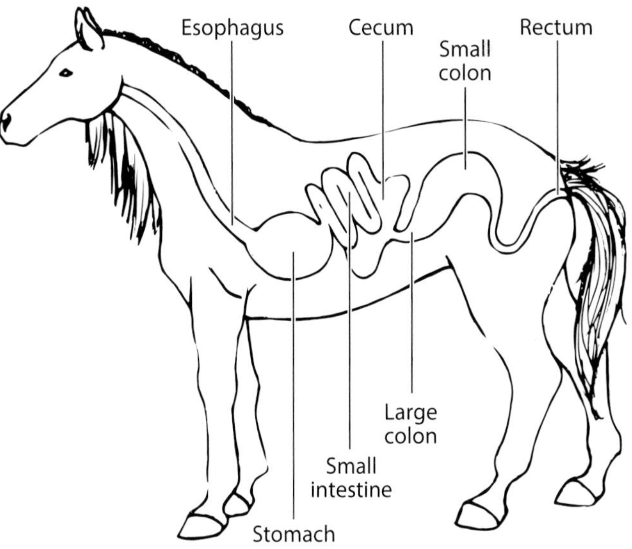

Figure 14.22. Comparison of the digestive tracts of a cow, a ruminant with a four-chambered stomach for foregut fermenting and a horse (a hindgut fermenter).

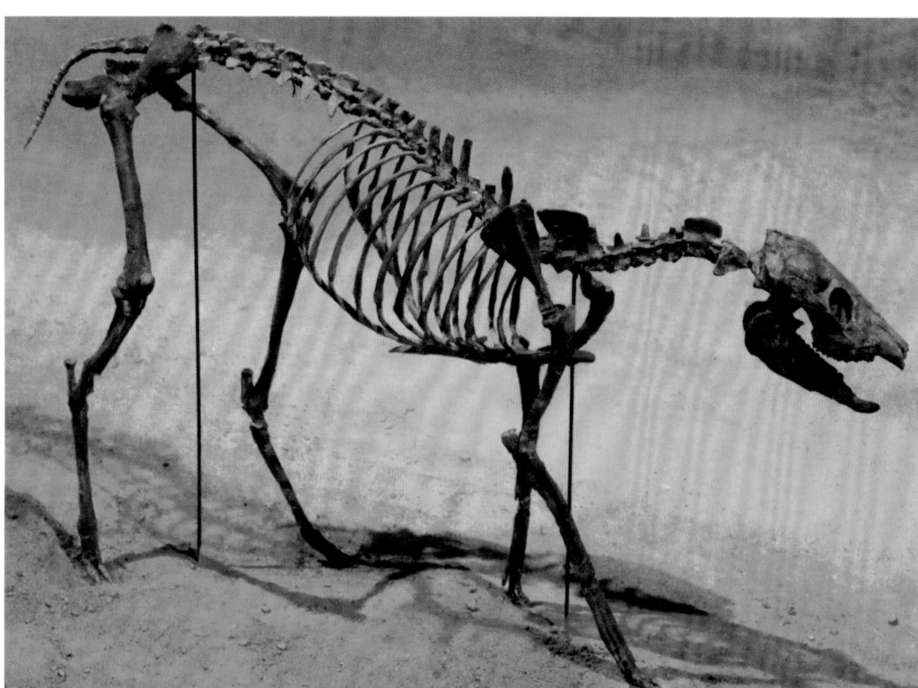

Figure 14.23. Skeleton of the primitive hornless ruminant *Leptomeryx*.

the males (bucks) but not in the does—except in reindeer, which have antlers in both sexes. Antlers, made entirely of bone, must be regrown each year and are then shed at the end of the breeding season in the winter. As these few examples show, there is a wide variety of different kinds of headgear in the different families of ruminants, and each family has its own distinctive pattern.

The earliest ruminants are known from teeth and skulls of mouse-size to rabbit-size creatures of the early middle Eocene (45 Ma) in eastern Asia. These include the specimens (including a beautiful skull) of *Archaeomeryx* from Mongolia and China as well as many other middle Eocene genera found in China (*Indomeryx, Miomeryx, Notomeryx, Lophiomeryx, Xinjangmeryx, Zhailimeryx*), Burma, and Thailand. By the late middle Eocene (40 Ma), very primitive rabbit-size ruminants such as *Leptomeryx* (Fig. 14.23), *Hypertragulus*, and *Hypisodus* were present in North America, and all three of these genera are common in the upper Eocene and lower Oligocene beds of the Big Badlands as well. By the early Oligocene, the endemic Eocene artiodactyls of Europe had vanished and been replaced by a wide variety of primitive hornless ruminants very similar to *Leptomeryx* in North America.

Family Tragulidae (Chevrotains, or "Mouse Deer")

The only living survivors of this great early Oligocene radiation of tiny hornless ruminants are the chevrotains, or "mouse deer" of tropical Africa and Asia (Fig. 14.24A). They have many features that are more primitive than the anatomical characteristics of most living ruminants. These include four functional toes on their forefeet, rather than two as in most ruminants, without any fusion of the central toes. Even though their stomach is four-chambered, the separation between the reticulum and abomasum is so weak that it is effectively three-chambered, like that in camels, hippos, and some pigs. In many ways, the chevrotains are "living fossils," anatomically between more primitive artiodactyls and the advanced ruminants, showing how the transition took place.

Living tragulids are tiny creatures, only about 50–100 cm (30–36 in) long, 20–36 cm (8–14 in) tall at the shoulder, and weighing 2–5 kg (5–11 lb). They have slender, hornless heads, but males have large canine tusks to compensate. Their eyes are large, to help them see their way in the dim light of the dense jungles of Africa and Asia where they live. They have very long, thin, delicate legs, allowing them to be relatively fast running in the dense undergrowth. Today there are only two genera and four species of tragulids: *Hyemoschus aquaticus* from tropical central Africa and three species of the genus *Tragulus* from India, Sri Lanka, and Southeast Asia.

Family Moschidae (Musk Deer)

Musk deer resemble chevrotains in some ways, as they are also small forest dwellers without horns but with large canine tusks in the males to compensate. They are slightly larger than chevrotains, weighing about 7–17 kg (15–38 lb), however, and in many other ways are more advanced than any primitive ruminant. Musk deer have a fully developed four-chambered stomach, and their forefeet and hind feet have only two main toes on each. The "musk" in their name refers to secretions from glands in the males' genital area that they use to mark their territory with scents that warn off other males. Unfortunately, in human commerce this musk is considered extremely valuable, as it is a source for the base of several kinds of perfume and is employed as well as in traditional Chinese "medicine" (yet it has no proven

Figure 14.24. Primitive ruminants. A. The living chevrotain, or "mouse deer," *Tragulus javanicus*. This is a female; males have long canine tusks. B–C. Skeleton and reconstruction of the extinct musk deer *Micromeryx*, from the Miocene of Europe.

medicinal properties). As a result, the animals that produce it are being hunted to extinction. Although only adult males produce musk—a single male may have only an ounce of musk in its gland—musk deer are hunted and trapped indiscriminately, including the females and young that lack musk glands.

The single living musk deer genus, *Moschus*, includes seven species that inhabit different forests in the mountains of Asia (China, Korea, Siberia, Mongolia, Tibet, and Nepal). However, the family Moschidae has a long history in the fossil record. The earliest fossil moschid is *Dremotherium* from the late Oligocene of Europe. During the early Miocene, moschids spread all over Eurasia, as evidenced by fossils of *Micromeryx* (Fig. 14.24B, C), *Hispanomeryx*, *Friburgomeryx*, *Pomelomeryx*, *Amphitragulus*, *Bedenomeryx*, *Oriomeryx*, and *Hydropotopsis*. They also spread to Africa, where a single genus (*Walangania*) occurred, although some scientists regard this fossil as a very primitive relative of the true deer (Cervidae). By the end of the early Miocene, most of these European moschids were extinct, except for *Micromeryx*, which managed to persist through the rest of the Miocene. The moschids survive in Asia, where today they are faced with extinction due to poaching, after a long and successful history lasting almost 30 m.y.

Also during the early Miocene moschids spread to North America, where they underwent their own local radiation, evolving into a subfamily known as the blastomerycines. Recent work on this group has reduced it from a taxonomic mess containing at least 22 species to just six genera and only eight species. These American musk deer largely resembled the living animals in most of their skeletal features, although late in their evolution they developed such specialized forms as *Longirostromeryx*, which had an unusually long snout compared to most moschids, containing highly reduced front cheek teeth. Blastomerycines vanished from North America at the end of the Miocene, along with rhinos, protoceratids, and so many other characteristic mammals of the Miocene savannas, probably as a result of the drying and cooling climate and the loss of forested habitats.

Family Giraffidae (Giraffes and Okapis)

Everyone is familiar with the living long-necked giraffes from zoos and nature documentaries, but giraffes and their relatives have a long and complex history in the fossil record as well. During most of their history giraffids had short necks and much more varied, and in some cases complicated, headgear than the living species (Fig. 14.25A). In fact the okapi, the only other living giraffid, is far more representative of the entire family than its familiar cousin (Fig. 14.25B). Okapis are large striped hoofed mammals with the characteristic head, teeth, and body of most giraffids but with a neck of normal length. They live in the dense jungles of the Congo Basin of central Africa and are so secretive and well camouflaged that they were not discovered by science until about 1900.

A wide spectrum of extinct giraffids resembling the okapi is known from the fossil record in Eurasia and Africa. Giraffids are first found as fossils in the early Miocene of Africa, in forms such *Canthumeryx*, a creature with a short neck and two long thin ossicones pointing sideways from above its eyes, and *Climacoceras*, with long thin branched ossicones that resembled deer antlers (Fig. 14.25A). The okapi, with its short stubby ossicones, can be considered a "living fossil" from this early Miocene radiation of giraffids.

During the middle and late Miocene, there was a great radiation of giraffids known as sivatheres in both Africa and southern Asia. *Giraffokeryx* had two sets of V-shaped ossicones over its eyes and its ears. *Brahmatherium* had thick conical ossicones atop its head. *Sivatherium* was a huge burly short-necked giraffid with broad palmate ossicones that resembled thick versions of moose antlers (Fig. 14.25C). This big animal was almost as tall as a living a giraffe, to 2.2 m (7.4 ft) at the shoulder and 3 m (10 ft) in total height, but was much more thick-necked and robust, weighing 500 kg (1100 lb). It apparently had a thick bulbous nose and upper lip resembling the snout of a modern moose. *Sivatherium* may have survived to as recently as 8,000 years ago, as it appears in some African cave paintings, and possibly on some ancient Sumerian figurines, from that period.

Another branch of the giraffids was the middle Miocene samotheres, which had two long ossicones that curved backward from the skull like a pair of bananas (Fig. 14.25A). Finally, in the middle Miocene, we find the earliest fossils (including *Bohlinia* of the middle Miocene of Greece and *Hunanotherium* from the late Miocene of China) that lead to the ancestors of the modern giraffe. Most of these fossil species had short necks, although a newly described specimen of *Samotherium major* (Fig. 14.25D)

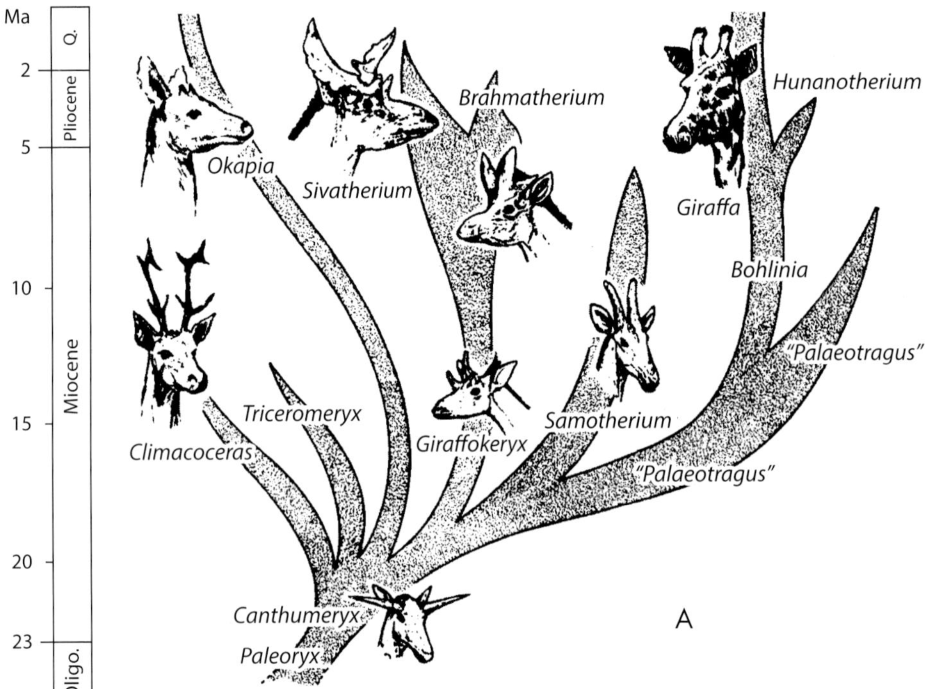

Figure 14.25. Giraffids. A. Family tree of the living and extinct giraffids since the Miocene–Pleistocene of Eurasia and Africa. Only the living giraffe has a long neck; its ancestors had many different kinds of horn-like ossicones.

Figure 14.25. Giraffids. B. The living okapi, a surviving short-necked giraffid typical of most of the group's history. C. The skull of the elephant-size giraffid *Sivatherium*. D. *Samotherium major* had a neck intermediate in length between the necks of a modern giraffe (top) and an okapi (bottom).

shows a neck that falls halfway between an okapi's neck and a modern giraffe's neck in length and structural development. This means the giraffid evolutionary record includes not only fossils representing a huge radiation of short-necked giraffes with many different kinds of headgear but also the transitional fossils that show how and when giraffes developed their long necks.

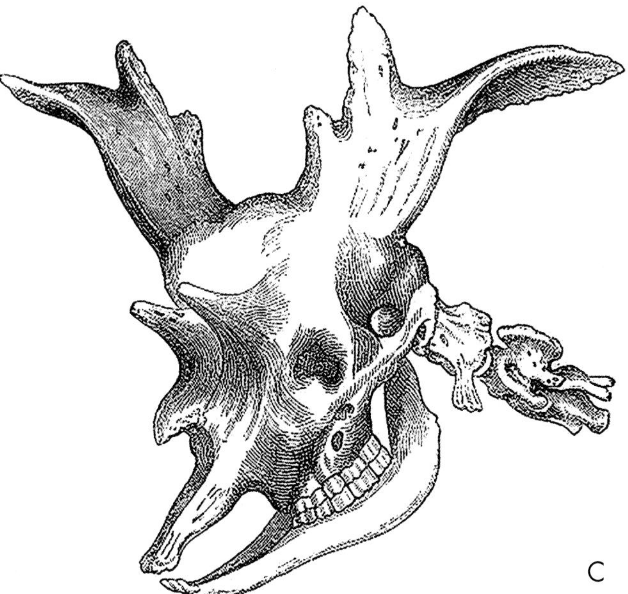

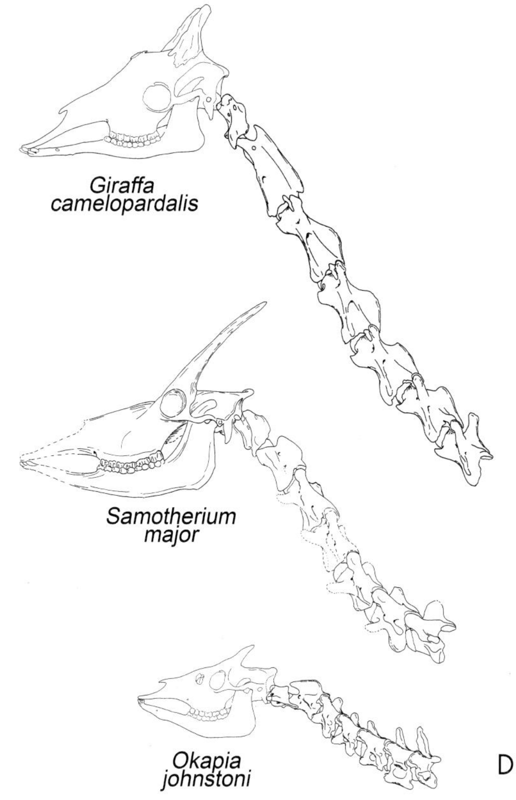

Family Antilocapridae (Pronghorns)

Roaming across the Rockies and Great Plains of North America is the single species of the family Antilocapridae, *Antilocapra americana*, the pronghorn. It is sometimes called the "American antelope," but this uniquely North American family has no close relations to the true antelopes of Africa and Asia (which are in the family Bovidae, along with cattle, sheep, and goats), so it is improper to call pronghorns "antelopes." Instead, they are great examples of parallel evolution, since they evolved to occupy the antelope niche in the American Miocene savannas and developed into a very diverse group with a wide variety of headgear, just as true antelopes did in Africa (Fig. 14.26).

The pronghorn is one of the fastest animals on land, running at speeds exceeding 88 km/h (55 mph). Despite extensive hunting, modern pronghorns still thrive on the grasslands of the western plains and Rocky Mountains, since they have no natural predators and can leap *through* (not over) the strands of a barbed-wire fence without breaking pace. In fact, some people have speculated that their great speed was developed due to the need to escape the Pleistocene American cheetah, which is now extinct.

The antilocaprid fossil record is impressive (Fig. 14.26). The family first appears in the early Miocene in North America, apparently evolving from a common ancestor with the giraffes that emigrated from Eurasia. The first radiation of pronghorns, during the early and middle Miocene, includes the small, delicate merycodonts. The best-known form, *Merycodus*, had short horns with two short branches, one pointed forward and one backward. Many of the merycodonts, such as *Ramoceros* and *Cosoryx*, had horns that branched into multiple tines (resembling deer antlers), while others had broad palmate horns (*Merriamoceros*), long straight horns with branched tips (*Paracosoryx*), and other variations. The more advanced antilocaprines diversified in the late Miocene and are represented by such bizarre forms as *Ilingoceros*, with long straight horns that have a corkscrew twist, and *Hexobelomeryx*, with six horns, three sprouting from each of its two horn bases. By the late Miocene, there were at least 12 different lineages of pronghorns, but most of these vanished during the end-Miocene extinction event. During the Pliocene and Pleistocene, the survivors included *Hayoceros*, bearing a pair of short two-tined horns in front and a pair of long straight upward-pointing horns behind; *Stockoceros*, which was much larger than the living pronghorn and had a pair of robust horns branched and pointing forward and back in a V pattern; and the tiny *Capromeryx*, with only a single short straight pointed horn over each eye. All these different pronghorns vanished in the great megafaunal extinction at the end of the Ice Ages in North America, leaving only the modern genus *Antilocapra*, the sole survivor of a great evolutionary radiation that dominated the herbivorous niches in North America.

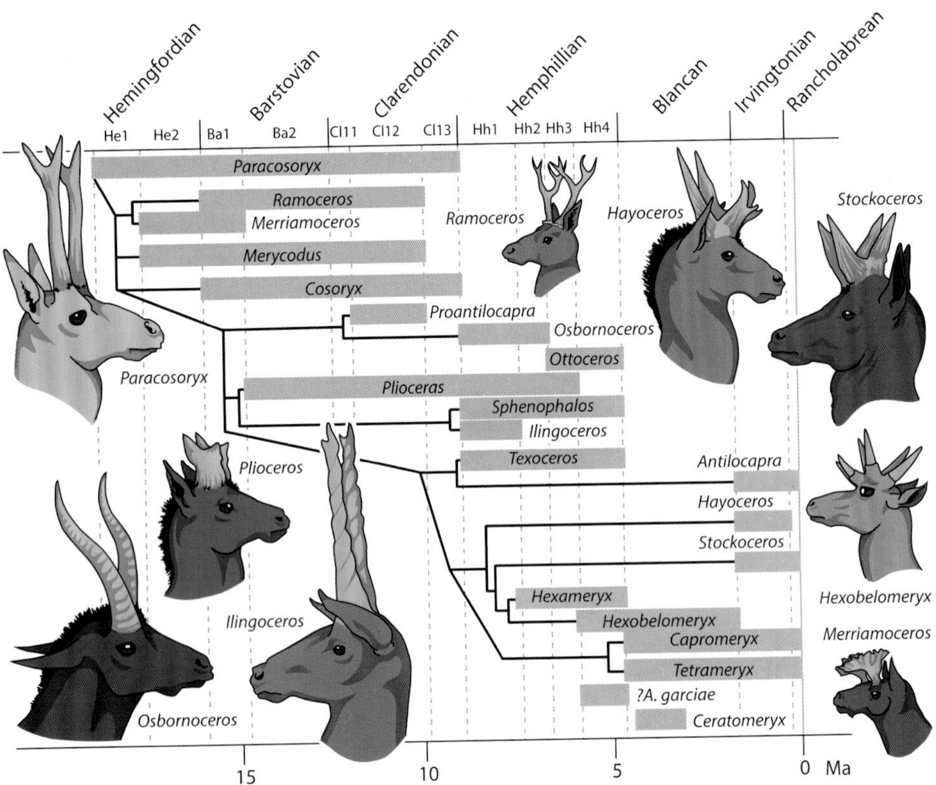

Figure 14.26. The evolution of the North American family Antilocapridae, the pronghorns, which are incorrectly called "antelopes."

Family Palaeomerycidae (Dromomerycids)

In the early Miocene of Eurasia an artiodactyl group appeared that was originally mistaken for true deer (Cervidae) or pronghorns (Antilocapridae). These ruminants have since been recognized as a distinct family, the Palaeomerycidae (Fig. 14.27). In many features and in their overall build they look quite deer-like, although they tend to have more primitive and lower-crowned teeth than cervids and probably ate only soft browse. But unlike deer (males), they did not have antlers, which are shed each fall and grown back in the spring. Instead the palaeomerycids had permanent bony horns that were never shed and apparently had a layer of skin around the bony horn core (similar to the condition in giraffes and okapis). This feature has made palaeomerycids controversial: Are they closer to giraffes, or are they closer to true deer? The anatomical evidence from the fossils suggests they are related to deer. Unfortunately, they are extinct, so we have no molecular evidence on the question.

Palaeomerycids first appeared in the early Miocene in European fossils such as *Ampelomeryx*, which have cranial appendages that make them look like a small moose. They migrated to North America about 18.5 Ma in three separate waves, each of which became the basis for a great evolutionary radiation of different subfamilies of palaeomerycids with a wide variety of cranial appendages. Many had relatively short limbs with broadly flaring hooves, which indicates they were probably browsers that lived in forests, like deer, and were not specialized for grazing or running as were the camels, pronghorns, horses, and many other denizens of the Miocene North American savannas.

One of these subfamilies, the Aletomerycinae, was a small group that included the short-horned but fast-running *Aletomeryx* (known from large quarry samples in western Nebraska) and the peculiar *Sinclairomeryx*, whose long cylindrical horns curved upward, swept back, and then pointed forward. Another subfamily, the Dromomerycinae, included the straight-horned *Dromomeryx*, the banana-horned *Rakomeryx*, and the moose-horned *Drepanomeryx*. The subfamily Cranioceratinae was the most diverse of the palaeomerycine groups. Beginning with the stubby-horned *Barbouromeryx*, the cranioceratines developed not only a straight upward-pointing horn over each eyebrow but a progressively longer horn that protruded upward and backward from the back of the skull, culminating in genera such as *Procranioceras*, *Cranioceras*, and *Pediomeryx*.

All of these North American palaeomerycines reached a peak in diversity in the middle Miocene, then declined rapidly. Only *Pediomeryx* survived to the end of the Miocene, when the group vanished in both Eurasia and North America. As with the North American rhinos, musk deer, protoceratids, and other groups that went extinct at that time, the palaeomerycids may have required more dense brush and a wetter climate than became available in the Pliocene world of cool, dry grasslands.

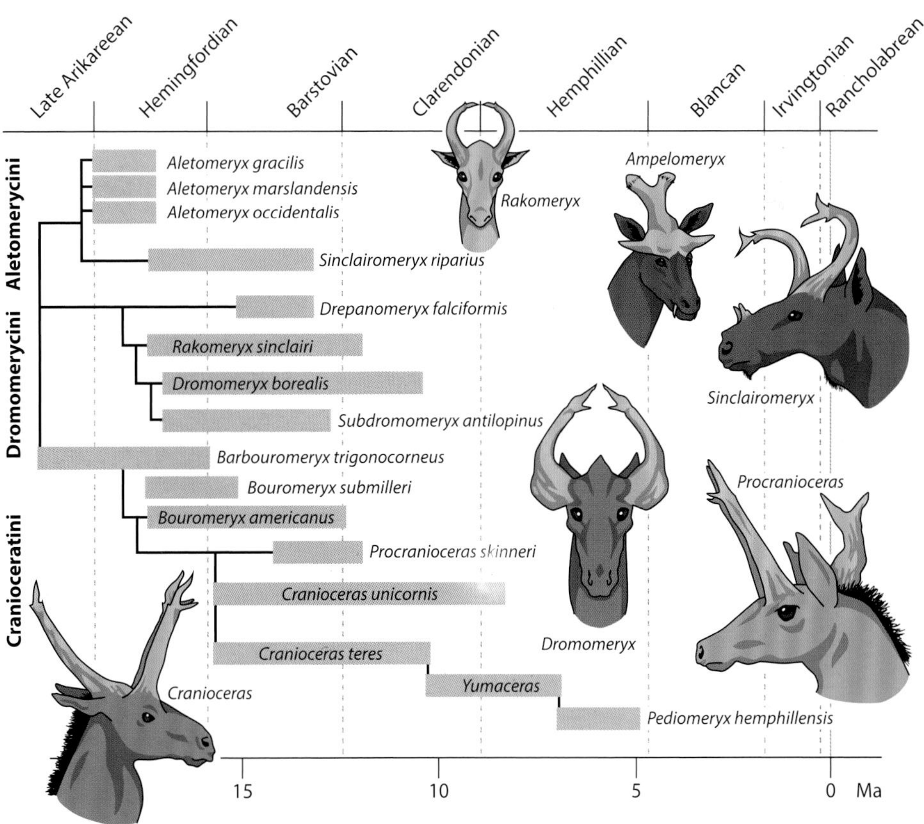

Figure 14.27. The evolution of the dromomerycine palaeomerycids.

In addition to these Northern Hemisphere fossils, a single jaw of a very primitive palaeomerycid, known as *Surameryx* and very similar to *Barbouromeryx*, has been just been described from beds laid down at least 10 Ma in the Peruvian Amazon. This fossil shows that palaeomerycids, along with tapirs, peccaries, and mastodonts, made it to South America long before the formation of the Panama land bridge about 3.5 Ma—but then vanished.

Family Cervidae (True Deer)

Among the living artiodactyls, the Cervidae is the second most diverse family alive and one of the most familiar. There are at least 20 genera and 90 deer species alive today, and hundreds more are known from the fossil record. The family Cervidae has numerous unique evolutionary specializations, the most familiar of which is the presence of antlers in males (bucks). Unlike horns in the bovids or ossicones in giraffes, antlers are not permanent; they are grown in a single year and then shed every winter. This makes them the fastest-growing type of bone known. In the early spring, the buck begin its antler growth, and for many months the growing bone is surrounded by a layer of fuzzy skin full of blood vessels called "velvet." When the antler is fully grown, by the fall breeding season, the buck rubs off the velvet to expose the bone and prepare for combat for breeding rights. As a buck matures, each season's antlers get larger, with more and more branches (tines), an indication of his size, strength, and maturity.

Deer first appear in the fossil record in the late Oligocene of Europe. A number of fossil deer evolved during the Miocene in Eurasia and North Africa, but most lacked the fully developed antlers found in the living cervids. Some middle Miocene cervids resemble the living deer with tiny antlers, such as the muntjacs, or forms with no antlers but large canine tusks, such as the Chinese water deer (*Hydropotes*). By the late Miocene, Eurasian deer had diversified into dozens of species in a number of genera, some with small antlers and others with large complex antlers.

One of the strangest of these was *Hoplitomeryx*, a bizarre form with permanent bony horns, rather than antlers (but still regarded as a deer relative, not a bovid). It had a long horn on its nose and a pair of pronged horns over its eyes (somewhat like the dinosaur *Triceratops*), plus long protruding canine tusks (Fig. 14.28). It is known only from cave deposits of the upper Miocene and lower Pliocene on the island of Gargano, off the southern tip of Italy. It is certainly one of the strangest creatures ever found.

BELOW: Figure 14.28. Skull of *Hoplitomeryx*, a strange Miocene deer with many different horns and tusks.

RIGHT: Figure 14.29. Skeleton of *Megaloceros*, the giant Ice Age deer incorrectly named the "Irish elk."

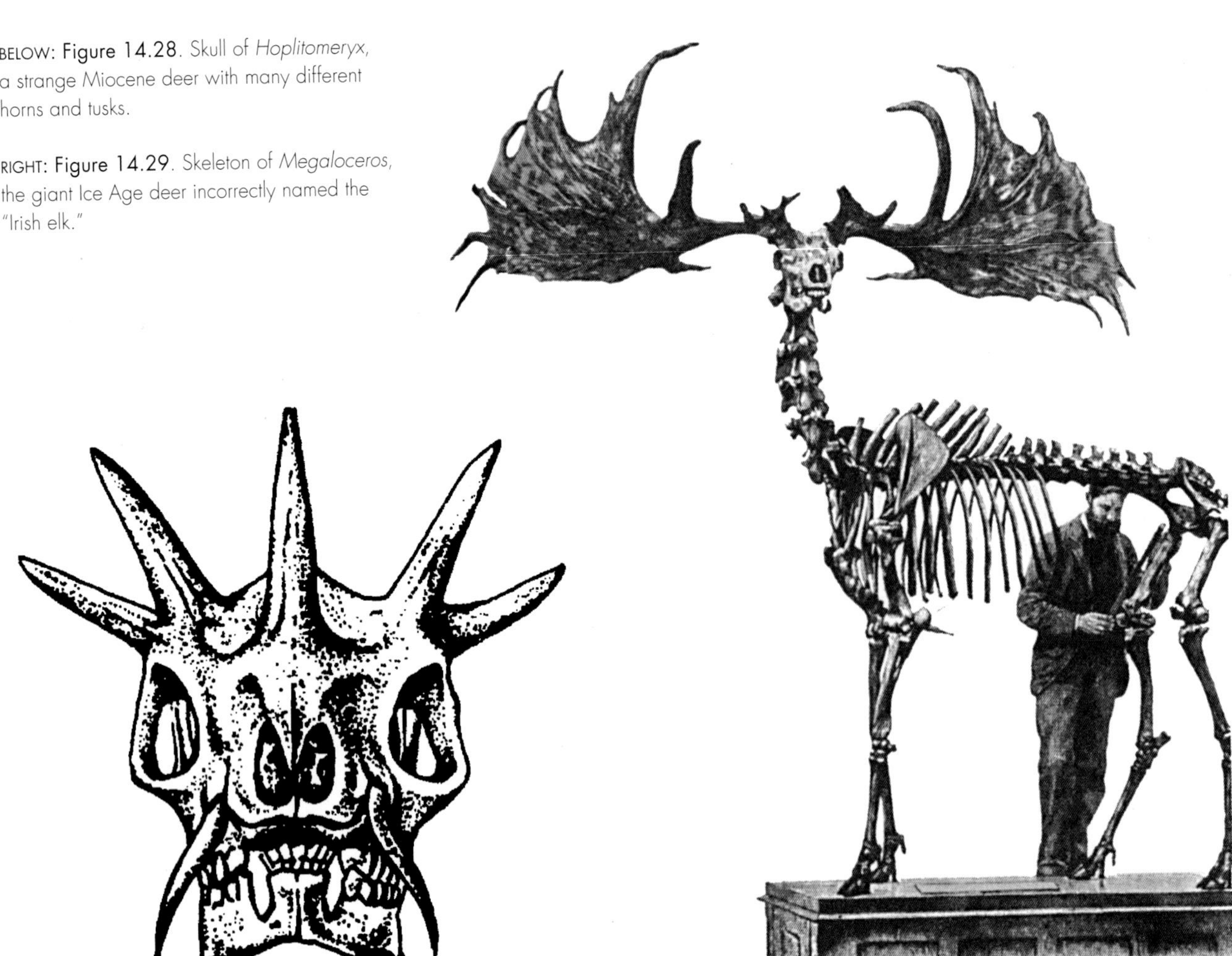

By the end of Miocene, cervids had emigrated from Eurasia to North America. The deer that evolved during the Pliocene and Pleistocene in North America include the extinct genus *Bretzia*, which had palmate moose-like antlers, and *Odocoileus*, the genus of the living white-tailed and mule deer. Some time before or since the formation of the Panama land bridge, about 3.5 Ma, deer migrated to South America along with the other invading mammals from the North American fauna. Once they were established there, an evolutionary radiation of uniquely South American deer occurred, leading to a modern diversity of six uniquely South American genera and dozens of species, including the brocket deer, pudus, huemuls (or guemals), marsh deer, and pampas deer.

The Eurasian continent has also yielded some truly spectacular fossil deer. *Eucladoceras* was the size of a moose and had huge antlers, over 1.7 m (5.6 ft) wide, with more than a dozen tines on each side. Most famous of the European mega-deer is the "Irish elk," *Megaloceros* (Fig. 14.29). Despite its name, it was not an elk nor was it exclusively Irish. Instead, it was a huge deer (related to fallow deer, not American elk or moose), bigger than the living moose, which was found in much of northern Europe during the Pleistocene. Its gigantic palmate antlers were somewhat like those of a moose, which in Europe is called "elk," hence the misnomer ("elk" in North America refers to the wapiti deer). Mature males of *Megaloceros* sported antlers that were over 3.5 m (11 ft) across and weighed about 45 kg (100 lb), an incredible amount of bone for a buck to grow each spring and carry on its head through the fall, then shed and regrow the next spring. Some scientists thought that their size meant that *Megaloceros* had evolved out of the control of natural selection, but that myth has been debunked. *Megaloceros* was the largest deer ever known, and its antlers are proportional to its huge size.

Family Bovidae (Cattle, Sheep, Goats, and Antelopes)

By far the largest family of artiodactyls is the Bovidae, which has 143 species in 60 genera and at least twice as many extinct species. Bovids are by far the most abundant large mammals on the planet, since they include not only wild cattle, goats, sheep, and antelopes, but also the huge herds of domesticated cattle, sheep, and goats. All bovids are distinguished by having a bony horn core around which there is a sheath of keratin, the hollow part of the horn. Unlike the antlers of cervids (shed every winter), or the horns of pronghorns (just the sheath is shed), the horns and sheaths of bovids are not shed, but accumulate growth layers throughout the life of the individual. Horns are particularly important in antelopes; the males use their more prominent horns not only for combat but especially for display, in recognizing their own species and impressing other males in competition for the females. Many bovids (especially cattle) use their horns for active fighting. In bighorns and many other sheep, the horns have shock-absorbing features so the animals do not suffer brain injury when they ram each other head to head.

Compared to other artiodactyls, bovids are a relatively recent addition to the landscape. Although there are a number of fragmentary jaws and teeth from the latest Oligocene and earliest Miocene that may be from bovids, the earliest well-known fossil bovid is *Eotragus*, from the late early Miocene (18 Ma) to middle Miocene in Eurasia and Africa. *Eotragus* weighed about 18 kg (40 lb), the same as a Thompson's gazelle. Males had simple straight conical horn cores about 8 cm (2.75 in) long, which with sheaths attached would have formed short straight horns. By the middle Miocene, there were about 15 genera, and the bovids had spread to India and China and become the dominant hoofed mammal group in the Old World and Africa. By the late Miocene the evolutionary explosion of bovids resulted in 70 new genera, a response to the expansion of dry grasslands in the tropics and subtropics (while cervids dominated the temperate latitudes). As the climate cooled and glaciers expanded, bovids found ways to adapt to the cold climates, so by the Pleistocene there were over 100 genera. They also spread to the New World in the latest Miocene, about 6 Ma. But the late Pleistocene extinctions pruned that down to 60 genera; North America was especially hard hit, although bison, muskoxen, mountain goats, and bighorn sheep survived there despite the losses of nearly all their relatives.

The interrelationships of the bovids are still controversial, as the older phylogenies based on fossils and anatomy give one set of results, and the newer molecular phylogenies give several different results. Part of the problem seems to be that the evolution radiation of the many branches of bovids happened so rapidly in the Miocene that the molecular differences between branches are not very great, so getting a clear-cut result is difficult. Alan Gentry of the Natural History Museum in London recognized two main branches based on fossils and living bones: the Boodontia (the cattle branch), which originated in Eurasia, and the Aegodontia (the antelopes, sheep, goats, and the rest), which diversified in Africa. The main tribes of bovids are all living today (Fig. 14.30), and most also have extensive fossil records.

Subfamily Boodontia, or Bovinae: Tribe Boselaphini contains the most primitive living bovids, the nilgai and the four-horned antelope, which are relicts of the Miocene bovid radiation and may be considered "living fossils." Even though these two species look similar to true antelopes and are sometimes called "antelopes," both molecular and anatomical data show that they are related to the cattle, as is the tribe Tragelaphini, or the spiral-horned antelopes (bongo, eland, kudu, nyala, sitatunga, and bushbuck).

The tribe Bovini, or the cattle, oxen, and their kin, is a large and very diverse group of animals, including the genera *Bubalus* (water buffalo, anoas), *Syncerus* (African Cape buffalo), *Bison* (American bison, wisents, and many extinct forms), and *Bos* (the familiar cattle, as well as yaks, banteng, gaur, kouprey, and the extinct Eurasian aurochs, thought to be ancestral to many of them). Bovines are among the most numerous and successful of all large land mammals, encompassing not only the wild herds

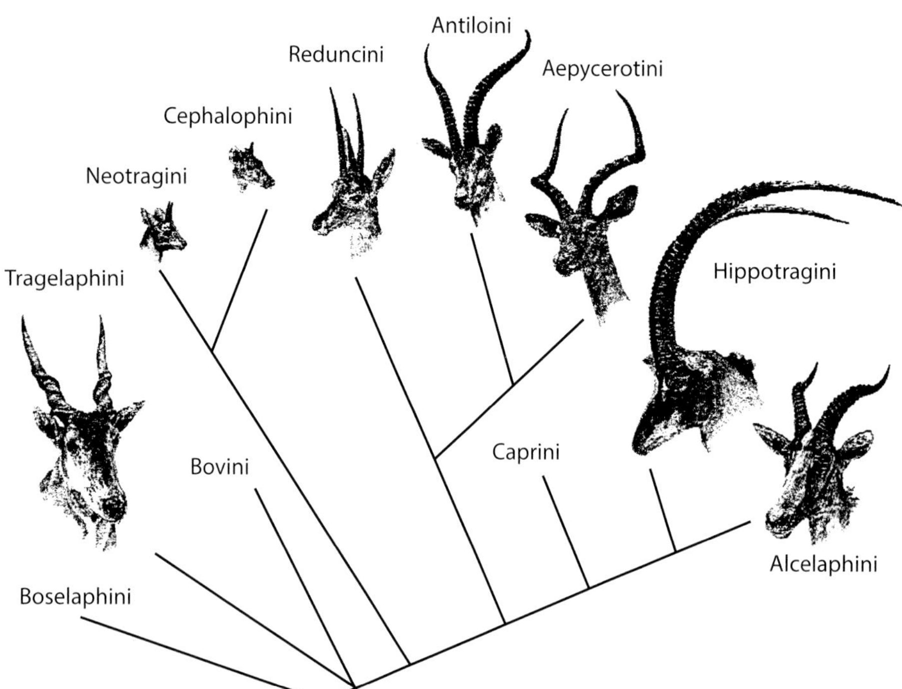

Figure 14.30. Family tree of the major branches of the bovids, including cattle and antelope. (Modified from Prothero and Schoch 2002.)

of water buffalo and African Cape buffalo and bison but also of course the huge numbers of cattle that humans have spread around the planet.

Bovine evolution is impressive. Gigantic bovines evolved many times, especially in the *Bison* genus. For example, there was a gigantic species from the Ice Age, *Bison latifrons*, that had horns spanning almost 2 m (6.5 ft) from tip to tip (Fig. 14.31). Several other large extinct bison are known, all of which were bigger than the living species. But *Bison latifrons* was not the largest extinct form of cattle to evolve. In the Pliocene and Pleistocene of Africa lived a genus of huge creatures named *Pelorovis* ("monster sheep"). Despite its name, *Pelorovis* is a gigantic bovine, not a sheep of any kind. This naming mistake was made with the discovery of some of the early forms from the fossil-hominid-bearing beds in Kenya and Tanzania, such as *Pelorovis oldowayensis* (Fig. 14.31), whose horns have a broad curve like those of a sheep (even though it is actually related to the bovine genus *Bos*). The huge Ice Age species *Pelorovis antiquus* (Fig. 14.31) wielded extremely long horns something like those of a giant longhorn bull—these revealed the species' true affiliation. *P. antiquus* weighed up to 2000 kg (4400 lb) and had a horn span about 4 m (13 ft) wide, but it differed from modern cattle in having longer legs and a longer face, looking more like a wildebeest than a buffalo. This amazing hunk of beef was actually a close relative of the African Cape buffalo, *Syncerus caffer*. There is even cave art in Libya that seems to be an image of this giant extinct buffalo.

Subfamily Aegodontia, the antelopes, sheep, and goats: Tribe Hippotragini, or the grazing antelopes (Fig. 14.30), includes the aquatic reduncins (waterbuck, reedbuck) and the horse-like hippotragins (sable and roan antelopes, oryx, gemsbok, and addax). The reduncins specialize in swampy habitats, where they wade and swim with great ease and love to hide among the tall reeds. The hippotragins are among the largest living antelopes; they have long horse-like faces and are specialized grazers. Many, such as the Arabian oryx and the gemsbok, are specialized for living in the regions around the Sahara desert and in the Arabian Peninsula. They can live on the sparse tough vegetation with their highly efficient ruminating stomachs, which extract all the water and nutrition they need. With their horse-like heads and long, almost straight horns, they look a bit like a unicorn in profile and may be responsible for the legends.

The tribe Antilopini, or true antelopes, includes the living antilopins (gazelles, blackbuck, springbok, and the long-necked gerenuk). All true antelopes are specialized for living in dense brush or for running on the plains of Africa. Almost all of them have distinctive curved horns in the males but none in the females. The gerenuks have extremely long slender necks and legs for rearing up and browsing on leaves out of the reach of most other herbivores, an example of convergence on the giraffe-like ecological niche. The South African mascot, the springbok, has evolved a peculiar behavior, called "pronking," wherein the males spring up vertically with stiff legs and raise a crest of white fur along their spines, which flashes to warn the rest of the herd of danger. After they land, the crest folds down and disappears, and they run away. Very different are the neotragins, or dwarf antelopes, including the tiny dik dik and klipspringers, the royal antelope, and steenbok. Klipspringers have short straight horns and are adapted for running and climbing in rocky outcrops with their sharply pointed hooves.

The members of the tribe Cephalophini are known as duikers (pronounced *DIKE-ers*), Afrikaans for "divers," because they will dive into the dense brush when threatened. There are 23 species of duikers in three genera, all relatively small deer-like forms, weighing between 3 and 70 kg (6.5–155 lb) and bearing short spike-like horns. Duikers are solitary most of their lives, collecting only to breed; females are larger than males. These tiny antelopes inhabit the dense tropical jungles of central and western Africa, where they rely on their speed and agility in the tangled undergrowth to escape predators. Sadly, all are being hunted nearly into extinction in the wild by starving Africans who kill them for bushmeat.

The tribe Alcelaphini includes large grazing herds of antelopes called wildebeests or gnus, plus the hartebeest, topis, and bonteboks. Familiar from the African savanna, these creatures live in huge numbers, especially during their annual migrations following the rains and green vegetation across the Serengeti. They have an extensive fossil record, and a number of extinct genera are known as well. Closely related is the tribe Aepycerotini, the impalas, which are highly specialized jumpers adapted for living in dense brush. They leap high above the brush when pursued, clearing at least 2.4 m (8 ft), and can make individual long jumps of as much as 9 m (30 ft). Studies have shown that alcelaphins and aepycerotins have very different patterns of evolution. Wildebeests and their relatives are very abundant, diverse, and speciose, displaying a wide variety of different horn shapes and adaptations through the past 5 m.y. However, they are also specialized grazers with high extinction rates over that period, as climate change in Africa has wiped out the grasslands and then brought them back. Impalas, on the other hand, are very unspecialized, wide-ranging browsers that can live on just about any kind of brush, so they are very low in diversity and species numbers; only one species of *Aepyceros* is known from the fossil record at a time, and only one species is alive today. These two closely related tribes of antelopes demonstrate very different responses to the same climatic and vegetational changes in Africa over the past 5 m.y.

Finally, the last major group of bovids is the tribe Caprini, which includes the goats, the sheep, and their relatives. One group, the saiga antelopes, is restricted to the Tibetan Plateau; they have bulbous noses that help filter out the dust in their parched habitat and warm the frigid air before it goes to the lungs. Another group, the rupicaprines, includes a variety of wild goats, such as the chamois of the Alps, the North American mountain goat, and the serows and gorals of Asia. The true caprines include the wild and domesticated sheep, goats, and the favorite of crossword-puzzle fans, the ibex, a denizen of mountains from the Pyrenees to the Alps to Siberia.

The ovibovines, or muskoxen, are restricted to the Arctic Circle today, but have an extensive fossil record in northern Eurasia and North America, due to their rugged distinctive skulls and horns. Their Pliocene ancestor *Euceratherium*, the "shrub-ox," was larger but more lightly built than the modern muskox, with shorter fur; it resembled a sheep with massive horns and lived in hilly grasslands. It crossed an early version of the Bering Land Bridge from Asia to North America about 2 Ma but vanished shortly thereafter. Its descendant *Soergelia* roamed Eurasia in the early Pleistocene all the way from Spain to Siberia, living mostly on the fringes of the advancing and retreating glaciers. It had horns that were intermediate between those of *Euceratherium* and modern muskoxen. It made a second crossing from Eurasia to North America in the early Pleistocene, about 1.8 Ma, then survived to about 10,000 years ago, when all the giant mammals vanished from this continent. From the middle and late Pleistocene we have an extensive fossil record of even more modern-looking muskoxen known as *Praeovibos*. By the middle Pleistocene the fossils of *Praeovibos* and the modern muskox (*Ovibos*) overlap in Eurasia. Sometime between 200,000 and 90,000 years ago, *Ovibos* migrated across the Bering Land Bridge (the third time for this group) and became established on the northern fringes of the glacial margins on this continent, where it still lives today, one of the few survivors of the great extinction of megamammals at the end of the Pleistocene.

Figure 14.31. Some of the more spectacular extinct bovines, including the enormous long-horned Ice Age *Bison latifrons*, and two huge cattle from the Pliocene and Pleistocene of Africa, *Pelorovis antiquus* (with the horns resembling those of a giant longhorn bull) and *Pelorovis oldowayensis* (with the down-curved sheep-like horns).

CHAPTER 15

LAURASIATHERIA: PERISSODACTYLA

"Odd-Toed" Hoofed Mammals: Horses, Rhinos, Tapirs, and Their Extinct Relatives

The order Perissodactyla, or the "odd-toed" hoofed mammals, is an important group of large herbivores. They are called "odd-toed" because the axis of their hands and feet (Fig. 14.2) runs through the middle digit (equivalent to your middle finger and middle toe), so they either have three toes (rhinos and tapirs) or only one toe (horses). The outer digits (thumb and little finger on the hand, big toe and pinkie on the foot) have been lost in most living hoofed mammals. This symmetry of the hands and

Figure 15.1. The gigantic hornless rhinoceros *Paraceratherium*, standing above a much smaller elephant. At its feet is the small running rhinoceros *Hyracodon*, from which it evolved.

feet, the reduction of side fingers and toes, and other details of their anatomy are unique to the Perissodactyla and distinguish them from all other groups of mammals.

This group was once very diverse, but only three families survive today. These include the family Rhinocerotidae (rhinoceroses), with five species in four genera; the family Tapiridae (tapirs), with one genus (*Tapirus*) and five species; and the family Equidae (horses), with one genus (*Equus*) and seven species of wild horses, asses, and zebras. But over the past 55 m.y. perissodactyls have been tremendously diverse, and the order has included numerous additional families that are now extinct, such as the gigantic horned brontotheres and the weird clawed chalicotheres. The largest land mammal that ever lived, the huge hornless rhinoceros *Paraceratherium*, was a perissodactyl (Fig. 15.1).

In the geologic past dozens of genera and hundreds of species of horses and rhinoceroses roamed all the northern continents, as did many genera of tapirs. During the Eocene horses, rhinoceroses, tapirs, and brontotheres were the most diverse and ecologically disparate group of large mammals on the northern continents, and they dominated the niches for large herbivores. Since the Eocene, however, the even-toed artiodactyls have gradually taken over, and today they are the bigger hoofed-mammal group, while only a tiny remnant of the original diversity of perissodactyls remains.

Perissodactyls apparently originated from a common ancestor with the phenacodonts (see Chapter 13) in the late Paleocene of Asia. The oldest known relative of the perissodactyls is *Radinskya*, discovered in the upper Paleocene rocks of China and described in 1989. It was named for the late paleontologist Leonard Radinsky, who did pioneering work on the early evolution of perissodactyls. *Radinskya* looks like an extremely primitive common ancestor of the earliest Eocene horses, tapirs, and rhinoceroses. By the early Eocene the most primitive known horses, rhinos, tapirs, and brontotheres were all established in North America and Asia, while Europe was inhabited its own endemic group known as the palaeotheres. These earliest horses, tapirs, and rhinos are so similar that it is almost impossible to tell them apart, except for subtle differences in their teeth and skulls that foreshadow their descendants. There is little indication of how different they would later become. Yet by the late early Eocene and middle Eocene, each lineage had evolved in quite different directions, and the skeletons and teeth of each are easily distinguished. Perissodactyls are an excellent example of evolution from a group of common ancestors that look extremely similar initially but then diverge, becoming a set of extremely distinctive lineages.

EQUOIDS

Family Equidae (Horses)

Horses are the most famous example of dramatic evolutionary change from a very primitive ancestral form to a wide variety of highly specialized descendants. In 1872 British biologist Thomas Henry Huxley (nicknamed "Darwin's bulldog" for his defense of evolution in debates and in the media) realized that the sequence of the fossils of horses from Europe seemed to form a series from Eocene palaeotheres (not true equids) to *Anchitherium* and *Hipparion* to modern *Equus*. In 1873 Russian paleontologist Vladimir Kowalewsky further developed the idea and proclaimed that he could show how horses evolved in Europe. But when Huxley visited the United States on a lecture tour in 1876, he visited O. C. Marsh at Yale University. When he saw Marsh's collections of horse fossils from the Rocky Mountains he realized that horse evolution had taken place primarily in North America, while only occasional horses had escaped to evolve in Eurasia.

By the early 20th century the sequence of horse evolution from tiny primitive Eocene forms no bigger than a medium-size dog (such as a beagle) to the modern horse, zebra, and donkey was well documented. In addition to the dramatic change in size, horses changed shape in many ways (Fig. 15.2A). Their snouts got longer, their brains grew larger, and their teeth transformed from low-crowned molars with simple crests and cusps to extremely high-crowned teeth capable of grinding down without wearing out as the animals ate gritty vegetation for their entire lifetime. Their limbs became dramatically longer for running, especially due to the lengthening of the bones of the hands and feet that became the lowest segment of the limbs. Meanwhile the four fingers and three toes of the earliest horses were reduced to three toes, with the middle digit expanded and the side toes reduced, and then to the single finger and toe bone, with the side toes reduced to tiny splints, in modern horses.

In the century since the original simple linear sequence of horses was published the story has gotten much more complex as more and more fossils have been found. Today the fossil record comprises dozens of horse genera and almost 100 different species. Instead of a single linear sequence steadily changing through time, the history of horses has a bushy, branching pattern, with many different lineages and many different specializations (Fig. 15.2B). The true horses originated in North America with a group of very primitive equids that have variously been lumped into *Eohippus* and more recently *Hyracotherium*. But research since 1989 shows that *Hyracotherium* (from the lower Eocene London Clay in England) is not a horse at all, but a palaeothere, so that name cannot be used for the true horses of North America. The name *Eohippus* was once used as a catchall name for every Eocene horse but today applies to only two of the many different species.

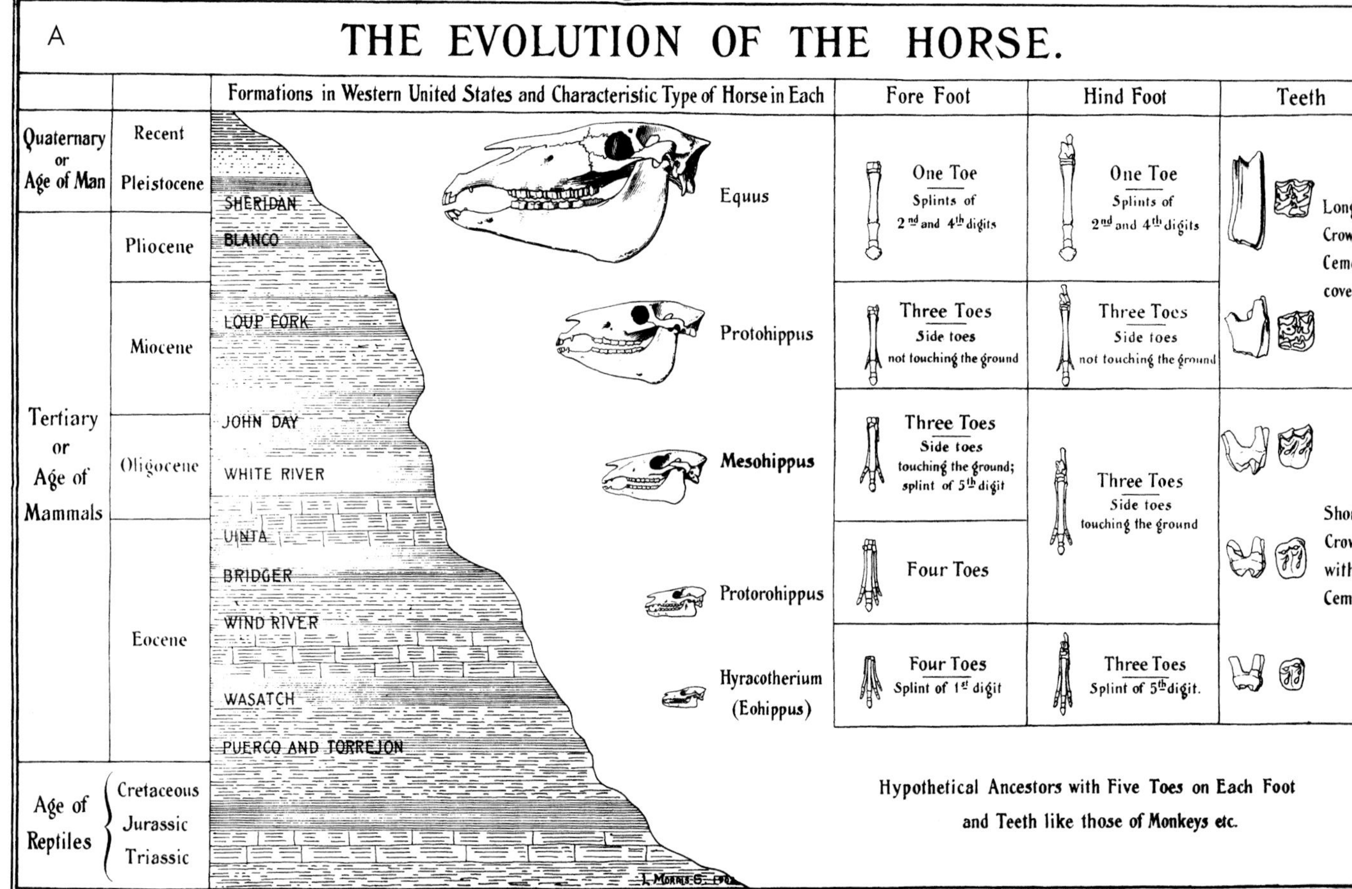

Figure 15.2. Horse evolution. A. The old linear notion of horse evolution from nearly a century ago, showing the generalized trend through time of larger body size, longer snout, longer legs, loss of side toes, and larger brain.

In addition to *Eohippus* there were many other genera of horses in the early Eocene, including *Protorohippus*, *Sifrhippus*, *Minippus*, *Arenahippus*, and some other names as well.

During the middle and late Eocene horse diversity reduced to just *Orohippus*, *Epihippus*, and *Haplohippus*. In the uppermost Eocene and lower Oligocene beds of the Big Badlands of South Dakota, horses are abundantly preserved. There we find *Mesohippus*, a genus with longer limbs, reduced side toes, and slightly higher-crowned teeth (Fig. 15.2). Three species of *Mesohippus* lived side by side with the slightly larger and more advanced *Miohippus* right into the late Oligocene, showing that horse evolution was bushy and branching throughout its entire history.

In the late Oligocene and early Miocene horses underwent a huge diversification. One group, the Anchitheriinae, retained primitive low-crowned teeth (and leaf-eating browsing diet) through almost the entire Miocene; one member, *Megahippus*, grew almost as large as a living horse. But the main horse lineage, the Equinae, became even more specialized. It evolved higher-crowned cheek teeth, ever-lengthening middle toes, and highly reduced side toes. There were numerous side branches, such as the hipparionine horses, which are recognized by the distinctive features in their teeth and the facial region of the skull. Not all of the extinct side branches showed the general increase in body size, and size reduction occurred independently in several horse lineages. Some groups, such as *Nannippus*, were dwarfed horses.

Miocene horses were so diverse that one quarry in Nebraska, about 12 m.y. in age, yielded fossils of 12 different species that all lived at the same time. The only surviving lineage of horses, however, began in the Pliocene, with *Dinohippus*, and by the Pleistocene only one genus survived, *Equus*. However, during the Ice Ages there were many different species of *Equus* spread across North America and Eurasia and eventually filtering into Africa and South America as well. Today only the three species of zebras, three species of asses and donkeys, and one species of wild horse still survive in nature, although domesticated horses are now among the most abundant hoofed mammals alive, thanks to humans.

Family Palaeotheriidae (Palaeotheres)

The Palaeotheriidae is an extinct group of horse-like mammals that lived in the dense forests of the island archipelago of Europe

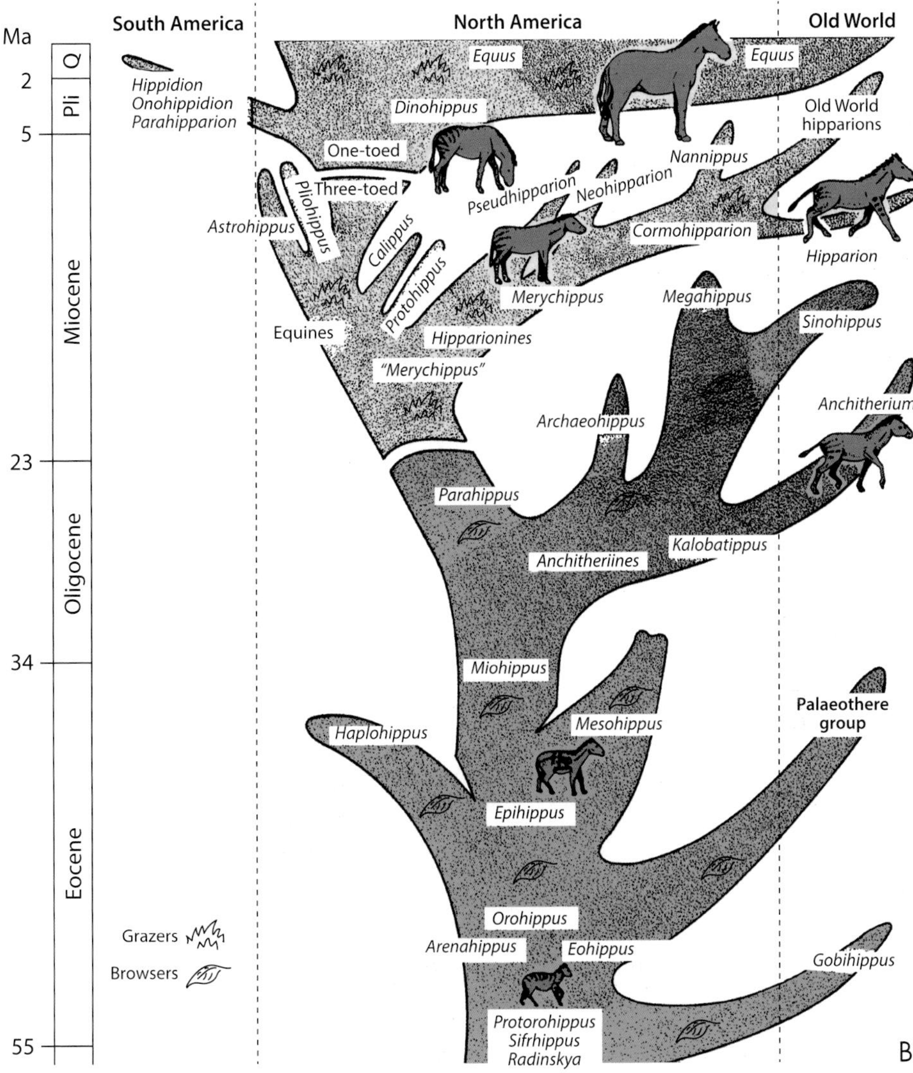

Figure 15.2. Horse evolution. B. A more modern view, showing the more branching and bushy history of horses, with multiple lineages coexisting at the same time and place. The primitive horses were all leaf-eating browsers (green pattern), but during the early Miocene the grass eaters (grazers) evolved in the Equinae (yellow pattern).

during the Eocene (Fig. 15.3). Palaeotheres have often been mistaken for horses, and a few paleontologists still call them horses, but a detailed analysis clearly shows they are not closely related the true horses, which were confined to North America during the Eocene.

Palaeotheres were among the first fossil mammals ever described scientifically. In 1804 the French naturalist Georges Cuvier (the founder of vertebrate paleontology and comparative anatomy) received a skull and skeleton that had come from the middle Eocene gypsum deposits in the Montmartre district of northern Paris. Cuvier examined the specimen, decided it came from an ancient extinct tapir, and gave it the name *Palaeotherium* ("ancient beast" in Greek). Later Cuvier pointed to *Palaeotherium*, along with the known mammoth and mastodont fossils, as proof that extinction had really occurred, an idea that was considered heretical by religious scholars at the time. They could not imagine that a God who watches after every sparrow would let any of his creations die off.

The largest *Palaeotherium* species, *P. magnum*, grew to be almost as large as a modern horse, although it had a much more tapir-like build and probably a short proboscis like a tapir as well (Fig. 15.3B). Since Cuvier's discovery, other palaeotheres were found all over northern Europe. In 1841 Richard Owen, often called "the British Cuvier," described a tiny lower Eocene fossil from the London Clay as *Hyracotherium leporinum* ("hyrax beast like a rabbit"). He thought it was related to the hyraxes, but in the mid-20th century scientists decided that *Hyracotherium* was the oldest valid name for the first horse. Then in 1989 Jeremy Hooker of the Natural History Museum in London showed that *Hyracotherium* was not a horse at all but the most primitive palaeothere known. (Any references that still call American early Eocene horses *Hyracotherium* are grossly out of date.) This sequence reminds us not only of how similar the earliest horses, rhinos, and tapirs were in the early Eocene but also that the palaeotheres are difficult to tell from the horses and tapirs of their time.

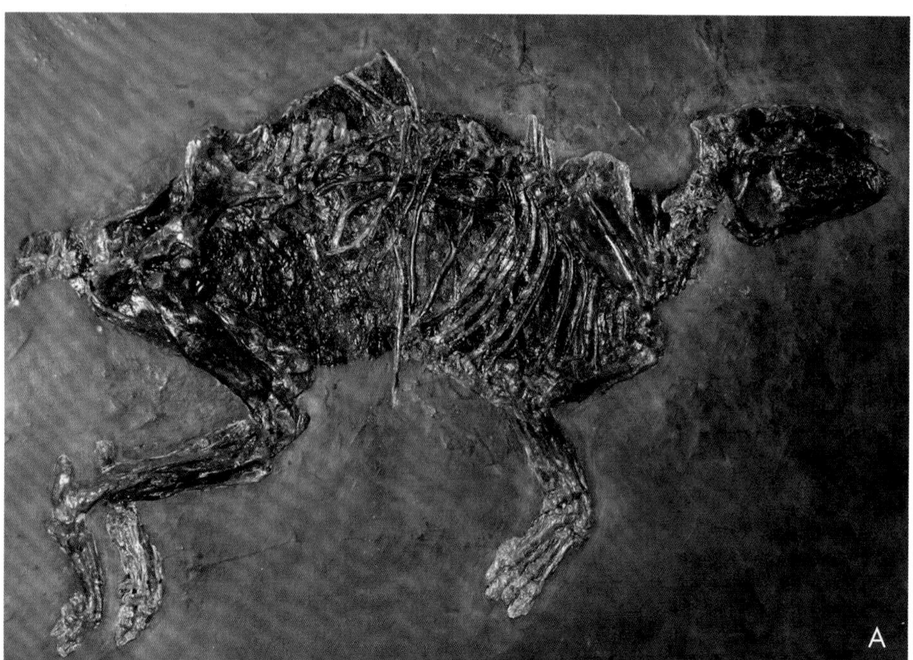

Figure 15.3. Palaeotheres were distant relations of horses found only on the Eocene islands of Europe, and they vanished when other mammals invaded from Asia in the Oligocene. A. Articulated skeleton of *Propalaeotherium* from the middle Eocene Messel lake beds of Germany, complete with body outline, hair impressions, and even stomach contents. B. Waterhouse Hawkins's reconstruction of *Palaeotherium*, showing its tapir-like snout. These sculptures were created for the Crystal Palace in 1851 and now reside in Sydenham Park in London.

The best-known palaeothere is *Propalaeotherium*, which is represented by complete articulate skeletons, famous for the extraordinary quality of their preservation, from the legendary middle Eocene lake shales of Messel, Germany (Fig. 15.3A). *Propalaeotherium* stood about 60 cm (24 in) tall at the shoulder and weighed about 10 kg (22 lb); it looked more like the tapirs than the early horses. The Messel specimens preserve hair impressions and even stomach contents, proving that these animals ate berries, leaves, and other vegetation picked from the lower stories of the dense Eocene forests.

Palaeotheres were severely impacted by the early Oligocene extinction event, caused by climate change, and also by competition due to the immigration of many Asian groups of mammals that ended the isolation of the European mammal faunas. Only *Palaeotherium* survived into the early Oligocene, but then the group vanished forever.

TAPIROIDS

Tapiroids is a large group of perissodactyls that includes not only the living tapirs but many extinct families, including the strange clawed chalicotheres. Often called "living fossils," the five surviving species of tapirs (genus *Tapirus*) are restricted to the tropics of Central and South America (four species) and the jungles of Southeast Asia (one species, the Malayan tapir). Living tapirs are pig-size creatures with a stocky body, short legs, and a thick, tough, smooth skin adapted for pushing through the jungle vegetation (Fig. 15.4). They have four hooves on their hands and three hooves on their feet, which help in walking in marshy ground, and they are excellent swimmers. The four New World species are reddish-brown, with bristly black hair on their manes, and the Malayan tapir has a pattern of broad black and white bands. The tapirs' most distinctive feature, however, is their long snout or proboscis, which is highly flexible and good not only for sniffing out danger but also for wrapping around leaves and stripping them from branches. The presence of such a proboscis is marked on their skulls by retraction of the nasal bone and an expanded open pit on the facial bones for the attachment of the complex proboscis muscles (Fig. 15.5). Tapir cheek teeth have a very distinctive pattern of two transverse cross-crests, a combination that is very efficient at shredding leafy vegetation. This kind of dentition is found in animals that are typically browsers (leaf eaters).

Figure 15.4. Some of the living species of tapirs. A. *Tapirus terrestris*, the South American tapir, found in the Pantanal wetlands of Brazil, Bolivia, and Uruguay. (See also next page.)

Figure 15.4. Some of the living species of tapirs. B. *Tapirus bairdii*, Baird's tapir. C. *Tapirus indicus*, the Malayan tapir. D. *Tapirus pinchaque*, the mountain tapir.

igure 15.5. The retracted nasal bones and wide asal incision of the tapir skull allow for the attachment of numerous muscles that control the animal's long exible proboscis. A. Diagram of the snout region of a living tapir compared to that of a rhinoceros (lower ght), showing how the proboscis muscles fit. B. Skull of *Hesperaletes* from the middle Eocene of San Diego, one of the first fossil tapirs to show the nasal retraction. C. Skull of *Megatapirus*, a rhino-size Ice Age tapir om China.

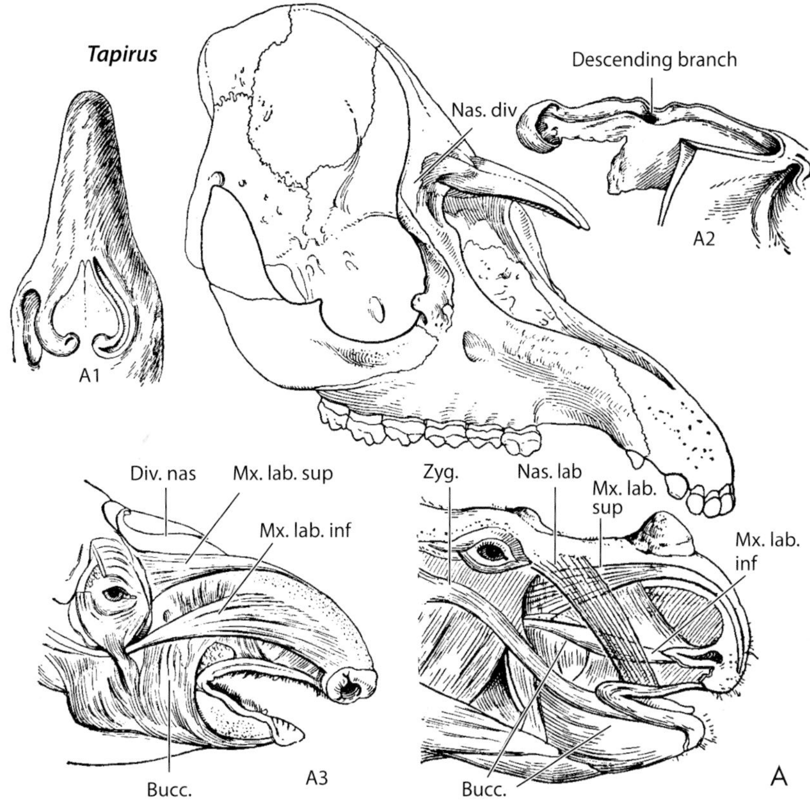

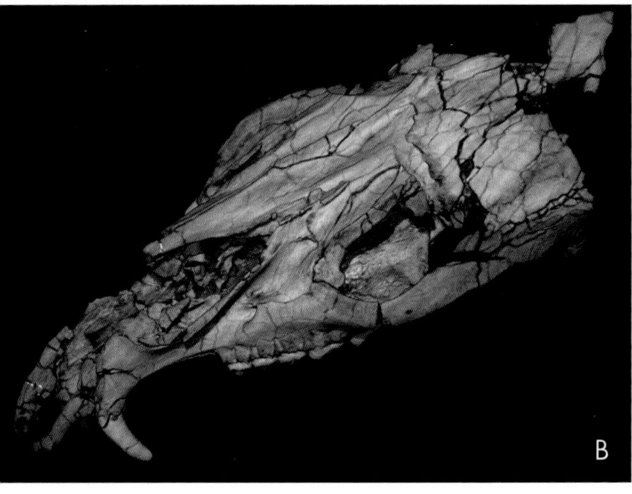

Family Tapiridae (Tapirs)

Tapirs have a fossil record going back 55 m.y. to the beginning of the Eocene in North America (Fig. 15.6). The earliest fossils are of *Homogalax*, beagle-size creatures with no proboscis that can barely be distinguished from the earliest horses and earliest rhinoceroses. Tapiroids then evolved and diversified in the middle Eocene, and moved freely back and forth from North America (Helaletidae, Isectolophidae) to Asia (Lophialetidae, Deperetellidae). The isolated islands of Europe had their own endemic tapiroids (Lophiodontidae) in the middle Eocene as well. After their great diversification in the middle and late Eocene of Asia and North America, tapiroids became much less diverse and much more scarce during the Oligocene. The long-lived Eocene–Oligocene genus *Colodon* was one of the last to show the primitive skull condition without a proboscis.

Upper middle Eocene deposits near San Diego, California, yielded *Hesperaletes*, the earliest tapiroid to show the distinctive retraction of the nasal bones and complex facial structures that supported a large proboscis (Fig. 15.5B). The retraction of the nasal bones is further developed in the Oligocene fossils

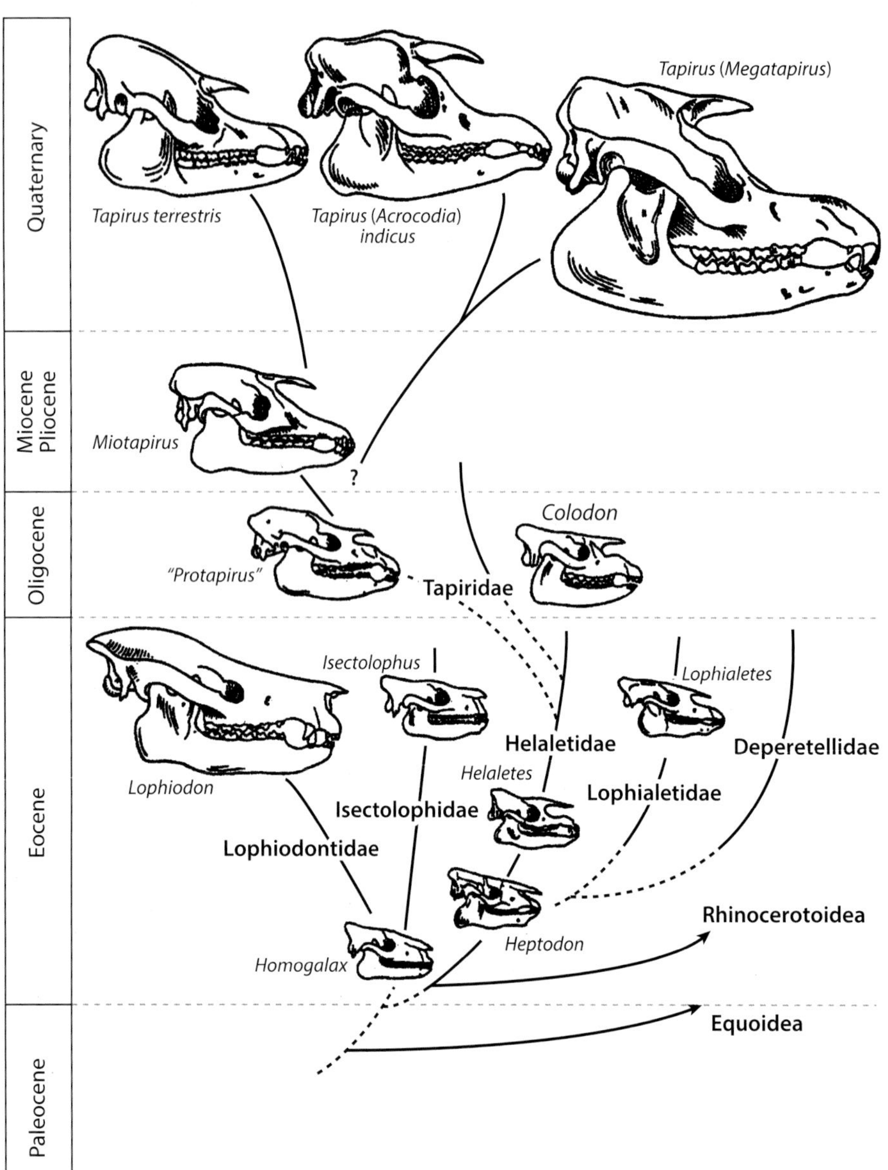

Figure 15.6. Evolution of the tapir skull, showing the increased enlargement of the nasal notch and retreat of the nasal bones as tapirs grew larger.

Protapirus and *Miotapirus* (Fig. 15.6). Tapirs continue to make rare appearances in the fossil record in Asia and North America through the rest of the Miocene and Pliocene. Their scarcity is probably due to their restriction to the forest habitats that were vanishing during the drying environments of the later Cenozoic. In the late Miocene (10 Ma) the first tapirs crossed the Central American island chain and arrived in the Amazon basin of South America (Fig. 5.4), where they evolved into four of the living species (Fig. 15.4A, B, C). Tapirs are occasionally found in the Ice Age deposits of North America, especially in regions like Florida, southern California, and coastal Texas, which were once more tropical, but they vanished from North America during the end-Pleistocene extinction. In Eurasia, Ice Age tapirs included the huge *Megatapirus* from China (Fig. 15.5C), which was almost the size of a modern rhinoceros. These vanished at the end of the Ice Age, leaving the Malayan tapir as the sol[e] surviving tapiroid in the Old World.

Family Chalicotheriidae (Chalicotheres)

By far the most peculiar of all perissodactyls were the chalicotheres. When the fossils of these creatures were first discovered by J. J. Kaup in Germany in 1833, scientists found that thei[r] skeleton resembled a large muscular horse with long arms. However, the same deposits that yielded these bones also included numerous curved claw bones that could not be associated with any other known animals from that quarry. The claws resembled those of some giant anteater or pangolin, and that is how the[y] were first identified. Finally, in the 1880s, the French paleontologist Henri de Filhol realized that the claws belonged to the strange creature named *Chalicotherium*, known for its horse-like ones from the same quarry. With this discovery, it was realized hat chalicotheres were one of the few hoofed mammals that had econdarily modified hooves into claws.

So what did these peculiar clawed ungulates do with their nique limbs? *Chalicotherium* itself (Fig 15.7) had very long fore-mbs and short hind limbs and apparently knuckle-walked, like a orilla or ground sloth, with the claws of its hands curved inward nd the knuckles touching the ground. The chalicothere *Moropus* Fig. 15.7), from the famous lower Miocene Agate Springs bone ed in western Nebraska, had less extreme limb proportions. cientists have made all sorts of suggestions about how chalico-heres used their claws, such as digging up roots or fighting off redators. Detailed studies of their arm muscles and the wear on heir claws rule these ideas out. The current scientific consensus that chalicotheres sat up on their haunches, leaned on trees, nd pulled down branches with their hooked claws, and then tripped leaves from the branches, much as ground sloths did.

Chalicothere teeth confirm this idea, since they are relatively ow-crowned, with simple W-shaped outside crests on the upper olars (much like the teeth of the leaf-eating brontotheres), nd their back teeth show little wear. When they became adults, halicotheres lost the front teeth in their upper jaw, suggesting hat they may have had a long prehensile tongue (as a giraffe has) hat also helped strip leaves off branches. This interpretation of heir diet has been confirmed by analysis of the microwear on he teeth, which indicates a diet of leaves, twigs, and bark.

Chalicotheres are thought to have been relatively rare crea-ures, since they were seldom fossilized, and only a few localities uch as Agate Springs Quarry and Morava Ranch Quarry in Ne-raska and a few places in Europe) have abundant chalicothere ossils. The group's earliest fossils, such as *Protomoropus*, *Pappo-oropus*, *Paleomoropus*, *Litolophus*, *Danjiangia*, *Lunania*, *Litolophus*, and *Grangeria* are found in the lower and middle Eocene rocks of both Asia and North America.

After the chalicotheres vanished from North America at the end of the middle Eocene they were restricted to Asia; forms such as *Schizotherium*, *Metaschizotherium*, *Phyllotillon*, and *Borissiakia* lived in small numbers in China, Mongolia, Pakistan, and Kazakhstan, and occasionally spread to Europe. By the early Miocene the gorilla-shaped *Chalicotherium* was widespread in Eurasia and Africa. Meanwhile, chalicotheres migrated back to North America (represented by *Moropus*) in the earliest Miocene (Agate Springs bone bed in Nebraska), about 23 Ma. During the early and middle Miocene of North America, there was a bizarre chalicothere named *Tylocephalonyx* with a large domed skull (Fig. 15.7). Both of these genera vanished by the late middle Miocene, while Asian chalicotheres such as *Ancylotherium* managed to migrate to Africa and persist until the early Pleistocene; along with the Asian *Nestoritherium* and the African native chalicotheres such as *Chemositia*, these were the last of their bizarre lineage. Some have suggested that these very late-surviving African chalicotheres might have been responsible for the legend of the "Nandi bear" in Africa (where humans had been evolving since 6 Ma).

Although paleontologists agree that chalicotheres are perissodactyls, the group's placement within the perissodactyls has long been controversial. Most analyses suggest that the chalicotheres are a subgroup within the tapir radiation, which Jeremy Hooker of the Natural History Museum in London called the Tapiromorpha, and Robert Schoch of Boston University called the Moropomorpha (tapirs, chalicotheres, and rhinos). This interpretation has been supported by several more recent studies, although others have placed chalicotheres as a cousin to a cluster of horses plus Ceratomorpha (the group including rhinos and tapirs).

gure 15.7. An assemblage of chalicotheres: the gorilla-like *Chalicotherium* (right background), the more horse-like *Moropus* (left foreground), the dome-ulled *Tylocephalonyx* (right foreground), and the large *Ancylotherium* (left background).

RHINOCEROTOIDS

Within the Perissodactyla is a subgroup that includes several families closely related to the modern rhinoceroses (family Rhinocerotidae). These include not only the living family but two extinct families, the Hyracodontidae and Amynodontidae (Fig. 15.8). They all have the distinctive upper molars but have many variations in size and proportions of the teeth. The three families are most easily distinguished by the shape of the crests of the last upper molar, which has different configurations in each family.

The entire rhinocerotoid radiation is derived from the widespread early middle Eocene genus *Hyrachyus* (Fig. 15.8), which has been found in North America, Jamaica, Europe, and even the Canadian Arctic. By later in the early middle Eocene, the three families of rhinocerotoids had diverged from *Hyrachyus* and were established in Asia and North America.

Family Amynodontidae (Amynodonts)

The amynodonts have been nicknamed the "hippo-like rh nos," because many of them have massive bodies with short r bust limbs and other hippo-like features. They also had broa mouths and widely flaring tusks, like modern hippos do. Eve the density of their bones suggests they were aquatic, and in th Big Badlands of South Dakota, they are usually fossilized in th sand deposits of the ancient river channels.

Amynodonts first appear in Asia during the middle Eocen and soon spread to North America. Among the best-known tax is *Metamynodon planifrons*, a hippo-size creature that roamed fro Asia to North America, where it is found in the lower Oligocen beds of the Big Badlands of South Dakota (Fig. 15.9). Also we known is *Zaisanamynodon protheroi*, a huge Asian rhino original

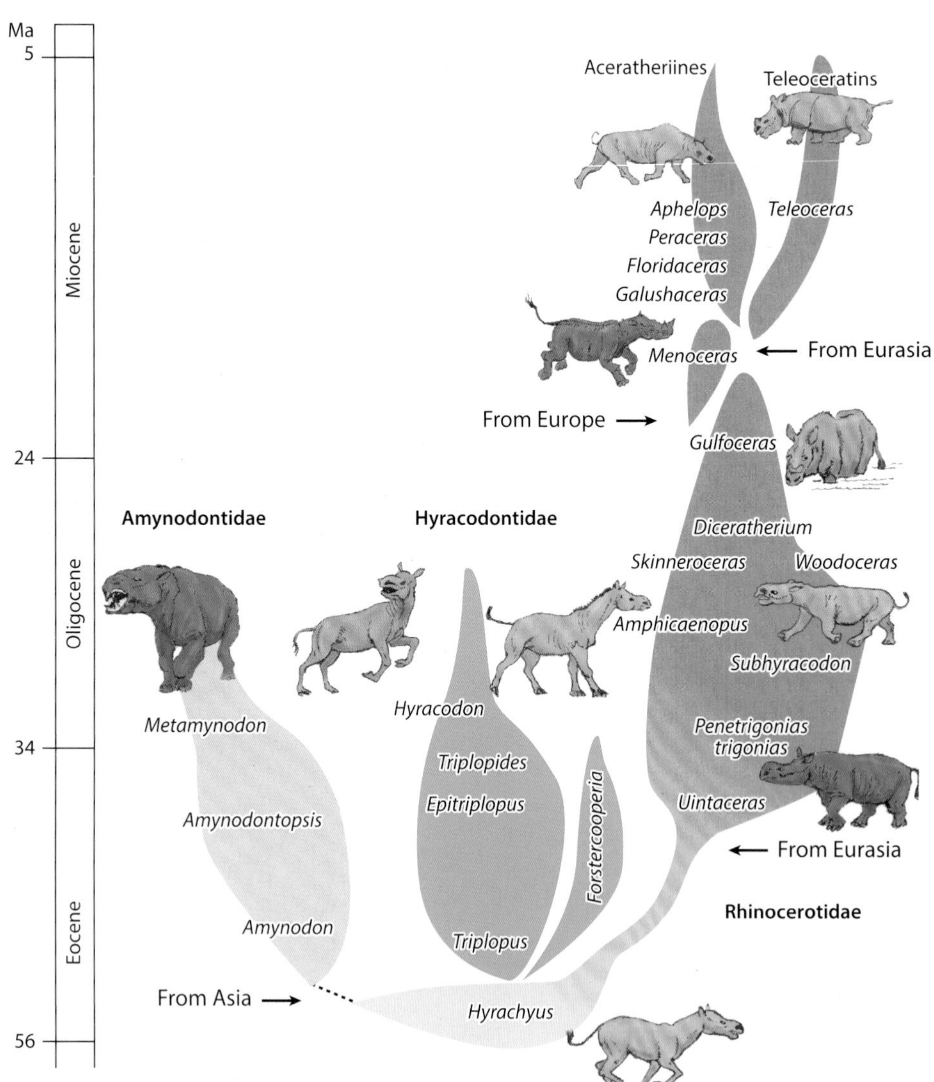

Figure 15.8. Three major families of rhinocerotoids, represented by dozens of genera and species, evolved in North America. These are the hippo-like amynodonts (left), the long-legged running hyracodonts (center), and the true rhinoceroses, family Rhinocerotidae The first two died out by the late Oligocene around the world, but the Rhinocerotidae survives today in Africa and Southeast Asia.

amed for the Zaisan Basin of Kazakhstan, where it was first ›und. It also occurs in the upper middle Eocene Clarno Formaon of central Oregon.

The other tribe of amynodonts was the cadurcodonts, known nly from Eurasia. The Mongolian and Chinese genus *Cadurodon* has highly retracted nasal bones and a large nasal opening, uggesting that it sported a small trunk or proboscis, like a tapir Fig. 15.9). The last of these was *Cadurcotherium*, which is found n upper Oligocene beds in Pakistan. Once *Cadurcotherium* vanshed about 23 Ma, at the end of the Oligocene (10 m.y. after he rest of its family died out), the family became extinct.

Family Hyracodontidae (Hyracodonts)

he second major family of rhinocerotoids was the hyracodonts. hey are sometimes nicknamed the "running rhinos," because heir legs are much longer and more slender than those of nost fossil rhinos. They first appeared in Asia in the early midle Eocene, and soon spread to the middle Eocene of North merica with the genus *Triplopus*. Their North American evoluion culminated with the genus *Hyracodon* itself, which is a very ommon fossil in the Big Badlands of South Dakota (Figs. 15.1, 5.9). *Hyracodon* was the size of a Great Dane and had very long lender limbs for a rhinoceros. Its teeth were simple and lowrowned, suggesting that it was a leaf eater, and it lived in the forests of the late Eocene and early Oligocene. *Hyracodon* first appeared about 36 Ma, as a much smaller form, then increased in size and in premolar complexity until the late Oligocene, about 26 Ma. It was the last of its family in North America.

However, the hyracodonts persisted in Asia and eventually became gigantic. The cow-size *Forstercooperia* is known from the late middle Eocene. In the late Eocene hyracodonts got larger and developed longer necks and legs, as in the camel-size *Juxia* from the late Eocene in China (Fig. 15.9). By the early Oligocene there were huge creatures, called *Urtinotherium*, but the culmination of this lineage, known as indricothere rhinoceroses, is the monstrous *Paraceratherium*, the largest land mammal that ever lived. It was 4.8 m (16 ft) at the shoulders, 8 m (26 ft) long, and heavier than any elephant or mastodont known (Fig. 15.1, 15.9). Original estimates put its weight at over 34 metric tons (37.5 short tons), but more recent methods of weight estimation place it around 20 metric tons (22 short tons), just a bit larger than the biggest elephant relatives that ever lived, the deinotheres.

Paraceratherium had a huge skull, over 2 m (6.5 ft) long, with a short proboscis or trunk (judging from the deep nasal opening on the skull), a prehensile lip for stripping vegetation from branches, and relatively primitive low-crowned teeth that were only suitable for leaf browsing, not for eating gritty grasses. It

igure 15.9. A stampede of rhinocerotoids. In the right foreground is the small ancestral *Hyrachyus*. On the extreme left is the running hyracodont rhino *Hyracodon*. Its relatives include the long-legged, long-necked hyracodont *Juxia* from China (far right) and, just behind it, in the background, the immense *Paraceratherium* from Asia, the largest land mammal known. To the left of the human are the hippo-like amynodont rhino *Metamynodon* and, just in front of it, the amynodont with the tapir-like snout known as *Cadurcodon*. In front of *Cadurcodon* is *Gaindatherium*; it is a member of the true rhino family, Rhinocerotidae, as are the remaining animals (described here left to right). The small rhino charging the human is *Meninatherium*. The huge one-horned hino in the center is *Elasmotherium*. To the right are the two-horned *Menoceras*, the earliest rhino to develop horns; the hippo-like rhino *Teleoceras*; and he woolly rhino *Coelodonta*.

had a very long neck that enabled it to reach treetops to browse, so in many ways it was a rhino trying to perform the role of a giraffe. But its most amazing feature is its feet. Despite the fact that it was heavier than any elephant, mammoth, or mastodont, it did not have the short stumpy toe bones flattened down from supporting their great weight that elephants do. Instead its toes were relatively long and slender, even though they supported more weight than any other land mammal and as much as many large sauropod dinosaurs (which also have short stumpy toes). This is another signature of the hyracodont ancestry of *Paraceratherium*. Even though it reached gigantic size, it never abandoned the long-toed blueprint of its ancestors.

Although names such as *Baluchitherium* and *Indricotherium* and others have been applied to different specimens of these giants (from the latter comes the group name "indricotheres"), the scientific consensus is that they are all the same as the genus *Paraceratherium*, the oldest name applied to the group. If they are analogous to elephants (the only living animals of similar body size), they would have had small populations spread widely over a huge area (another reason to believe there was probably no more than one genus in all of Asia). They would have walked slowly, about 20 km/h (12 mph), but with their long legs would have covered a lot of ground. They would probably have fed at night and early morning, and rested during the heat of the day, since their large body size means they would have overheated very easily. Such large animals have slow pulse rates (30 beats per minute) and also long life spans (over 50 years). They were far larger than any predator during the Oligocene, so they did not have to fear predation except when their calves were small.

Paraceratherium lasted through the entire Oligocene in Asia (China, Mongolia, Pakistan, Turkey, Kazakhstan), but in the earliest Miocene, about 20 Ma, it vanished. The cause of its extinction is unknown, but its disappearance coincides with the escape of mastodonts from Africa (which could have destroyed its forest habitats, as elephants do today). In addition, the African predators of mastodonts, such as the giant bear dog *Amphicyon* and the huge creodont *Hyainoilouros*, also escaped from Africa at this time and might have been the first predators that could actually take down such huge beasts. With the extinction of *Paraceratherium*, the hyracodont family vanished forever.

Family Rhinocerotidae (True Rhinoceroses)

When people hear the name "rhinoceros" they think of a dangerous thick-skinned beast with horns and bad eyesight, originally from Africa or Asia. Most do not know that rhinoceroses have a rich fossil record, not only from Africa and Asia but also from Europe and especially from North America. The first true rhinoceroses, or rhinocerotids (family Rhinocerotidae), appeared in both Asia and North America in the late Eocene (about 40 Ma) and diversified on both continents for the next 35 m.y. (Fig. 15.8). The earliest fossil rhinoceroses had no horns; in fact, based on the fossil evidence, most rhinoceroses through time have lacked horns. Unlike the bony horns of ruminant artiodactyls (such as goats, sheep, cattle and antelopes), or the bony antlers of deer, rhinoceros horns are made of densely packed hair-like fibers glued together, and have no bone in them at all. They are usually not preserved in fossil rhinoceroses (except in mummified woolly rhinoceroses from the Ice Ages), so we must infer their position and size by the roughened areas on the top of the skull where they were attached.

During their long history, rhinoceroses have evolved into a wide variety of body sizes and ecological niches. There were small dog-size running rhinos, including the earliest rhinoceroses that can barely be distinguished from the earliest tapirs or horses; barrel-chested short-legged hippo-like rhinos, like the Teleoceratini; rhinoceroses with long prehensile snouts for stripping leafy vegetation (the Aceratheriinae); and gigantic rhinoceroses of the Ice Ages with horns on their foreheads (the Elasmotherini). While most rhinoceroses have been hornless, some have had horns in a variety of configurations: a single horn on the nose (the Teleoceratini and the living Asian rhinos); a pair of horns mounted in tandem on the nose and forehead (the Dicerotini, or the African rhinos); a single gigantic horn on the forehead (the Elasmotherini); paired horns mounted on side-by-side knobs on the nose (the Menoceratini) or on paired flanges on the nose (the Diceratheriini). In some rhinos only the males had horns (e.g., the Diceratheriini and the Menoceratini), but in most others there is no significant difference in horns or body size between males and females. However, we can distinguish the two genders by their front teeth. Most male rhinoceroses had a large pair of tusk-like lower incisors in their jaw (used for fighting and slashing rivals), while the females had much smaller lower incisors with a blunt conical point.

It is also through the teeth that paleontologists recognize fossils as belonging to the rhinoceros lineage. The cheek teeth of all rhinoceroses have a distinctive arrangement of crests that looks like the Greek letter π (Fig. 1.9). This is unique to the family. In addition to these features, there are many other distinctive aspects of the rhinoceros skull and skeleton that allow us to identify almost any rhinoceros fossil, whether it had a horn or not.

Rhinoceroses went through a remarkable pattern of diversification, intercontinental migration, local extinction, and then final restriction to just a few places in the tropics of Africa and Asia (Fig. 15.8). They originated in eastern Asia (the earliest fossils come from China and Mongolia) from a common ancestor with hyracodonts at the end of the middle Eocene (40 Ma). The earliest rhinoceroses then migrated to North America (the oldest known fossil is *Teletaceras*) in the late Eocene, when they were the size of German shepherd dog and not very different from their early tapir or horse relatives. They then grew in body size, exemplified by the cow-size hornless *Subhyracodon* and *Trigonias* in the late Eocene–early Oligocene of North America (Fig. 15.8). Rhinoceroses are unknown in Europe during the Eocene, but rhinos such as *Ronzotherium* and *Epiaceratherium* flooded into western Europe during the immigration event known as the Grande Coupure in the early Oligocene, driving out many of the native Eocene ungulates, such as palaeotheres and the artiodactyls (discussed in Chapter 14).

After brontotheres vanished at the end of the Eocene, rhinoceroses became the largest land mammals in both Eurasia and North America and remained so until mastodonts escaped from Africa about 20 Ma. During the 14 m.y. interval of "rhinoceros rule" from 34 to 20 Ma, North America was dominated by the descendants of *Subhyracodon*, diceratheriine rhinoceroses with paired flanges on their noses. Meanwhile, Eurasia was ruled by a number of genera of primitive rhinos, including *Protaceratherium*, *Mesaceratherium*, *Brachydiceratherium*, and *Aceratherium*. In the late Oligocene another paired-horn group, the Menoceratini, emerged in Europe and then spread to North America about 23 Ma.

At the same time several other rhinoceros groups—the hippo-like Teleoceratini, the prehensile-lipped Aceratherini, plus the Rhinocerotini, which includes the living Indian and Javan rhinos—also evolved in Eurasia, and the first two of these groups migrated to North America in the early Miocene (about 19 Ma). The Teleoceratini and Aceratheriini (Figs. 15.8, 15.9) came to dominate most Eurasian and North American faunas in the middle and late Miocene, usually occurring in pairings of a leaf eater or browser (typically an aceratherine) with a grazer or grass eater (typically a teleoceratine).

Near the end of the Miocene the earliest members of the Javan rhinoceros group (the Dicerorhini), the Indian and Javan rhinoceros group (the Rhinocerotini), and the giant Elasmotherini evolved in Eurasia (Fig. 15.9); and the tandem-horned rhinoceros group (the Dicerotini) appeared in Africa, becoming the living white rhino and black rhino. Some of the relict archaic groups (Teleoceratini, Aceratherini) vanished during the great extinction event at the end of the Miocene, leaving North America without rhinoceroses. This extinction was possibly triggered by climatic changes to colder and drier conditions due to the drying of the Mediterranean.

Meanwhile, the Dicerorhini diversified during the Pliocene and Pleistocene, evolving into the woolly rhinoceros of the Ice Ages of Eurasia and represented by the living Sumatran rhinoceros. The Elasmotherini also diversified across Eurasia, from Spain to China, culminating in the elephant-size Eurasian form *Elasmotherium*, which had a single huge horn on its forehead (Fig. 15.9). These rhinoceroses vanished before the end of the Pleistocene, leaving only the Dicerotini in Africa and the Rhinocerotini in southern Asia. This long history of success and survival is coming to an unfortunate end. Rhinoceros horn is now so valuable (worth more per gram than cocaine or gold) for use as a "medicine" (though it is completely ineffective) in China and Vietnam that rhinoceroses are being poached to extinction in the wild, and there is very little that can be done to prevent it.

BRONTOTHERES, OR TITANOTHERES

The largest and most spectacular land mammals during the Eocene were the creatures known as brontotheres (family Brontotheriidae), whose name means "thunder beast"; they are also often called titanotheres, meaning "titanic beast." When these huge animals ran, it might have sounded like a thundering stampede of elephants. Although the last of the brontotheres had huge blunt paired horns on their noses and were the size of rhinos or elephants, they are only distantly related to rhinos, horses, and tapirs. Starting as dog-size animals in the early Eocene (*Lambdotherium*), through the remaining 20 m.y. of the Eocene brontotheres evolved and diversified into many genera and species, tending to become larger and larger, until the brontotheres of the latest Eocene (formerly thought to be early Oligocene) reached elephant size and bore huge blunt nasal horns on their massive skulls (Fig. 15.10).

A fragmentary brontothere jaw with a few teeth found in 1846 in the Big Badlands of South Dakota was among the first fossil mammals discovered in America west of the Mississippi River. It was so incomplete that it was first called a "giant *Palaeotherium*" (referring to a European fossil distantly related to horses). This and other specimens led to the first "bone rush" of explorers and scientists trying to find fossils and map the geology of the American West. Numerous partial skulls and jaws were recovered in the 1850s, which were sent to the paleontologists Joseph Leidy of the University of Pennsylvania in Philadelphia and O. C. Marsh at Yale. Leidy gave them names such as *Titanotherium* and *Megacerops*, and shortly thereafter Marsh named some of his first good specimens *Brontotherium*, *Allops*, *Menops*, *Titanops*, *Brontops*, and other names. The huge skulls and bones in the upper Eocene beds at the base of the strata of the Big Badlands proved irresistible to collectors, so soon the major eastern museums, especially Yale's Peabody Museum, the American Museum of Natural History in New York, and the Smithsonian, had large collections of the huge skulls. In 1884 Marsh sent the young John Bell Hatcher out to find more specimens. Hatcher soon proved to be a master at fieldwork, collecting and sending 118 boxes of brontothere bones weighing 24,136 lb (10,948 kg) to Yale in 1886 alone!

Marsh and Hatcher both died before they could devote time to publishing records of all their specimens, so it fell to Henry Fairfield Osborn to work on them. Like many taxonomic "splitters" of his time, Osborn named many new species and genera, usually based on slight differences in the skulls or horns. He gave little consideration to the possibilities

LEFT: Figure 15.10. Brontothere evolution portrayed as a simple linear process, advocated by Henry Fairfield Osborn in the early 20th century.

OPPOSITE: Figure 15.11. A modern view of brontothere evolution, showing that the group was much more diverse than Osborn thought and had a branched and bushy family tree, like that of most animals. Blue circles and ovals indicate American taxa, and red circles and ovals are Asian taxa.

that these differences might be due to the variability of a natural population or differences between males and females or distortion due to the stresses on the specimen after it was buried. In the 1920s Osborn also received numerous brontothere specimens from the American Museum of Natural History's Central Asiatic Expeditions to Mongolia. Eventually Osborn's work was published in an immense pair of volumes, spanning 951 pages, full of outdated names, inadequate comparisons, and outmoded ideas about evolution. This gigantic book discouraged anyone else from working with brontotheres for the next 60 years. Although paleontologists recognized that Osborn's work was incompetent and outdated, they knew it would be a major job to correct all the mistakes. In 1989 Bryn Mader made a first attempt to address the problems, but our current understanding of brontotheres comes from the 2008 work by Matthew Mihlbachler of the New York Institute of Technology. Osborn recognized more than 30 genera and dozens of species in North America alone, but in 2008 Mihlbachler recognized only 15 valid genera and 22 species in the entire world (Fig. 15.11). For example, Osborn had assigned at least eight different generic names (*Brontotherium*, *Titanotherium*, *Brontops*, *Allops*, *Menops*, *Menodus*, *Symborodon*, *Diploclonus*) and 37 different species to the late Eocene (37–34 Ma) brontotheres, while today Mihlbachler, Mader, and others recognize in this group only one valid genus, *Megacerops*, with two species, *M. coloradensis* and *M. kuwagatarhinus*.

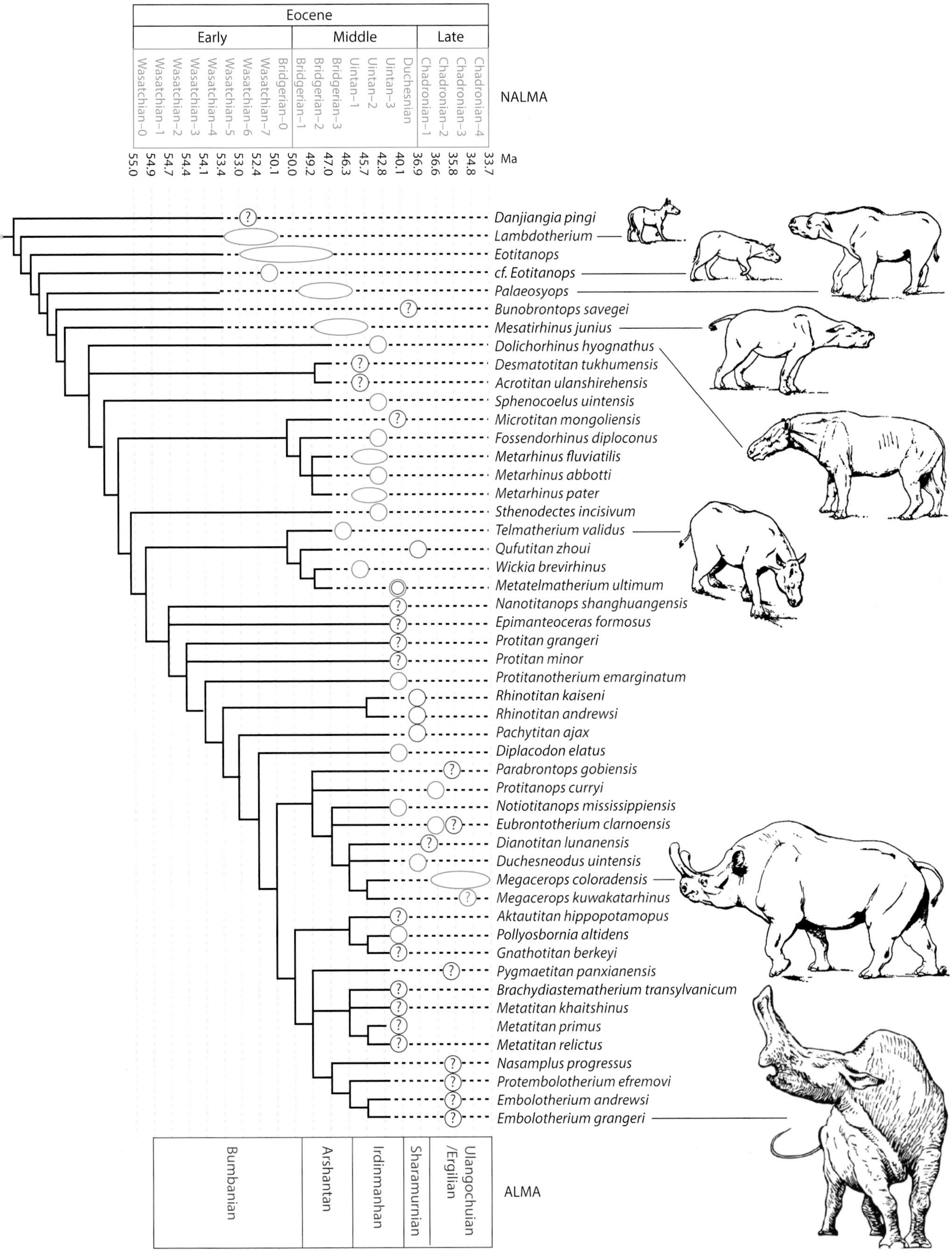

Eocene
Early | Middle | Late

NALMA
Wasatchian-0
Wasatchian-1
Wasatchian-2
Wasatchian-3
Wasatchian-4
Wasatchian-5
Wasatchian-6
Wasatchian-7
Bridgerian-0
Bridgerian-1
Bridgerian-2
Bridgerian-3
Uintan-1
Uintan-2
Uintan-3
Duchesnian
Chadronian-1
Chadronian-2
Chadronian-3
Chadronian-4

Ma
55.0 54.9 54.7 54.4 54.1 53.4 53.0 52.4 50.1 50.0 49.2 47.0 46.3 45.7 42.8 40.1 36.9 36.6 35.8 34.8 33.7

Danjiangia pingi
Lambdotherium
Eotitanops
cf. Eotitanops
Palaeosyops
Bunobrontops savegei
Mesatirhinus junius
Dolichorhinus hyognathus
Desmatotitan tukhumensis
Acrotitan ulanshirehensis
Sphenocoelus uintensis
Microtitan mongoliensis
Fossendorhinus diploconus
Metarhinus fluviatilis
Metarhinus abbotti
Metarhinus pater
Sthenodectes incisivum
Telmatherium validus
Qufutitan zhoui
Wickia brevirhinus
Metatelmatherium ultimum
Nanotitanops shanghuangensis
Epimanteoceras formosus
Protitan grangeri
Protitan minor
Protitanotherium emarginatum
Rhinotitan kaiseni
Rhinotitan andrewsi
Pachytitan ajax
Diplacodon elatus
Parabrontops gobiensis
Protitanops curryi
Notiotitanops mississippiensis
Eubrontotherium clarnoensis
Dianotitan lunanensis
Duchesneodus uintensis
Megacerops coloradensis
Megacerops kuwakatarhinus
Aktautitan hippopotamopus
Pollyosbornia altidens
Gnathotitan berkeyi
Pygmaetitan panxianensis
Brachydiastematherium transylvanicum
Metatitan khaitshinus
Metatitan primus
Metatitan relictus
Nasamplus progressus
Protembolotherium efremovi
Embolotherium andrewsi
Embolotherium grangeri

ALMA
Bumbanian
Arshantan
Irdinmanhan
Sharamurnian
Ulangochuian /Ergilian

Brontotheres show a clear sequence of fossils (Fig. 15.10) that grew larger through the 20 m.y. of the Eocene, from 54 to 34 Ma. As they became larger, the horns on their skulls developed from hornless nasal bones to tiny nubbin-like horns to huge battering rams over 1 m (3.3 ft) long. However, brontothere evolution has been misinterpreted as well. Osborn thought the animals demonstrated evolution out of control and beyond the restraints of natural selection, as their horns grew larger and larger. This conjecture is false, since the horns were clearly functional for more than 20 m.y., and the largest horned brontothere species lasted for 2–3 m.y. without any problems from their horns. Osborn also tended to show brontothere evolution as a single lineage through time (Fig. 15.10), when we know now that there were many parallel lineages through time (Fig. 15.11), both in Asia and in North America, and that they migrated freely back and forth during the middle and late Eocene.

Much of what has been written about brontothere paleobiology is also wrong. Numerous accounts have speculated that brontotheres engaged in violent high-impact head-to-head butting with their horns, as modern bighorn sheep do today. However, this would have been impossible, since the bone supporting the base of the horns is thin and spongy and would have shattered under a high-stress impact. Instead, they probably used the horns for wrestling and head-to-head pushing and swinging, as many antelopes do today. They may have also hit each other side to side. There is one brontothere skeleton on display at the American Museum of Natural History (Fig. 15.12) that shows a broken rib that had rehealed. The healed break is high enough that only another brontothere could have inflicted such a wound, with a sideways blow from its horns.

Another myth about brontotheres concerns their age and extinction. Due to changes in the geologic time scale, the last of the brontotheres are now known to have vanished at the end of the Eocene. Specimens labeled "early Oligocene" in older publications and outdated sources are now all considered late Eocene in age. Osborn argued that brontotheres suffered from "racial senescence"—their lineage had evolved for so long and in a nonadaptive direction that they become "senile" and died out due to "old age." This silly notion is no longer taken seriously. The reasons for their extinction are not completely understood, but brontotheres vanished during a mass extinction near the end of the late Eocene that also wiped out many other primitive groups of mammals that had thrived in the Eocene forests. In the Oligocene those forests vanished, and drier scrublands prevailed. Brontotheres, like many other victims of the late Eocene extinctions, had very low-crowned teeth adapted for eating soft vegetation, which apparently became scarce in the harsher, drier climate of the early Oligocene. In 2002, Matt Mihlbachler and Nikos Solounias of the New York Institute of Technology studied the wear on the teeth of brontotheres and concluded that they ate mostly soft leaves.

Once the brontotheres vanished, about 34 Ma, there were no land mammals that reached such a large size until the middle Oligocene, when the huge indricothere rhinos like *Paraceratherium* evolved in Asia. But in North America, the rhinos never got this large, and such large mammals did not roam the Americas again until the first mastodonts appeared.

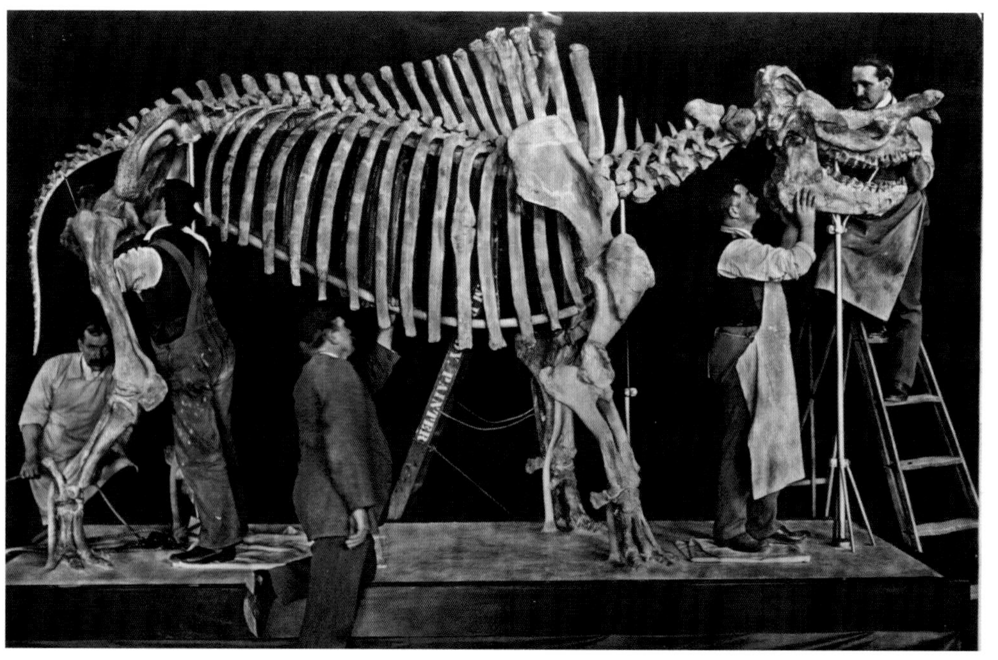

Figure 15.12. An old photograph of the mount of the huge late Eocene brontothere *Megacerops* at the American Museum of Natural History in New York, where it is still on display. This specimen is particularly revealing, because there is a broken rib (rehealed) just behind the middle of the shoulder blade that was probably a result of an impact from side-to-side wrestling with another brontothere

CHAPTER 16

LAURASIATHERIA: MERIDIUNGULATA

South American Hoofed Mammals

During most of the Cenozoic, South America was an island continent, isolated from mammalian evolution elsewhere in the world. As it broke away from Gondwana in the Late Cretaceous, it inherited a strange mixture of unusual animals: marsupials (which evolved to fill the niches of carnivores; see Chapter 3); xenarthrans (sloths, armadillos, and anteaters; see Chapter 5); and a suite of different native hoofed mammals not closely related to hoofed mammals on the other continents (Figs. 16.1, 16.2). Since they had no competition from perissodactyls, artiodactyls, or mastodonts, these native hoofed mammals evolved

Figure 16.1. An array of unusual South American hoofed mammals. In the lower left is the tiny horse-like litoptern *Thoatherium*. In the left background is the large litoptern *Macrauchenia*, with the long proboscis. The large beast between them with the tusks and proboscis is *Astrapotherium*. On the right middle is another beast with a trunk, *Pyrotherium*. The rest of the creatures are notoungulates. The tiny rabbit-like creature on the lower right is *Pachyrukhos*. The animal behind it is *Thomashuxleya*. The large creature to the right of the human is *Homalodotherium*. The horned toxodont to the left is *Trigodon*, and the big hippo-like animal in the background is *Toxodon*.

to fill the vacant ecological niches. Many converged on body forms that occur in familiar mammals from other parts of the world. In some cases, however, the South American hoofed mammals evolved strange body forms that resembled nothing seen in other parts of the world. In a few cases they surpassed their ecological counterparts in evolutionary development; the South American equivalent of a horse, for example, was more completely one-toed than true horses have ever been.

Most of these creatures were first discovered and collecte in the late 19th–early 20th century by Argentinean paleontolo gist Carlos Ameghino and then described and named by hi brother, Florentino Ameghino. Together these men launche South American paleontology. But they worked in isolatio (both geographically and intellectually) from paleontologists i the Northern Hemisphere. Florentino was fooled by the extrem examples of evolutionary convergence he saw in his fossils. H

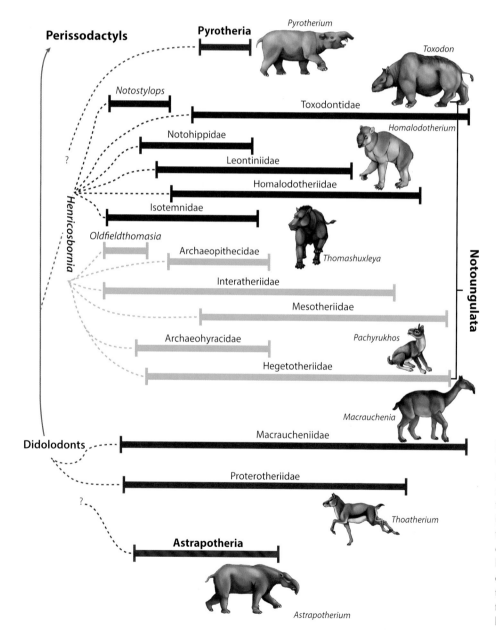

Figure 16.2. The evolutionary history of the South American native hoofed mammals. Most of the lineages are notoungulates, subdivided into the toxodonts (blue) and the typotheres (green). The second largest group is the litopterns (pink), subdivided into the camel-like macraucheniids and the horse-like proterotheriids. Both of these groups appear to be closely related, based on their protein sequences. The relationships of the mastodon-like pyrotheres and the tapir-like astrapotheres are less well established.

often assigned them to groups found on other continents and argued that they were ancestral to all the mammals from elsewhere. Those mistaken notions were largely corrected by paleontologists such as George Gaylord Simpson in the mid-20th century.

Because of this confusion, many of the South American fossil beasts have inaccurate names that suggest they are members of Northern Hemisphere groups, often with the prefix *noto-*, meaning "southern" (e.g., *Notohippus* means "southern horse"). In addition, Florentino Ameghino began to run out of ideas for conventional names derived from Greek or Latin, so he became fond of naming new fossils after famous naturalists of the time. These include genera such as *Henricosbornia* (after Henry Fairfield Osborn), *Thomashuxleya* (after Thomas Henry Huxley), *Asmithwoodwardia* (after paleontologist A. Smith Woodward), *Oldfieldthomasia* (after British anatomist Oldfield Thomas), *Ricardolydekkeria* (after British mammalogist Richard Lydekker), *Guilielmofloweria* (after British anatomist William Flower), *Carolodarwinia* (after Charles Darwin), *Ricardowenia* (after Richard Owen), and many others.

Since the days of the Ameghinos paleontologists have collected and described hundreds of additional South American species and sorted out some of their relationships. However, there are still four major orders of South American hoofed mammals (Notoungulata, Litopterna, Astrapotheria, and Pyrotheria) that cannot be easily related to any hoofed mammal from elsewhere. Paleontologists have tried to ally certain groups of South American hoofed mammals with specific North American fossils, but so far all such efforts have been inconclusive. However, in 2015 an analysis of the proteins from well-preserved late Ice Age fossils suggested that both the notoungulate *Toxodon* and the litoptern *Macrauchenia* might be distantly related to the perissodactyls. If this is true, then South American hoofed mammals descended from an early ungulate relative in North America before the perissodactyls split off (Fig. 16.2). The best candidates for the animals that link South American ungulates to other hoofed mammals are the didolodonts of the Paleocene and Eocene of Argentina, which have been compared to the North American archaic ungulates known as hyopsodonts (Chapter 13). Some are considered to be ancestral to the litopterns (particularly the transitional didolodont known as *Victorlemoinea*).

McKenna (1975) clustered South American native ungulates into a larger group he dubbed "Meridiungulata," but so far there is no strong evidence they are all closely related to one another. Thus, we will look at each group individually and point out the amazing examples of evolutionary convergence they demonstrate.

ORDER NOTOUNGULATA (SOUTHERN UNGULATES)

The largest and most diverse group of South American hoofed mammals is the Notoungulata (Fig. 16.2). Fourteen families are now known, encompassing at least 100 genera and several hundred species, recorded as fossils from the late Paleocene right up to the mass extinctions at the end of the Pleistocene. However, most of them vanished when the flood of North American competitors came over in the mid-Pliocene with the formation of the Panama land bridge (Fig. 5.4). Some of the highlights of notoungulate evolution follow.

Suborder Toxodontia

Shown in blue in Figure 16.2: The genus *Toxodon* itself (Fig. 16.3) survived into the late Pleistocene, becoming one of the last South American hoofed mammals to vanish. It was discovered by none other than Charles Darwin on his *Beagle* voyage in 1833 and brought back to England. Darwin gave it to Richard Owen, who described it and noted that it resembled living rhinoceroses and hippos in body shape and proportions. Yet its teeth are very high-crowned, suitable for eating grasses, and have broad bow-shaped curved crests (*Toxodon* means "bow tooth" in Greek) not seen in any Northern Hemisphere mammal.

Not all toxodonts resembled hippos. They came in a variety of sizes and shapes (Fig. 16.2). The late Miocene fossil *Trigodon* was built a bit like a rhino (Fig. 16.1) and even had a short bony horn on its forehead (although true rhino horns are made of cemented hairs, not bone). The sheep-size *Adinotherium* also had a horn on its forehead. *Homalodotherium* converged remarkably with the perissodactyls known as chalicotheres (Figs. 15.7, 16.1, 16.2). Like those extinct creatures, it had long robust limbs with claws instead of hooves, possibly for pulling down branches to eat leaves. *Rhynchippus* was very much like primitive horses in its size and proportions, and even its teeth look superficially horselike. *Thomashuxleya*, on the other hand, resembled a warthog, except its teeth look nothing like those of a pig (Figs. 16.1, 16.2).

Suborder Typotheria

Shown in green in Figure 16.2: Another branch of notoungulates, the typotheres, had an equally diverse array of shapes

Figure 16.3. Skeleton of the huge rhino-like notoungulate known as *Toxodon*. Even though its body size and proportions resemble those of a hippopotamus or rhinoceros, the details of the teeth and skull are uniquely notoungulate.

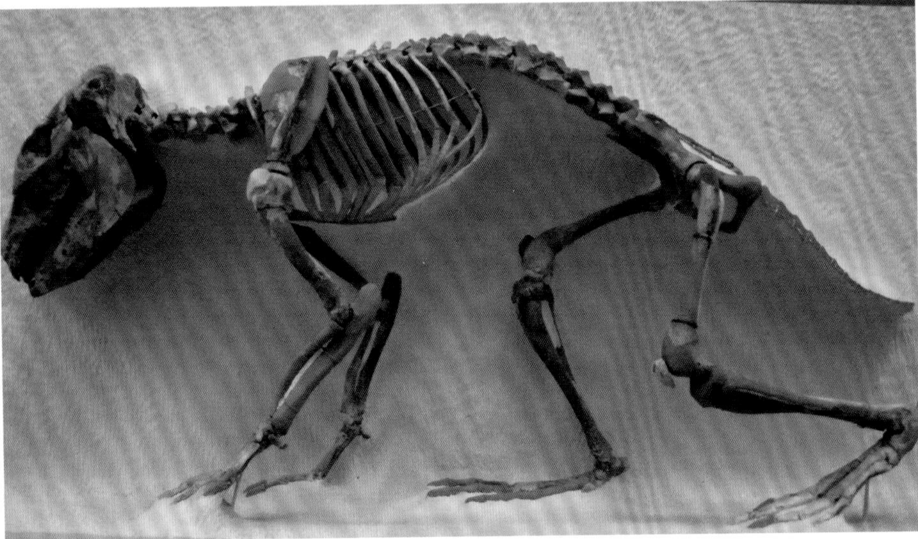

Figure 16.4. Skeleton of the rabbit-like notoungulate known as *Pachyrukhos*.

and adaptations. They include the sheep-like interatheres, and the mesotheres, which had chisel-like front teeth like those of a rodent. The archaeohyracids, as their name implies, vaguely resembled the hyraxes of Africa. The hegetotheres were remarkably convergent with rodents and rabbits. One of them, *Pachyrukhos*, even had large ears, long legs for hopping, large eyes for nocturnal vision, and chisel-like front teeth (Figs. 16.1, 16.4).

ORDER PYROTHERIA ("FIRE BEASTS")

A very peculiar South American hoofed mammal was *Pyrotherium* ("fire beast"), so named because its fossils were found in volcanic ash deposits of Oligocene age. It was about the size and shape of a rhinoceros or mastodont and had short slender legs, a long barrel-shaped body, and a skull with short upper and lower tusks (Fig. 16.1). The nasal bones on top of the head are shifted backward, suggesting that it also had a short trunk like that of a mastodont or tapir as well (Fig. 16.5). Its molars have well-developed cross-crests, similar to those of tapirs and some mastodonts, suggesting a leaf-browsing diet.

Only six genera of pyrotheres are known, scattered from the Paleocene to the Oligocene, and their relationships are

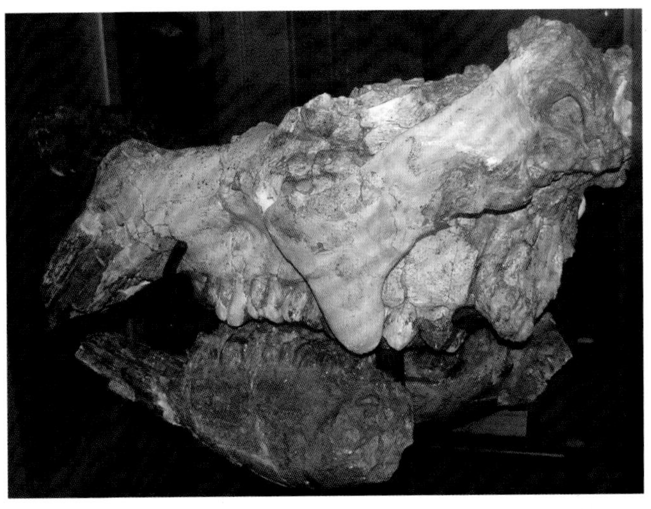

gure 16.5. Skull of *Pyrotherium*, which had prominent upper and lower sks and a big opening in the nasal region for the attachment of a trunk proboscis.

ontroversial. Despite the tusks and proboscis that fooled early ientists like the Ameghinos, the pyrotheres are clearly not lated to Proboscidea (Chapter 6). The details of their teeth ave been used to suggest they are related to uintatheres (Chapter 7), but their ankles resemble those of the arsinoitheres of the ligocene of Africa (Chapter 6). Other analyses place them as ose relatives of the notoungulates. Unfortunately, there are not nough good anatomical characters in the specimens known to solve the issue with confidence. None were known after the ligocene, so we cannot get DNA or proteins from them (as we id with notoungulates and litopterns to establish that those vo groups were related to perissodactyls).

ORDER ASTRAPOTHERIA ("LIGHTNING BEASTS")

ven more bizarre-looking than pyrotheres were the astrapotheres. Thirteen genera are known, ranging from the late Paleocene to the middle Miocene. The strangest of all of them was *strapotherium* itself, which was the size of a rhino or mastodont about 3 m/10 ft long) and had a very long trunk and spindly gs, features that suggest it was amphibious (Fig. 16.1). Its skull truly amazing, with large hippo-like curved upper and lower isks protruding from its mouth and a deeply retracted nasal one suggesting a longer proboscis or trunk (Fig. 16.6). Once gain we have an example of a South American hoofed mammal nimicking hippos or mastodonts, complete with the trunk and isks of the latter group.

The relationships of astrapotheres are even less clear than n the other South American hoofed mammals (Fig. 16.2). hey have been allied with many other groups of mammals, ut the similarities to hippos and mastodonts are due to onvergence.

Figure 16.6. Skull of the bizarre beast known as *Astrapotherium*. It had peculiar upper and lower tusks unlike those seen in any other mammal, and the nasal notch was retracted far back on the skull, suggesting a large proboscis or trunk. Yet in the teeth and every other detail of the skull, it looks nothing like an elephant or mastodont, so its apparent trunk is an amazing example of evolutionary convergence.

ORDER LITOPTERNA (LITOPTERNS, OR "SMOOTH HEELS")

he fourth order of South American hoofed mammals was the itopterna (Fig. 16.1, shown in pink in Fig. 16.2). Composed of t least five families and dozens of genera, it is the second most iverse group of hoofed mammals on the island continent after otoungulates and also came in a variety of shapes. One of the nost striking litoptern fossils is *Macrauchenia* (Figs. 16.1, 16.7), also discovered in 1833 by Charles Darwin on the *Beagle* voyage and described by Richard Owen. *Macrauchenia* was about 3 m (10 ft) long, weighed over 1000 kg (2200 lb), and had a camel-like body and long neck, but it apparently had a long proboscis similar to that of a tapir. Its long legs were adapted for fast running and it had rigid ankles for rapid cornering and

Figure 16.7. Skeleton of *Macrauchenia*, the Ice Age litoptern that had an elepha like proboscis but was built like a camel. In the left foreground is the archaic Nort American ungulate *Phenacodus* (see Chapter 13).

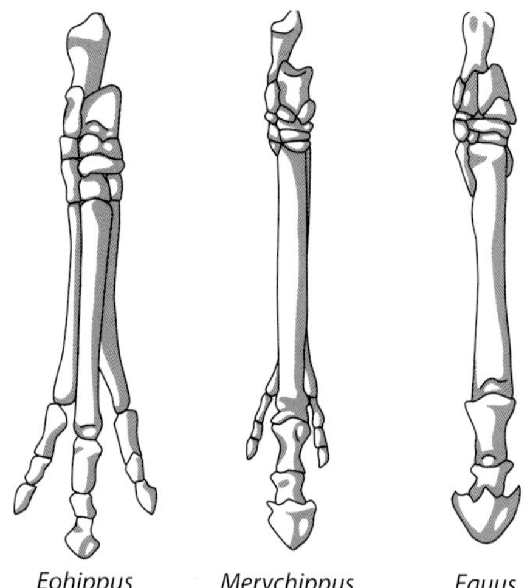

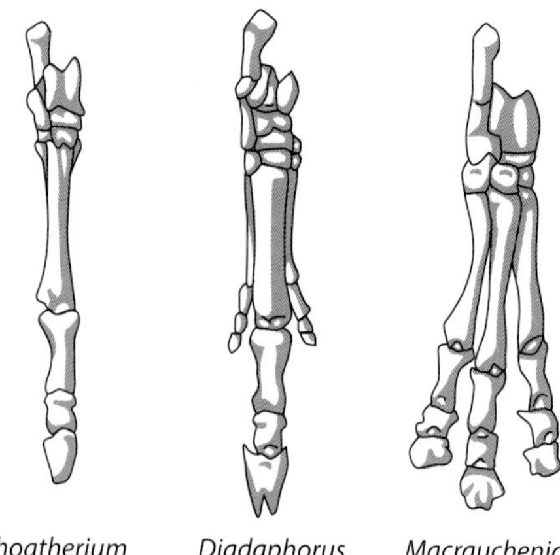

Figure 16.8. Comparison of convergence in the feet of three-toed and one-toed horses (top) and the feet of some litopterns. *Macrauchenia* retains the robust three-toed foot seen in the earliest horses like *Eohippus*. *Diadaphorus* resembles the three-toed horses like *Merychippus* or *Mesohippus*. *Thoatherium* was completely one-toed, losing even more of its side toes than did the living horse *Equus*. This is a remarkable example of evolutionary convergence, since in the rest of the skeleton (especially th details of the teeth and skulls), litopterns looked nothing like a horse.

turning. Its teeth were relatively low-crowned compared to thos of other South American mammals, and their chemistry su gests that *Macrauchenia* was mostly a leaf browser, using its lon neck and trunk to strip the leaves from taller trees. It first ap peared about 7 Ma and vanished near the end of the last Ic Age, about 20,000 to 10,000 years ago.

Another distinctive type of litoptern was the proterothere (Fig. 16.2), including the three-toed *Diadaphorus* (Fig. 16.8 which resembled three-toed horses of the Oligocene and M ocene of North America, and the one-toed *Thoatherium* (Fig 16.1, 16.8). *Thoatherium* looked much like the late Miocen horses that lived in North America, but it was an even mor specialized runner than any northern horse. While even mo ern horses have tiny splints of their reduced side toes alongsid their main toe, *Thoatherium* had only one central toe on eac foot, with no trace of side toes. Once again, South America hoofed mammals demonstrate amazing convergence with u related mammals from other continents. But in this case *Th atherium* has reduced its side toes even more than has the livin horse, so it was a more completely one-toed horse than any tru horse ever was!

CHAPTER 17

UINTATHERES, PANTODONTS, TAENIODONTS, AND TILLODONTS

this chapter, we will look at four early Cenozoic groups of tinct mammals whose relationships to the rest of the mammals e not well established, and which cannot be included in any of e previous chapters. Most were large-bodied and had hooves, they are sometimes treated as ungulates, but there is no clear evidence to support this. All four groups were among the largest mammals on the planet during their time, and all died out in the late Paleocene or Eocene as climate changed and the dense jungle habitats of the globe vanished.

ORDER DINOCERATA (UINTATHERES)

ne of the strangest-looking groups of large mammals in the ocene was the uintatheres (Fig. 17.1), order Dinocerata (the name means "terrible horns"). Although the North American uintatheres are better known, the group evolved in Asia as well.

gure 17.1. The immense horned beasts known as uintatheres were the largest land mammals of the middle Eocene. The earliest is *Prodinoceras* (right eground), which is followed by the huge tusked *Bathyopsis* (right background). The largest was *Eobasileus* (left background). An aberrant uintathere th a bulbous nose was *Gobiatherium* (left foreground).

They were among the largest land animals on the planet during the middle Eocene in both Asia and North America. The uintatheres evolved from pig-size *Prodinoceras* from the late Paleocene and early Eocene of Asia, which weighed about 175 kg (385 lb). They culminated in the elephant-size *Uintatherium* and *Eobasileus*, which weighed as much as 4500 kg (9900 lb). *Eobasileus* ("dawn emperor") was the most spectacular, with six long paired knobs protruding from its nose, forehead, and back of its head, and long upper tusks that were protected by a bony flange on the lower jaw (Fig. 17.2). Yet despite their huge body size, all uintatheres had remarkably simple, small cheek teeth with crown patterns that make them resemble a giant version of the teeth of the earliest relatives of rabbits (which also appeared in Mongolia and China at the same time as the earliest uintatheres). Many scientists have noted that the teeth of *Eobasileus* seem much too small to grind the amount of vegetation needed to sustain such huge bodies.

Perhaps the strangest-looking of the uintatheres was *Gobiatherium*, a rhino-size beast found in the middle Eocene of the Gobi Desert of Mongolia. Instead of the many knobs on the top of the skull or the huge canine teeth displayed by *Eobasileus*, *Gobiatherium* sported a huge bulbous nose supported by bone (Figs. 17.1, 17.3) and broad flaring cheekbones. What this large nose w needed for is anyone's guess, although lots of paleontologi have speculated about its function.

When the first uintatheres were discovered, their large biza skulls were important trophies in the "bone wars" between p neering paleontologists Edward Drinker Cope, Joseph Lei and O. C. Marsh. In the summer of 1872, all three men a their crews were working in the middle Eocene beds of t Bridger and Washakie basins of Wyoming. They found sk after skull of the large and impressive uintatheres, yet they we out in very remote parts of the wilderness with limited opp tunity to publish their finds and get their name in print fir Nonetheless, that did not stop them from trying. All three m would go to the nearest railway depot to ship off their spe mens and send telegraph messages east to have their scienti name and description published in their local journal to g credit. Although he was not normally one to rush to publis Leidy got in print first with his description of *Uintatherium bustum* on August 1, 1872. On August 17, Cope's telegram fro Black Buttes in the Washakie Basin arrived at its destination Philadelphia badly garbled, so his intended name *Loxolophod* ended up printed as *Lephalophodon*. The next day another noti

Figure 17.2. The last and largest of the uintatheres, *Eobasileus*. A. The huge mounted skeleton of *Eobasileus*.

ɜHT: Figure 17.2. The last and largest the uintatheres, *Eobasileus*. B. The zarre skull of *Eobasileus*, with three pairs knobs on the top of its skull and large nine tusks.

ʟOW RIGHT: Figure 17.3. The ɪd-looking, bulbous-nosed skull of *obiatherium*.

rived in Philadelphia, which Cope had sent earlier, and named ope's new species *Eobasileus cornutus*. (Although they did not ıow it at the time, *Eobasileus* would be a valid name for the last ıd biggest of the uintatheres.) On August 22 Cope corrected ʀe garbled telegram with another that restored the spelling of *ɔxolophodon*, although it was an invalid name, because Cope ad already recklessly used it on a different animal. Meanwhile, ı August 20, Marsh sent a telegram naming his new specimens *inoceras* and *Tinoceras* (these names are now considered synoyms of Leidy's *Uintatherium*.) When all three scientists came ack east that fall, they found out that they had created a mess ith their competing names for the same taxa, but each insisted ıat his names were valid. This dispute eventually drove Leidy, ɔnsidered the founder of vertebrate paleontology in the United tates, out of the field altogether. Cope and Marsh fought it out ı the journals, publishing 16 more articles on uintatheres beveen August 1872 and June 1873, ignoring their rival's names ɴd Leidy's work as well. The taxonomy of uintatheres reached state of chaos, with dozens of invalid names for what are now ːcognized as just a handful of genera and species. Eventually ıe uintathere wars led into battles over other fossils. By 1884 ɪarsh had enough fossils to publish a huge 237-page volume ɴtitled *Dinocerata: A Monograph of an Extinct Order of Gigantic ɪammals*, with giant folio pages and lavish illustrations. Not ɴtil almost a century later, in 1961, did paleontologist Walːr Wheeler unravel the mess and establish the valid names for ʌmerican uintatheres.

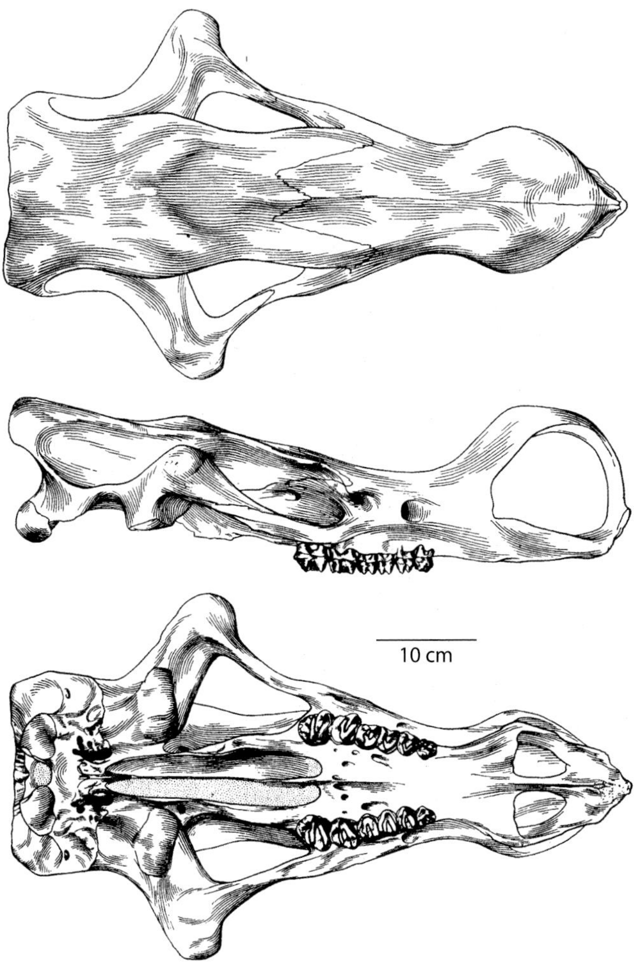

There has also been much argument about the paleoecology f uintatheres. Cope thought they had small trunks and were ːlated to elephants. William Turnbull of the Field Museum in

Chicago pointed out that uintatheres have a very deep, broad rib cage and huge hip region, suggesting that they had a large gut fermentation chamber to digest plants (which could not have been thoroughly chewed up with such small molars). Uintatheres have slightly retracted nasal bones, possibly indicating a prehensile lip, and a large spout-like front end of the lower jaw for a long prehensile tongue. A number of authors suggested that their best analogue might have been hippos, since the uintatheres are recovered from river-channel deposits in ancient jungles of the middle Eocene and have the stature of a semi-aquatic plant eater.

Uintatheres were so large that they had almost no threat from predators except when their calves were young and vulnerable. They lived during the early and middle Eocene, when the largest predators were wolf-size creodonts, which were no threat to adults. So the spectacular horns and upper canine tusks were probably not for fighting off predators but for display and combat wi members of their own species. There seems to be a distinctic between males and females (sexual dimorphism), with presum male skulls being larger and having longer horns. This suppor the idea that the horns were for intraspecific competition.

The relationships of uintatheres are still unresolved. A though their large body size tends to give them features th resemble those of other large hoofed mammals (an opinic that Earl Manning and I published in 1988), they have no tru unique specializations of the true ungulates. Spencer Lucas the New Mexico Museum of Nature and Science and others su gested that their teeth most resemble the Mongolian rabbit rel tive *Pseudictops*, only much larger. In their view, uintatheres a "giant horned bunnies." Given the lack of strong evidence f any relationship, most scientists regard the relationships of tl Dinocerata with any other order of mammals as inconclusive.

ORDER PANTODONTA (PANTODONTS)

One of the earliest groups of mammals to evolve large body size was the extinct order Pantodonta (Fig. 17.4). Pantodonts were stocky creatures with a primitive skeleton and robust limb often with blunt claws or hoof-like structures. Later pantodon

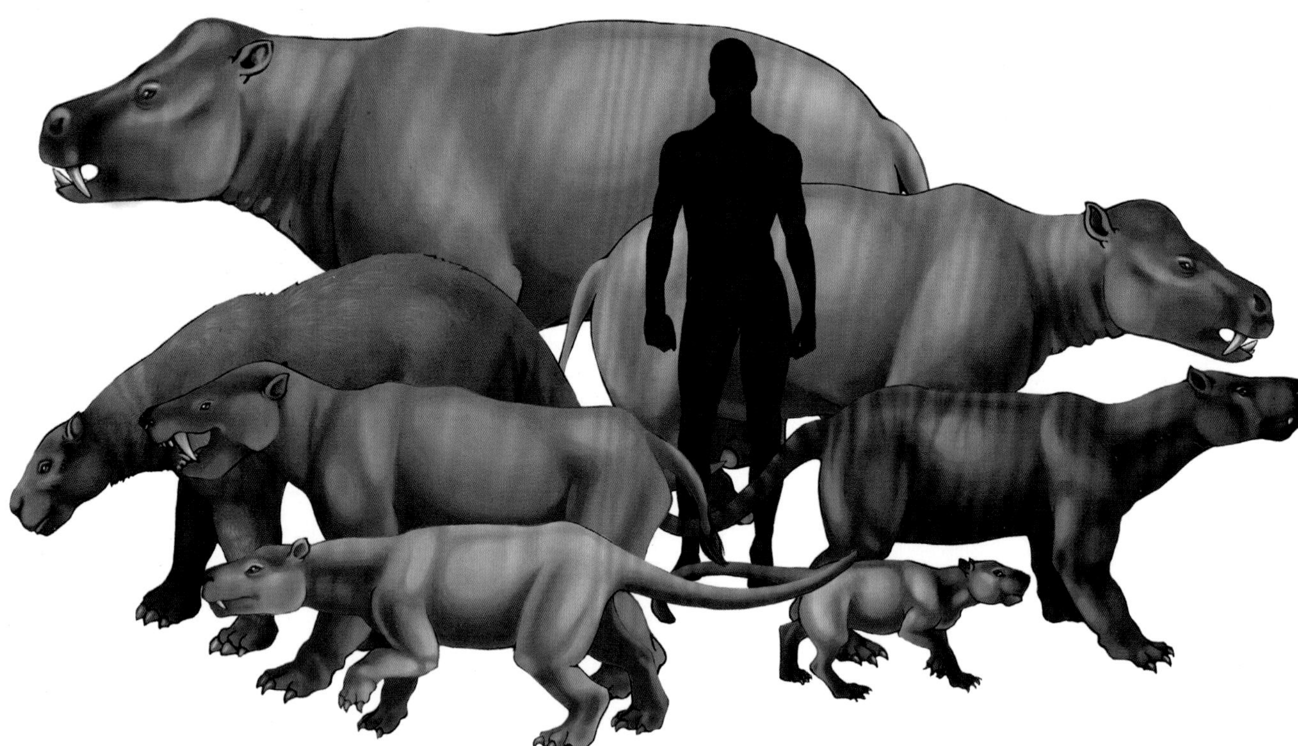

Figure 17.4. Pantodont diversity. In the right foreground is the earliest and smallest form, *Pantolambda bathmodon*, and on the left is *Pantolambda cavirictus*. In the middle left is *Titanoides*, and in the middle right is *Caenolambda*. In the left background is *Barylambda*, and on the right is *Coryphodon* In the background is the largest and last of the pantodonts, the rhino-size *Hypercoryphodon*.

d a hippo-like build that suggested they were partially aquatic. ther smaller pantodonts had long tails and slender limbs ggesting they could have been tree dwellers.

The most distinctive feature of pantodonts is their cheek eth, which had a distinctive pattern of crests resembling a V the Greek letter lambda (Λ). This is why many pantodont nera have the root "lambda" in their names. The primitive ntodont tooth pattern is similar to that of modern tapirs, d is well suited to eating soft leaves and other browse from e Paleocene and Eocene jungles where they lived. Pantodonts re never a very diverse group (fewer than two dozen genera d species are known), but they were widespread in North merica and eastern Asia (China and Mongolia) in the Pale-ene and early Eocene. They also lived in both the Canadian rctic and on Svalbard Island in the European Arctic, and iefly in Europe.

One lineage even managed to cross into the otherwise iso-ted island continent of South America as well. Known only om teeth and jaws from the early Paleocene of Bolivia, *Alcid-orbignya* was a very small, primitive pantodont. Named after e famous early 19th-century French paleontologist Alcide 'Orbigny, *Alcidedorbignya* demonstrates that there was some ay during its time for a few mammals to move from the iso-tion of South America to North America and vice versa. Its tremely primitive features and small size raise the possibility at pantodonts originated in South America and migrated to orth America and on to Asia, although most scientists regard pantodonts as an Asian group that migrated to the Americas and elsewhere across the Arctic to Europe.

From the beginning of the Paleocene, pantodonts stood out due to their relative large body size at a time when most other mammals were rat-size to cat-size (Fig. 17.4). In the early Pale-ocene, the earliest known pantodont in Asia (*Bemalambda*) was the size of a large dog. Sheep-size *Pantolambda* was the first pan-todont to migrate from Asia to North America, which it did during the middle Paleocene. By the late Paleocene the North American pantodonts had evolved to pony-size *Barylambda*, which was 2.5 m (8 ft) long and weighed about 650 kg (1430 lb). It had a heavy skeleton with long claws and a robust tail like that of a ground sloth, so it may have reared up on its hind feet and tail to reach higher vegetation. *Barylambda* was bigger than any other mammal in North America of its time (Figs. 17.4, 17.5A). In the late Paleocene of North America there was an even larger pantodont, the bear-size *Titanoides*. It was up to 3 m (10 ft) long and weighed about 150 kg (330 lb). Its broad robust snout sported sharp canine tusks.

By the early Eocene the last of the American pantodonts, the genus *Coryphodon*, had reached the size of a small rhinoc-eros (Figs. 17.4, 17.5B). Small species were about 2.25 m (7.4 ft) long, and weighed about 500 kg (1100 lb), but the largest species weighed as much as 700 kg (1500 lb) and was significantly larger than any other land mammal in the early Eocene. The genus migrated from Asia to North America, replacing *Barylambda*, and even spread to Europe and the Canadian Arctic before vanishing

igure 17.5. Pantodonts. A. Skeleton of *arylambda*, showing the robust limbs nd heavy tail. (See also next page.)

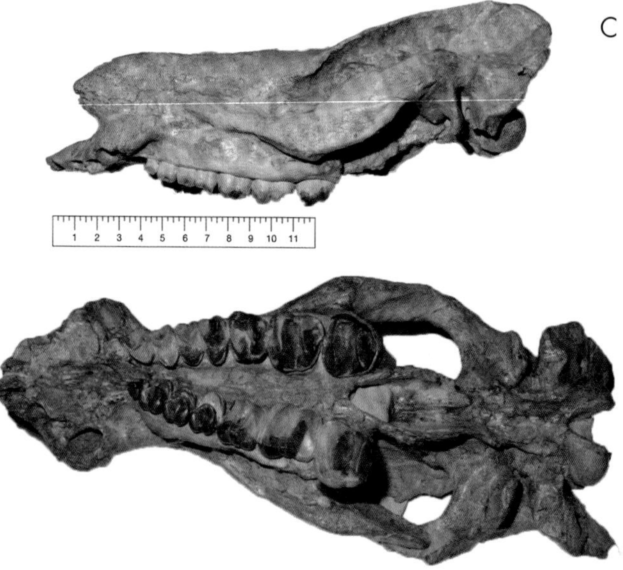

Figure 17.5. Pantodonts. B. The last and largest of the American pantodonts, *Coryphodon*, which had a hippo-like skeleton with large flaring upper and lower tusks. C. The immense skull of the rhino-size Asian pantodont *Hypercoryphodon*, the last and the largest of the order (scale bar in inches).

at the beginning of the middle Eocene. *Coryphodon* had very robust stocky limbs and was built like a hippopotamus, suggesting it had an aquatic lifestyle. It had a stocky muscular neck, head, and body, a broad muzzle, and prominent canine tusks for battling with other animals and rooting up vegetation. There is evidence that the tusks were larger in males than in females. Through most of their range, *Coryphodon* species lived off leaves and soft browse, but those that lived in the Canadian Arctic show evidence in their tooth wear of having switched to a diet of leaf litter, twigs, and evergreen needles during the six months of darkness when plants did not grow.

The very last of the pantodonts, *Hypercoryphodon*, from Mongolia (Figs. 17.4, 17.5C), was the size of a rhinoceros and survived until the end of the middle Eocene (about 40 Ma). It is known primarily from a huge skull found in Mongolia in the 1920s by the American Museum of Natural History paleontologist Walt Granger, and described by Granger and Henry Fairfield Osbor in 1932.

The relationships of pantodonts remain controversial. Ma paleontologists have lumped them with other heavy-bodied ear Cenozoic mammals, especially hoofed mammals, based on the primitive skeletal features, but shared evolutionary novelties d not provide strong evidence of relationship. Some have note the similarity of the V-shaped crests on their cheek teeth to thos of primitive insectivorous mammals of the Cretaceous and su gested a relationship, but such a simple tooth pattern could ea ily be due to convergent evolution. For now their relationshi to the other groups of mammals are unresolved, since there no strong evidence or consensus that links them with any pa ticular group.

ORDER TAENIODONTA (TAENIODONTS

The taeniodonts were archaic mammals from the Paleocene and early Eocene that became highly specialized as large-bodied diggers (Figs. 17.6, 17.7). They were never very diverse, with only 10 genera over their entire history, all in North America. The first taeniodont was *Schowalteria*, known from an incomplete fossil from the Late Cretaceous of Canada. Much better known, from a nearly complete skeleton, is *Onychodectes* from the early Paleocene (Fig. 17.7C). It was only rat-size, with about the same general proportions as a rat or a cat. Most of the rest of the taeniodonts from the early and middle Paleocene were similarly small, although *Wortmania* (Fig. 17.7B) from the early Paleocene weighed about 20 kg (44 lb) and was one of the largest mammals to evolve right after the dinosaurs became extinct.

The middle late Paleocene taeniodont *Psittacotherium* ("parro beast") was the size of a large dog (Fig. 17.6). It reached abo 1.1 m (3.5 ft) long and weighed about 50 kg (110 lb), making one of the largest mammals of the late Paleocene. It had massiv robust jaws that featured large canine tusks, chisel-like front inc sors, and broad grinding cheek teeth with a pattern unlike tha found in any other mammal. The forelimbs were especially mu cular and robust and had large heavy claws, possibly for digging These trends of robust skull with chisel-like incisors and larg canine tusks were further developed in the late Paleocene–earl Eocene taeniodont *Ectoganus*.

These trends culminated with the last and the largest of th taeniodonts, *Stylinodon* (Figs. 17.6, 17.7A, 17.8), which arose i

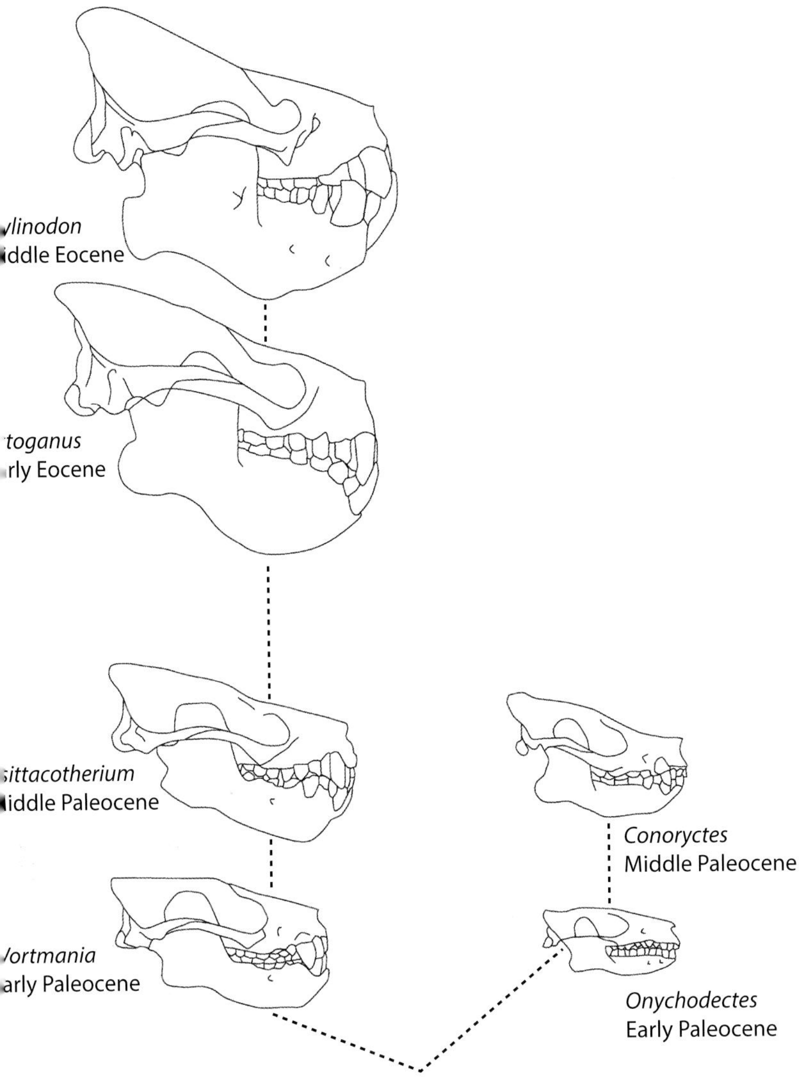

ylinodon
iddle Eocene

toganus
rly Eocene

ittacotherium
iddle Paleocene

Conoryctes
Middle Paleocene

/ortmania
arly Paleocene

Onychodectes
Early Paleocene

gure 17.6. Taeniodont evolutionary history.

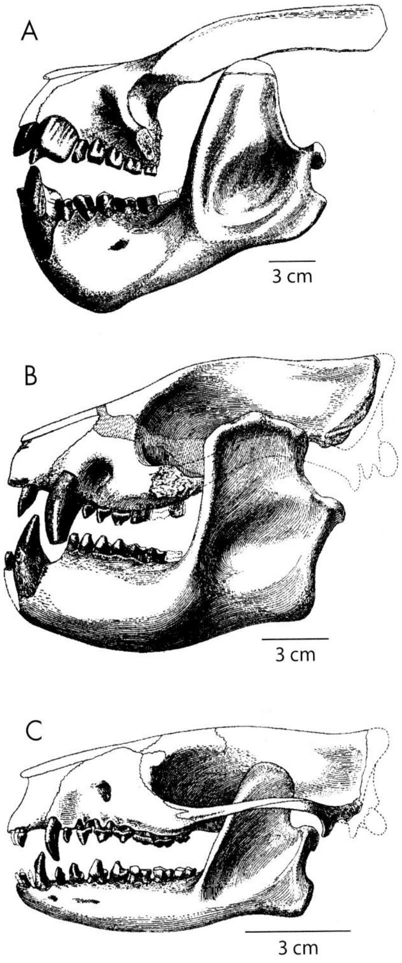

A

3 cm

B

3 cm

C

3 cm

Figure 17.7. Skulls of taeniodonts. A. *Stylinodon*. B. *Wortmania*. C. *Onychodectes*.

igure 17.8. Reconstruction of the aeniodont *Stylinodon*.

the early Eocene (53 Ma), and persisted to the very end of the middle Eocene (about 40 Ma), a duration exceeding 13 m.y. It was the size of a large pig, weighing as much as 80–110 kg (180–242 lb). Although *Stylinodon* had massive robust forelimbs with many strong muscles and short stout claws, it was apparently not a digger, since its teeth do not show wear and scratches from a gritty diet. The skull of *Stylinodon* was massive and muscular, with deep powerful jaws. Its canines had evolved into large rootless incisor-like teeth with a chisel-like edge to match the incisors. They had a thin ribbon of enamel on the outer surface to make them self-sharpening. The molar teeth were also high-crowned, so they would not wear down too quickly, because the diet apparently created a lot of wear in many individuals of *Stylinodon*. The molars of most *Stylinodon* specimens are so worn that none of the original crown remains. Later taeniodonts ac ally had rootless, ever-growing cheek teeth, so they would nev wear down. Taeniodonts may have fed like ground sloths, pu ing down branches with their massive claws and forelimbs bring leaves within reach.

As with pantodonts and tillodonts (described next), the ne est relatives of taeniodonts are a mystery. Some scientists ha compared taeniodonts to primitive hoofed mammals, althou they have no clear specializations of ungulates. Others ha noted the similarities of their teeth to Cretaceous insectivoro mammals such as *Cimolestes*, although there appears to be a l of convergent evolution in such simple teeth. Since there is r consensus or strong evidence either way, the relationships taeniodonts are still unresolved.

ORDER TILLODONTIA (TILLODONTS)

The tillodonts (Fig. 17.9) are an order of extinct mammals known from the earliest Paleocene to middle Eocene of North America, Asia (China, Mongolia, and Pakistan), and briefly Europe. Never very diverse—about 20 genera are known—they were generally rare except during the latest Paleocene and early Eocene, from which time *Esthonyx* is a common fossil. Tillodont skulls show a strange mixture of features of different mammals. They have ever-growing chisel-shaped incisors in the front (like those found in rodents, but tillodonts are clearly not related to rodents) and tiny canines (or none at all), yet their molars have V-shaped crests that look vaguely like those of pantodonts. The skull has huge areas for the jaw muscles used for chewing, but the braincase is short and small.

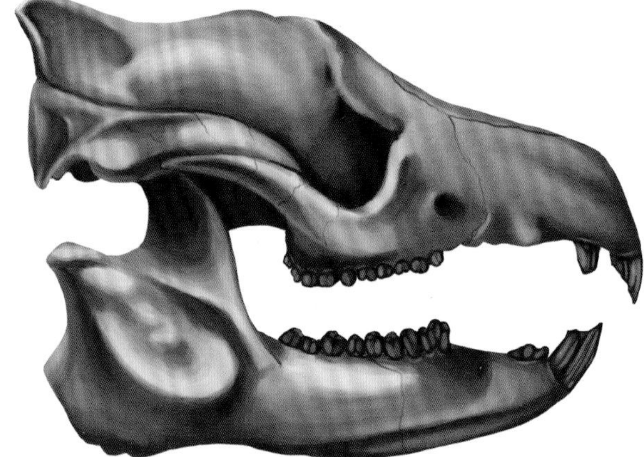

Figure 17.9. The tillodont *Trogosus*. A. Reconstruction. B. Skull.

Tillodonts began with tiny forms emerging in the earliest Paleocene, with skulls only 5 cm (2 in) long, and went on to become one of the first groups of large-bodied mammals of the early Cenozoic. This trend culminated with *Trogosus* (Fig. 17.9), a bear-size creature weighing about 150 kg (330 lb), with a skull 35 cm (14 in) long. The last of the American tillodonts, *Trogosus* lived from 53 to 47 Ma, dying out in the early middle Eocene. It is the only tillodont known from not just jaws and teeth but also its skeleton, which is heavily built and compact, apparently supporting a very muscular animal. It had stout forelimbs and long claws, and together with the chisel-like incisors these features suggest a habit of digging up roots and tubers for food. Their teeth often show very heavy wear, consistent with the idea that they consumed a lot of gritty food that wore down their teeth rapidly. However, grooves on the sides of their incisors suggest that they stripped leaves off of branches as well.

Tillodonts originated in the earliest Paleocene of Asia (China, Mongolia, and possibly India), with small-bodied forms such as *...enaius*, *Meiostylodon*, and *Huananius*. Tillodonts were always more diverse in Asia than anywhere else and persisted the longest there, vanishing near the end of the middle Eocene (40 Ma). The tillodont *Esthonyx* was the first to migrate from Asia to North America in the late Paleocene, about 56 Ma, where it was among the more common mammals. Four other genera of tillodonts are found in North America through the early and middle Eocene, until they vanished when *Trogosus* became extinct. Two genera of tillodonts, *Franchaius* and *Plesiesthonyx*, appeared in Europe during the early Eocene but soon vanished.

The relationships of tillodonts are controversial. Their chisel-like incisors initially suggested a relationship with rodents, but the similarity is clearly due to convergent evolution. Their heavy bodies and plant-grinding molars have often led scientists to link them with hoofed mammals, but they have no truly unique anatomical features of that group. The V-shaped crests on their molars suggested relationships with pantodonts, but this is probably another case of convergent evolution due to similar diets. There is simply not enough evidence to determine which group of mammals is most closely related to tillodonts.

Figure 17.10. Composite image of reconstructions of some of the more remarkable extinct mammals. Behind the group on land is the giant hornless rhinoceros *Paraceratherium*, the giant ground sloth *Megatherium*, and the American mastodon. Across the front row (from left to right): the fin-backed *...metrodon*; the South American mastodon-mimic *Astrapotherium*; the giant creodont *Megistotherium*; the broad-cheeked peccary *Skinnerhyus*; the saber-toothed marsupial *Thylacosmilus*; human for scale; the giant otter *Enhydriodon*; the taeniodont *Stylinodon*; the giant rodent *Josephoartigasia*; the horned pig *Kubanochoerus*; and the South American hippo-mimic *Toxodon*. In front of the mastodon is the gorilla-like horse relative *Chalicotherium*. Behind the human is the giant short-faced kangaroo *Procoptodon*; the South American camel mimic *Macrauchenia*; and the six-horned uintathere *Eobasileus*. In front of the giant rhinoceros is the huge *Bison latifrons*, and the giant pantodont *Hypercoryphodon*. In the water (but not to scale) are the huge predatory sperm whale *Livyatan*, eating a baleen whale; the swimming sloth *Thalassocnus*, and the whale-sized Steller's sea cow.

CHAPTER 18

MAMMALIAN EVOLUTION AND EXTINCTION

Whenever people learn about the amazing and spectacular extinct beasts of the past, certain questions always seem to pop up. There is not room in this book to answer all of them, and some require a detailed technical understanding of paleobiology t explain, but some questions demand a discussion, if not a fin conclusive answer.

WHY WERE PREHISTORIC MAMMALS SO BIG?

The question of what controls mammalian body size is a complex one, with many different factors in operation, so there is no simple answer to why mammals reached certain body sizes. The pioneering American paleontologist Edward Drinker Cope (see Chapter 17) was one of the first to collect and describe many of the fossil mammals discussed in this book. As early as 1885 he noticed that there was a trend for the species in most mammalian lineages to grow larger over their evolutionary history, which is now called "Cope's rule." (A similar trend was noticed and published by the French paleontologist Charles Depéret.)

"Cope's rule" and the factors that control mammalian body-size trends have been studied and debated by many different generations of paleontologists. Although body-size increase is a common trend, it is not universal. Many mammalian lineages show no size trends at all, while others show both increases and decreases in body size. (For example, many lineages of horses grew larger through time, others stayed constant, and a few became dwarfed—see Fig. 15.2B).

There are many examples of dwarfing and size reduction in the fossil record as well. This is particularly common with larger mammals like elephants, rhinos, hippos, and with other hoofed mammals isolated on small islands. Ecological theory suggests that once they no longer had to cope with large predators, and also had to live on the much smaller base of plant resources of an island, large mammals shrank in body size because there were few advantages to being huge. It is a very wasteful and expensive evolutionary strategy to keep a large body size when it is no longer essential to survival.

Growing a large body size has its trade-offs. Large body siz makes it easier to fight off predators and competitors (especiall large males competing for mating rights). For predators, it allows them to capture bigger prey, fight off other members c their own species for a carcass, and also fight off scavengers an smaller predators trying to steal their food. Larger animals typ cally have the fat reserves to survive bad times and can trave longer distances in search of food and water. Large animals als are more thermally stable and less vulnerable to climate change and extreme fluctuations in weather conditions. As a collatera benefit, larger animals tend to live longer (body size is correlate with life span) and to have greater intelligence.

But there are downsides to being larger. A larger animal requires more food and water resources. Carrying a large bab to term is a much greater burden on the mother, so most larg animals have only one or a few babies each breeding cycle. The must invest a lot of parental energy and resources protectin their young. (By contrast, most small mammals have lots of babies at once, and only a few strong ones are expected to survive.) Large mammals also tend to grow slowly, and they hav a long period when they are vulnerable juveniles. They mus cope with many seasons of drought and famine to survive t adulthood. Large animals also tend to be highly specialized int certain niches, so they are more vulnerable when major extinction events happen, compared to smaller and more ecologicall flexible animals.

Once land animals reach certain body sizes, there are mechanical limits to how much weight normal bones can support no matter how robust those bones are. There are problems with how the circulatory system can pump blood up to their heads, high above the ground, and many other issues. Marine vertebrates such as whales and fossil marine reptiles can grow significantly larger because a dense medium like water is supportive of great size and weight, while land mammals must fight gravity with only the support of air. However, most discussions of body size trends focus on land animals.

To some extent the increase of body size through time is an artifact of the reality that most groups first start out as smaller creatures, and the only direction to go is up. The earliest dinosaurs in the Late Triassic were chicken-size, so as they evolved they could only stay the same size or get larger. Even individual lineages, such as the huge sauropod dinosaurs, the duckbills, and the horned dinosaurs like *Triceratops*, started with relatively small ancestors. As they diversified, it was inevitable that they got larger as they tried different body shapes.

This trend of starting small and eventually getting big is particularly true in the evolution of land mammals. For the first two-thirds of their history (about 134 m.y., from 200 to 66 Ma), mammals were mostly rat-size or smaller (see Chapter 2) because they were constrained by the dinosaurian overlords who ruled all the large body-size niches. After the non-bird dinosaurs vanished 66 Ma, all the lineages that would give rise to Cenozoic mammals eventually would have to grow large enough to occupy all the vacant large land animal niches vacated by the dinosaurs. Still, by the beginning of the Eocene 54 Ma, the largest mammals were only cow-size (e.g., *Coryphodon*, Fig. 17.5B), and by the end of the middle Eocene and late Eocene (40–34 Ma), only the brontotheres and uintatheres were as large as a rhinoceros. By the Oligocene there were the huge indricothere rhinos (Figs. 15.1, 15.9), but they were the exceptions in their habitat; most mammals were sheep-size or smaller. During the Miocene a number of proboscidean lineages got to be huge, especially the deinotheres (Fig. 6.4C, D), and a number of elephants and mammoths were big. But by and large, most of the mammals of the Neogene were relatively small. Only a few mammals took the risks and accepted the trade-offs to become huge, and it happened on and off in only a few lineages through time.

The impression that all extinct mammals were large is also an artifact of the bias in what gets presented to the general public. Big, fierce, extinct animals are glamorous and eye-catching, and have the "wow" factor that gets public attention. When the American Museum of Natural History sponsored a traveling exhibit a few years ago called *Extreme Mammals*, it featured mostly the largest mammals. But the vast majority of extinct mammals were small and unspectacular, so the public seldom hears about them. Thus people get the impression that *all* extinct mammals were huge, when it is really only what the media have chosen to publicize. (Even in this book I have had to bow to the pressure to give more space to the big and bizarre, while glossing over the entire order Rodentia in one chapter, even though it contains almost half the diversity of the Mammalia).

There is also another factor to consider: Up until 10,000 years ago mammoths and mastodonts and huge bison, deer, and ground sloths roamed the Americas and Asia, and gigantic wombats and kangaroos were found in Australia. Giant mammals were the norm for most of the middle and late Cenozoic, and their absence in the last 10,000 years is a relatively recent phenomenon. They were here not long ago, but most have been wiped out by the great extinctions at the end of the Ice Ages. This leads us to our second question.

WHERE HAVE ALL THE MEGAMAMMALS GONE?

A visit to La Brea Tar Pits and Museum or the Cincinnati Museum of Natural History and Science, or a number of other natural history museums around the Northern Hemisphere, puts the viewer face to face with the enormous skeletons of mammals of the Ice Ages: huge mammoths and mastodonts, enormous bison, deer, and ground sloths, bear-size beavers, and huge predators like giant short-faced bears, Ice Age lions, and saber-toothed cats. One cannot escape with the impression that the earth has long been the domain of megamammals (mammals greater than 40 kg/88 lb in body weight; some put the threshold at greater than 100 kg/220 lb). Today only a pitiful remnant of the megamammals survives, mainly in tropical Africa and Asia, with elephants, rhinos, hippos, and giraffes.

The losses of megamammals happened on nearly every continent. North America lost 33 of 45 genera of mammals that weighed more than 40 kg, including mammoths, mastodonts, giant bison, several species of horses, camels, deer, peccaries, tapirs, giant beavers, ground sloths, glyptodonts, and pampatheres, and in addition most of its large predators: giant short-faced bears, cave bears, saber-toothed cats, Ice Age lions, the American cheetah, and dire wolves. Today, only the American bison, the moose and several other kinds of deer, plus bears, and cougars remain of the North American megafauna.

South America lost 46 of 58 large mammal genera. These included not only immigrant groups like mastodonts, horses, saber-toothed cats, and giant bears, but also many of the continent's native giants, including several genera of ground sloths, pampatheres, glyptodonts, and native hoofed mammals such as toxodonts (*Toxodon*) and litopterns (*Macrauchenia*).

Australia lost 15 of its 16 large mammals, including rhino-size diprotodonts, huge kangaroos, tapir-like palorchestids, and the marsupial "lion" *Thylacoleo*. In addition to the mammals, the extinctions affected the enormous flightless dromornithid birds, huge crocodiles, a 6 m (20 ft) long python called *Wonambi*, and *Megalania*, the monster Komodo dragon exceeding 6–7 m (20–23 ft) in length.

The extinctions in Eurasia were less severe. Europe lost only seven of 23 genera (mainly mammoths, rhinos, hippos, giant Ice Age deer and cattle, cave lions, cave bears, and saber-toothed cats). Sub-Saharan Africa escaped with its megamammals largely unscathed, with only two of 44 genera going extinct. Thus the extinction pattern is not simple and not consistent from one continent to the next (Fig. 18.1).

Ever since the huge Ice Age beasts were first discovered, people have speculated about why they vanished. Most scholars before 1800 rejected the entire notion of extinction, since they believed that a God who watched over even the smallest sparrow would not allow one of his creatures to go extinct. Thomas Jefferson was not only the author of the Declaration of Independence, our first secretary of state, our third president, and the founder of the University of Virginia, he was also a scholar and naturalist. When he received some large fossil claw bones from the Appalachian region, he thought they came from a giant lion. (The fossils turned out to belong to the giant ground sloth now known as *Megalonyx jeffersoni*.) Jefferson wrote in 1799:

> The bones exist: therefore the animal has existed. The movements of nature are in a never-ending circle. The animal species which has once been put into a train of motion, is probably still moving in that train. For if one link in nature's chain might be lost, another and another might be lost, till this whole system of things should vanish by piece-meal. ... If this animal had once existed, it is probable on this general view of the movement of nature that it still exists.

He was convinced that these strange animals were still living out in the unexplored Northwest. Jefferson even instructed Meriwether Lewis and William Clark to be on the lookout for a gigantic lion and mammoths during their expedition to the West. He wrote in 1781:

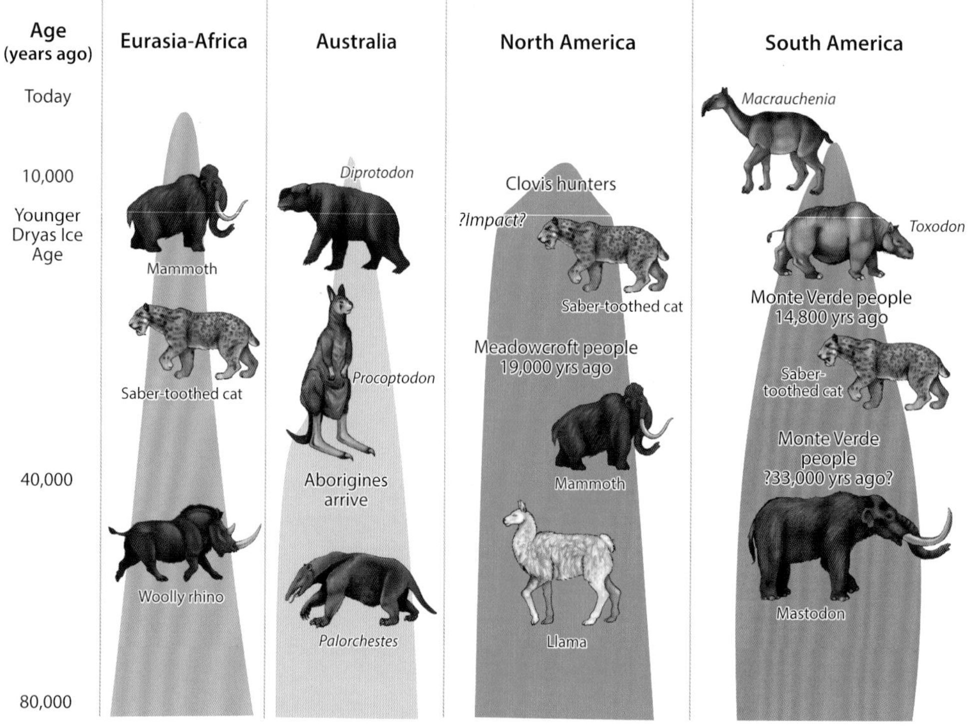

Figure 18.1. The pattern of extinction of large mammals during the end of the last Ice Age. Large mammals gradually vanished in Eurasia until about 10,000 years ago, when the climate changed dramatically and the ice sheets retreated. However, the last mammoth survived on Arctic islands as late as 3,700 years ago. Large mammals gradually vanished from Australia after the arrival of Aborigines and their dingoes about 60,000–40,000 years ago, but did not completely vanish until about 10,000 years ago. Large mammals were thriving in North America until about 11,000–10,000 years ago when the climate changed and hunters using Clovis points arrived, but humans had reached North America from Asia at least 19,000 years ago (Meadowcroft site in Pennsylvania). There are claims of a comet impact 12,900 years ago, but if it occurred it had almost no effect on human cultures nor did it cause extinction of megamammals. According to South American paleontologists, their native large mammals went through pulses of extinction (the "zigzag" model), until the last megammals vanished 9,000 years ago, later than those on most other continents. Humans had reached South America (Monte Verde site, Chile) at least 14,800 years ago and possibly as early as 33,000 years ago.

It may be asked why I [list] the mammoth as if it still existed? I ask in return, why should I omit it, as if it did not exist? Such is the economy of nature, that no instance can be produced of her having permitted any one race of her animals to become extinct; of her having formed any link in her great work so weak as to be broken. To add to this, the traditional testimony of the Indians that this animal still exists in the northern and western parts of America, would be adding the light of a taper to that of the meridian sun.

But in the early 1800s, the founder of vertebrate paleontology and comparative anatomy Georges Cuvier published the first scientific descriptions of extinct mammoths, mastodonts, and other bizarre creatures such as the mosasaur and *Palaeotherium* (Fig. 15.3). The world was well enough explored by that time that he was confident that no such huge beasts as mammoths and mastodonts still roamed the planet. Thus Cuvier made the conclusive case that some creatures had gone extinct.

But what could have caused the extinction of the Ice Age megamammals? Many ideas have been proposed, but some are merely speculation that cannot be tested by any scientific evidence. Most of the arguments boil down to just a few likely factors:

The "overkill" or "blitzkrieg" hypothesis: Many scientists point to the wave of extinctions between 12,000 and 10,000 years ago in North America, and the arrival of people with Clovis spear points at the same time, and argue that sophisticated human hunters "blitzed" through the American mammals with no previous experience of human hunting (Fig. 18.1). Although the coincidence in timing seems persuasive, there are lots of counterarguments. For one thing, we only have direct evidence that humans hunted mammoth, bison, horse, and just a few other species, so if they wiped out the megafauna, much of it was probably indirectly via habitat destruction, or wiping out the prey species on which large predators depended. Some animals we know they hunted, like deer and bison, did just fine and are still with us.

In addition, the Clovis people were not the first hunters in North America. Sites such as the Meadowcroft rock shelter in Pennsylvania have produced artifacts of a pre-Clovis culture at least 16,000 and perhaps as many as 19,000 years ago (Fig. 18.1). If humans were such devastating hunters, why didn't the pre-Clovis people decimate the Ice Age megamammals some 8,000 years earlier? Thus, the idea that the first arrival of hunters from outside automatically results in mass extinction does not work. The only way to salvage the overkill explanation is to say that the Clovis cultures were much more sophisticated and rapacious hunters than the pre-Clovis peoples.

South America has a similar story. There the last of the Ice Age mammals vanished about 9,000 years ago, slightly later than in North America, and some scientists have argued that human hunters were the primary cause. However, as with the Meadowcroft site left by pre-Clovis peoples in North America, several sites in Chile, including Monte Verde, date back to at least 14,800 years ago, and possibly to 33,000 years ago, before the peak of the last Ice Age (Fig. 18.1). Monte Verde probably holds the oldest evidence of human habitation in the Americas. Surprisingly, the oldest sites are not close to where humans came across from Asia but the farthest possible distance; it is a long way from the Bering Land Bridge to the southern tip of Chile. Whatever the pre-Clovis peoples of the Americas were doing, they showed no signs of overhunting the large mammals for thousands of years.

Australia presents a different story. Aboriginal humans and their dingo dogs arrived at least 48,000 years ago, and possibly as early as 60,000 years ago. Some of the Australian fauna began to vanish then, but some may have persisted until as late as 7,000 years ago. If the Aborigines overhunted the native mammals, it was a gradual decline, rather than the rapid disappearance seen in the Americas. Others have pointed to how Aborigines even today use fire to change the landscape and make their hunting easier. Pollen evidence from lake sediments suggested that humans destroying the dense brushy habitat that protected many mammals (including those not even hunted for food) may have been just as important as direct hunting.

A totally different picture is found in Eurasia and Africa, where most of the large mammals survived over millions of years alongside many species of humans as they evolved hunting skills. Most of these animals were never overhunted, while some vanished at times when there is no evidence of overhunting. Overkill advocates argue that since these beasts evolved side by side with hunters, they were used to these attacks and not as vulnerable as American and Australian mammals that had evolved in the absence of human predation.

Of course, in the past few hundred years the arrival of humans in many places, such as Madagascar, New Zealand, and other islands in the Indian and Pacific Oceans and the Caribbean Sea, has clearly been the major cause of the extinction of the native wildlife. But are these extinctions of vulnerable naïve animals with small populations on islands really comparable to the extinction of large populations of mammals on continents? This point is very controversial.

The climate-change hypothesis: The end of the last Ice Age was also a time of rapid climatic and vegetational change around the world. The planet went from the peak of the last glacial maximum about 20,000–18,000 years ago, when large ice caps covered both poles, to our present interglacial beginning about 10,000 years ago. There are many signs from the population structures and other odd features of the mammalian communities shortly before their extinction that suggest that they had reached maximum stress due to climate change, especially in the loss of their preferred habitats and vegetation (in the case of herbivores).

Critics of the climate-change model point out that no similar mass extinction happened at the previous glacial-interglacial transition about 125,000 years ago. But there are many signs that the most recent glacial-interglacial transition was unusually severe. For one thing, there was an extremely rapid cold event about 11,700 years ago, known as the Younger Dryas event. The entire planet, which had nearly finished warming up from the last glacial maximum, was suddenly (in less than a decade) plunged into a new glacial period. Although large mammals had adjusted to the slow transition from glacial to interglacial climate over thousands of years, perhaps the rapid switch back to another glacial period was too much for many of them.

The impact hypothesis: The fad for blaming all mass extinctions (such as happened at the end of the Cretaceous when the dinosaurs vanished) on impacts of objects from space was extended to the Pleistocene in 2007. That year a group of scientists proposed that the North American extinctions were due to a comet or meteorite impact over the Carolinas near the beginning of the Younger Dryas event, about 12,900 years ago. The original evidence for this supposed impact was a "black mat" of organic material in many Clovis sites, plus microscopic nanodiamonds in deep-sea cores, and rare platinum group metals in Greenland ice cores from around 12,900 years ago.

This idea was trendy for a few years, but now the evidence against it has become overwhelming. Most of the nanodiamonds and the "black mat" and other supposed impact indicators have been explained by other causes. More important, if it occurred at all, it was about 1,000 to 2,000 years too early for most of the North American extinctions, which were largely between 11,000 and 9,000 years ago. Nor is there any sign that American Paleo-Indian cultures suffered any effects that would be expected if a huge impact had occurred. Finally, an impact over the Carolinas cannot explain extinction on any other continent, most of which had no significant extinctions at 12,900 years ago anyway.

Other ideas: There are no limits to the ingenuity and imagination of scientists trying to solve the Pleistocene extinction puzzle. One important idea is the **keystone species** hypothesis. The advocates of this model have demonstrated that a "keystone" species, like the African elephant today, is essential to breaking up dense forest and providing a diversity of habitats from brushy thickets to patchy grasslands. If humans or climate had wiped out mammoths and mastodonts, their absence could have changed the overall pattern of vegetation and triggered extinction in many other animals who lost their preferred habitat.

Another trendy idea is the **hyperdisease** proposal. Accordir to this model the spread of exotic mammals from Eurasia t North America introduced new diseases that wiped out who communities of animals. There are many problems with th notion, especially the fact that most diseases are highly specif to one or two closely related species, and there are none th attack almost all species at the same time. In addition, it is ve hard to find evidence to test the hypothesis that some unusu virus or bacterium wiped out an entire fauna of animals once. Occasionally we can find evidence of specific diseases an bone deformations in a few pathological fossils but not of som virulent disease that wiped out lots of animals quickly.

There are other factors to consider as well. Although most the megamammals vanished between 11,000 and 9,000 yea ago in the Americas, and earlier elsewhere, there were som that survived into the Holocene. Mammoths still survived unt 3,700 years ago on polar islands, such as Wrangel Island, nort of the Bering Sea between Alaska and Siberia. Mastodon survived in North America until about 4,000 years ago. Groun sloths persisted on Caribbean islands until 4,700 years ago.

Finally, the fossil bone record may not be complete enough t give a reliable answer. In 2009 scientists analyzed samples of so from Alaska's deeply buried permafrost. They found DNA fror extinct horse and mammoth in permafrost that was 2,000 year younger than the youngest bones of these animals! Even thoug there are no fossilized bones to show it, apparently horse and mammoths survived for thousands of years after huma hunters arrived, long after the major changes in climate and th supposed "blitzkrieg" by human hunters.

People like simple yes or no answers when they ask questior like, "What killed the Ice Age megamammals?" But nature i not simple, nor is it prone to clear-cut answers. The real stor probably had many different interacting causes, as is typica for any complex event and most natural systems. Certainl humans contributed to the extinction of megamammals, bu so apparently did climate and vegetational change and possibl other factors as well. People do not like to deal with complexit or ambiguity, but reality is not simple. We just have to accep that the answer is complex and that we may never know th complete story, even for an extinction occurring only 10,00 years ago.

HOW DID MAMMALS DIVERSIFY AFTER THE DINOSAURS VANISHED?

One of the more interesting topics that mammalian paleobiologists have studied in recent decades is how the mammals diversified after the extinction of the non-bird dinosaurs 66 Ma. As we have already mentioned, in the earliest Paleocene mammals were still the rat-size creatures they had been for over 130 m.y., during the reign of the large dinosaurs. Once there were n large dinosaurs to rule above them, mammals diversified rapidl into many body sizes and shapes. By the early Eocene the larges mammals had reached the size of a cow, and by the late middl Eocene there were brontotheres and uintatheres about the siz

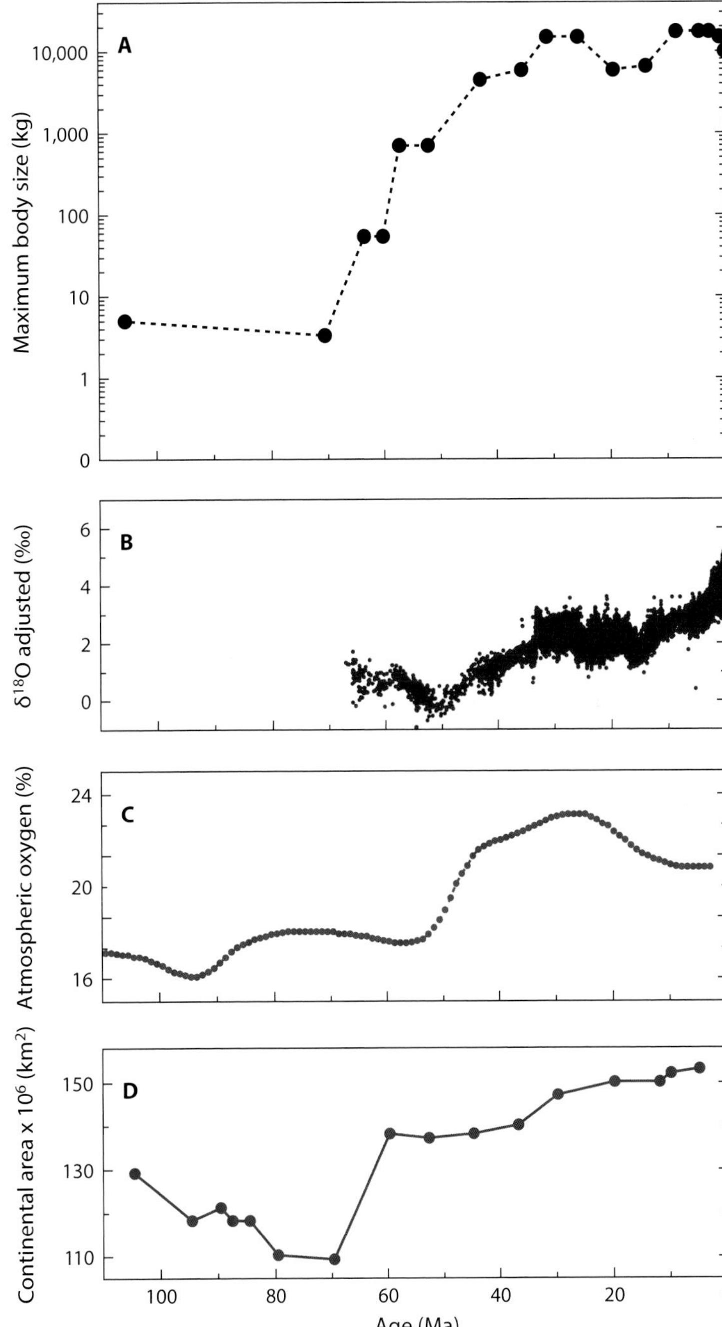

ɟure 18.2. Plot of the maximum body size of land mammals ough the Cretaceous and Cenozoic, showing the peak ·d then plateau about 40 Ma, in the middle Eocene. This nd is compared to the global ocean temperature trend (as own by the oxygen chemistry of the ocean waters, or $\delta^{18}O$), e percentage atmospheric oxygen (which peaked in the ·cene), and the exposed land area as seas retreated in the ·er Cretaceous.

: elephants or rhinos. By 50 Ma the first whales were ·tablished in the oceans (Chapter 14), the first bats ere flying in the skies (Chapter 10), and there were ·veral types of burrowing mammals and ant-eating ıammals—so mammals had colonized nearly every iche on land, air, and water that they occupy today.

A number of paleontologists have plotted the trends ı maximum body size after the end of the dinosau- an reign (Fig. 18.2). As expected, body size increased ·pidly through the Paleocene until it peaked in the ıiddle Eocene. Some scientists have pointed out that ıe peak corresponds to the peak in global temperature : 40 Ma, as well as peaks in oxygen levels and exposed .nd area, although these data are so broad that it is ard to decide whether this correlation means much.

Another way to analyze the explosive evolutionary ıdiation of Cenozoic mammals (Fig. 4.3) is to count ow many species or genera or evolutionary lineages ere present at each time interval. This has been done y a number of scientists over the years, but the results re generally the same. The best data come from North .merica, which has the most complete record from the aleocene to present and also a large number of speci- ıens and of scientists analyzing them. A typical plot ⁻ig. 18.3A) shows a rapid increase in number of line- ges through the Paleocene, from fewer than 10 in the ıtest Cretaceous to almost 60 by the late Paleocene. 'here are several ups and downs, with the first really big eak of about 80 lineages by the middle Eocene, and ıen a big decline through the late Eocene and early)ligocene, followed by another peak in the late Oligo- ene and early Miocene. There is the extinction event t the end of the Miocene, when rhinos, protoceratids, ıusk deer, palaeomerycids, horned rodents, and several other roups vanished from North America, and many other groups horses, camels, peccaries) were greatly reduced in diversity.

The records from Africa, South America, and Australia are oo incomplete (especially for the early Cenozoic) to record in a neaningful plot like Figure 18.3A. However, in Eurasia there is similarity to North America in the overall trends in generic di- ersity, with peaks of diversity in the middle Eocene and middle Aiocene, and lows in the Paleocene, Oligocene, and Pliocene Fig. 18.4).

So what causes these ups and downs in diversity through the Cenozoic? And why are there intervals of major extinction? These are puzzles that have seen lots of different analyses over the years, and many attempts to rigorously quantify and explain the pattern seen in Figure 18.3A. The most obvious possible correlation would be with global temperature and climate. A typical Cenozoic temperature curve, based on the chemistry of the shells of the plankton in the open ocean, is shown in Figure 18.3B. There are some rough similarities that scientists have noted many times. For example, the peak of warming in the early middle Eocene seems to match the

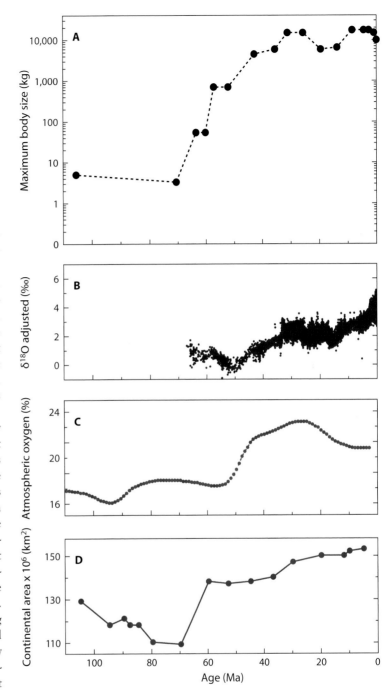

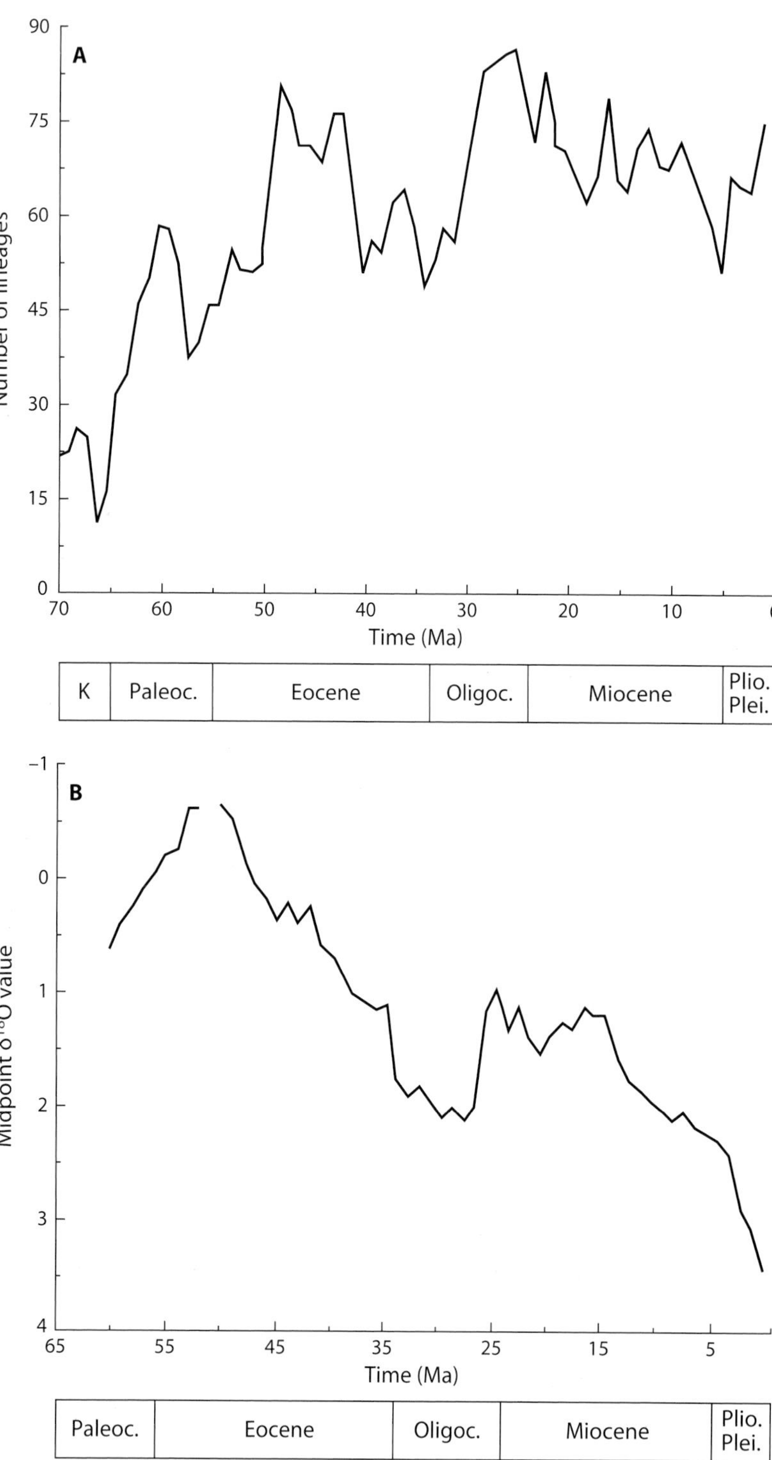

Figure 18.3. Mammal diversity and climate. A. Divers of mammalian lineages through the Cretaceous and Cenozoic. B. Global ocean temperature through the Cenozoic, as measured by the oxygen isotope chemis of the seawater. Analyses of these two data sets show statistically significant correlation between temperature and diversity.

diversity peak in the middle Eocene, and the decline through the cooling in the late Eocene–Oligocene matches the drop in diversity at that time. However, the rest of the Oligocene and Miocene temperature curve does not predict the diversity all that well. Rigorous statistical analysis has shown that there is no true statistical correlation between temperature and diversity of North American mammals, so there are clearly other factors at work.

Some scientists were not satisfied with this ty of analysis and decided to break down the co position of the Cenozoic mammalian faunas finer detail. Using a different method, they fou that there were certain groups that clustered t gether, which they called "faunas" or "subfa ily taxonomic units" (SFTUs). These clusters mammalian groups seemed to each form a pu of diversification, only to be replaced by and ov lap with the next "fauna" or SFTU (Fig. 18. And each replacement of one fauna with t next seemed to correspond to a major climat event, such as a cooling or warming trend, or peak in warming ("climatic optimum"). For e ample the "Paleocene fauna" consisting of t dominant Paleocene groups (such as "viverravi carnivorans, pantodonts, and archaic ungulat such as hyopsodonts, arctocyonids, and pery tychids) forms a discrete cluster in the Paleocen This group declines rapidly during the early E cene, when the classic early middle Eocene faur (consisting of "miacid" carnivorans, creodont mesonychids, primitive rhinos, horses, and a tiodactyls, plus taeniodonts and archaic ung lates like hyopsodonts and meniscotheres) tak over during the extreme warming of the Eoce (EECO, "early Eocene climatic optimum"). N included in this analysis was the huge diversi of smaller mammals, especially the tree-dwellin plesiadapids, primates, multituberculates, and i sectivorans, which make up the bulk of the fossi of Paleocene–early Eocene mammals.

Analysis of land plant fossils, ancient soil and global climatic records shows that the Pal ocene and especially the early middle Eocene w a warm "greenhouse" world, with high levels c carbon dioxide in the atmosphere. There w no snow or ice anywhere, and temperate-climat plants and mammals and even crocodilians lived above th Arctic and Antarctic Circles, where they experienced a war climate even though it was dark six months a year. Land plan from places like North Dakota and Wyoming, which today exp rience blizzards and weeks of subzero winter temperatures, ind cate that these places looked like the jungles of Central Americ

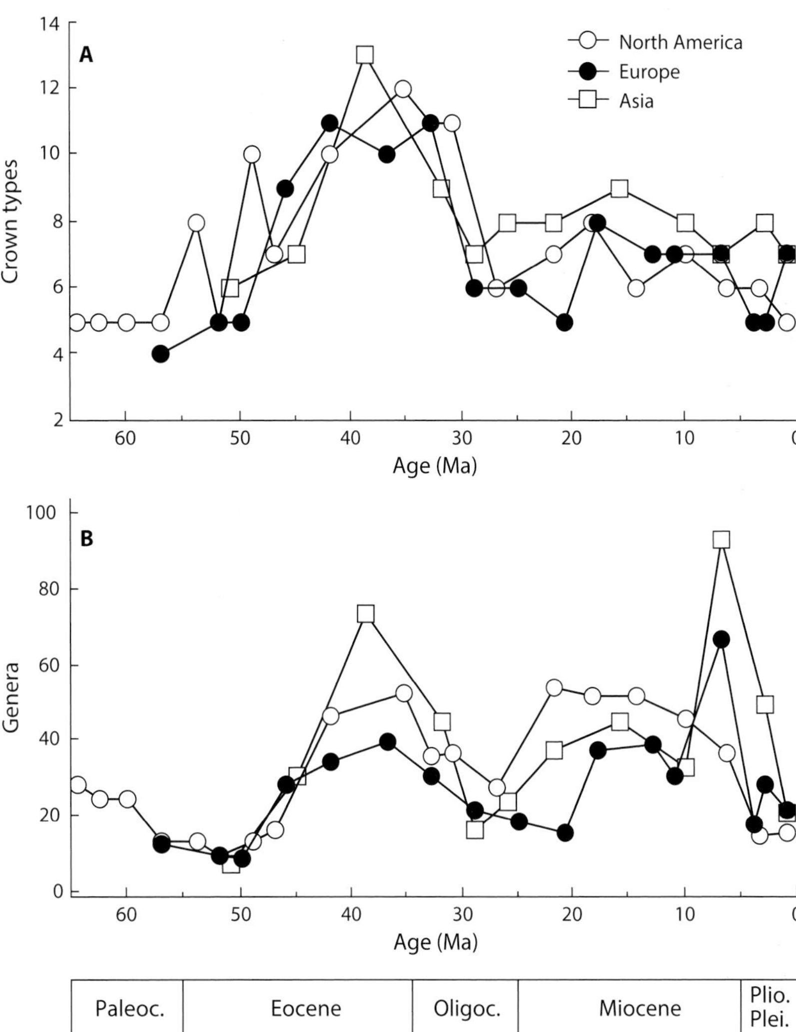

gure 18.4. Plot of tooth shape diversity (top agram) and generic diversity (bottom diagram) Europe, Asia, and North America through the enozoic. There seems to be a common peak in the ddle Eocene, and possibly another peak in the ddle Miocene, with lower values in the Paleocene, ligocene, and Pliocene.

ack about 60–40 Ma. Not only do the land plants suggest this, e high diversity and abundance of crocodilians, snakes, tures, and other climatically sensitive reptiles confirm it.

This global climate may have been triggered by a sudden blast f methane from the sediments of the deep ocean at the beginning of the Eocene, known as the Paleocene-Eocene Thermal laximum (PETM). The methane became a potent greenhouse as, warming the planet abruptly. It was then converted to caron dioxide, which remained for millions of years and kept the lanet warm right through the end of the middle Eocene, about 7 Ma.

Temperature then began to decline abruptly at the middle late Eocene transition and again in the early Oligocene ig. 18.3B). The causes for this global cooling event are complex nd controversial, but they probably were largely an effect of the rmation of the Antarctic Circumpolar Current as Australia nd South America pulled away from Antarctica. This current erves to thermally isolate Antarctica from the warmer waters at used to reach it in the Eocene. Once the isolation was comlete in the early Oligocene, Antarctica showed the first signs of glaciation at about 33 Ma, the "greenhouse" of the Age of Dinosaurs ended, and our modern "icehouse" world began.

Many of these mammalian groups were then replaced by a typical middle late Eocene assemblage. These included "miacid" and "viverravid" carnivorans, hyaenodont creodonts, plus primitive artiodactyls (protoreodonts, oromerycids, leptotraguline protoceratids, helohyines, and antiacodontines), and primitive perissodactyls (brontotheres, amynodont rhinos, hyracodontid rhinos, primitive rhinocerotids and tapiroids, and horses). These faunal differences would be even more dramatic if the analysis had included small mammals like the rodents, primates, and multituberculates, which show even more dramatic changes in the Paleocene and Eocene.

Both the early middle Eocene fauna and the middle late Eocene fauna then died out rapidly during the extreme cooling event of the late Eocene–early Oligocene (Fig. 18.5). They were replaced by a fauna adapted to the cooler, drier conditions of the Oligocene, such as that found in the Big Badlands of South Dakota and related beds of the White River Group in North Dakota, Nebraska, Wyoming, and Colorado. Ancient soils and leaf

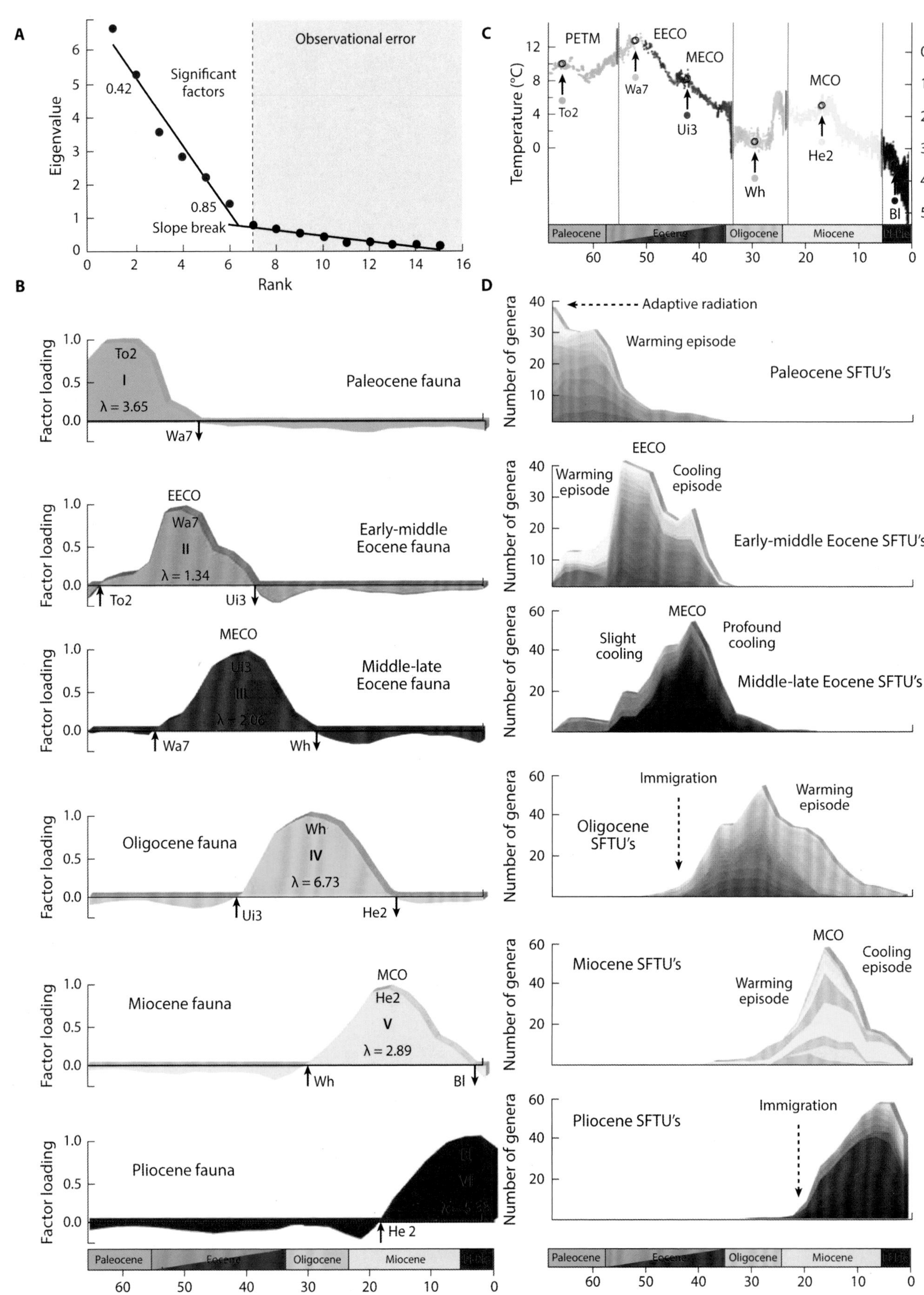

A
Observational error
Significant factors
0.42
0.85
Slope break
Eigenvalue
Rank
B
Factor loading
To2
I
λ = 3.65
Paleocene fauna
Wa7
EECO
Wa7
II
λ = 1.34
Early-middle Eocene fauna
To2
Ui3
MECO
Ui3
III
λ = 2.06
Middle-late Eocene fauna
Wa7
Wh
Oligocene fauna
Wh
IV
λ = 6.73
Ui3
He2
MCO
He2
V
λ = 2.89
Miocene fauna
Wh
Bl
Pliocene fauna
Bl
VI
He 2
Paleocene
Eocene
Oligocene
Miocene
Pl-Ple
C
Temperature (°C)
PETM
EECO
MECO
MCO
To2
Wa7
Ui3
Wh
He2
Bl
D
Number of genera
Adaptive radiation
Warming episode
Paleocene SFTU's
EECO
Warming episode
Cooling episode
Early-middle Eocene SFTU's
MECO
Slight cooling
Profound cooling
Middle-late Eocene SFTU's
Immigration
Warming episode
Oligocene SFTU's
MCO
Miocene SFTU's
Warming episode
Cooling episode
Pliocene SFTU's
Immigration

ure 18.5. Detailed dissection of Cenozoic land mammal diversity North America based on clusters of groups that co-occurred and panded and vanished together, known as "faunas" or "subfamily onomic units" (SFTU). See text for discussion.

ssils, plus land snails and reptiles, show that the region went om dense forests in the late Eocene to open, dry scrublands the early Oligocene. This "Oligocene SFTU" is composed typical Badlands mammals: hesperocyonine and early boroagine dogs, nimravid "false cats," primitive bear dogs, plus a hole new assemblage of artiodactyls (oreodonts, hypertragulid d leptomerycid ruminants, anthracotheres, entelodonts, cam-) and perissodactyls (tapirids, primitive horses, and primitive inocerotoids).

After the long cooling of the early middle Oligocene there s an abrupt warming in the late Oligocene–early Miocene, hich peaked at the Miocene Climatic Optimum (MCO). In dition, North America received many new immigrant groups om Asia. These mammals make up the "Miocene SFTU" ig. 18.5), which dominated the hot dry grassy savannas of orth America through most of the Miocene. This assemblage s dominated by borophagine dogs, early raccoons, huge bear gs, and a distinctive suite of artiodactyls (the last oreodonts, us musk deer, protoceratids, miolabine and protolabine cam-s, dromomerycines, hesperhyine peccaries) and perissodactyls nchitheriine and equine horses, tapirs, and aceratherine and eoceratine rhinos).

Finally, the cooling at the end of the Miocene resulted in e extinction of many typical savanna-adapted mammals, as entioned earlier. The causes of this late Miocene cooling and tinction are not well understood, but they are probably linked to the drying up of the Mediterranean, which became a gigantic salt basin like Dead Sea almost 1500 m (5000 ft) below sea level during the end of the Miocene. This event was caused when the Strait of Gibraltar closed due to mountain uplift of the Atlas Mountains and a global drop in sea level when the Antarctic glaciers expanded.

These typically Miocene groups were replaced by the cold steppe-climate-adapted "Pliocene SFTU," with all the mammalian groups found in typical Pliocene localities in the Great Plains and Rockies, from the Texas Panhandle to Hagerman Fossil Beds in Idaho. These include advanced canine dogs, weasels, raccoons, skunks, saber-toothed machairodontines, and true cats, along with their prey: advanced cameline camels, pronghorns, equine horses, tayassuine peccaries, deer, cattle, and the gomphothere mastodonts. Many of these groups were also immigrants from Asia, and they persisted through most of the Ice Ages as well, until the events at the end of the Pleistocene discussed earlier.

Such analyses dissect the details of Cenozoic mammalian history much better than the broad-brush statistical approaches shown in Figures 18.3 and 18.4, but they are still very coarse in resolution. These analyses use subfamilies and genera, clustering them over large time increments. When we actually look at a crucial climatic change event in great detail, a different pattern emerges. For example, in my own research on the dramatic cooling event at the Eocene-Oligocene boundary I found a remarkable paradox. There are plenty of climatic indicators, such as ancient soils, plant fossils, changes in the land snails and turtles and crocodilian faunas, that clearly indicate that the late Eocene in places like the Big Badlands of South Dakota was covered in a dense forest like that of present-day Nicaragua, but the early Oligocene habitats just a million years later were dry scrublands like those of Baja California (Fig. 18.6).

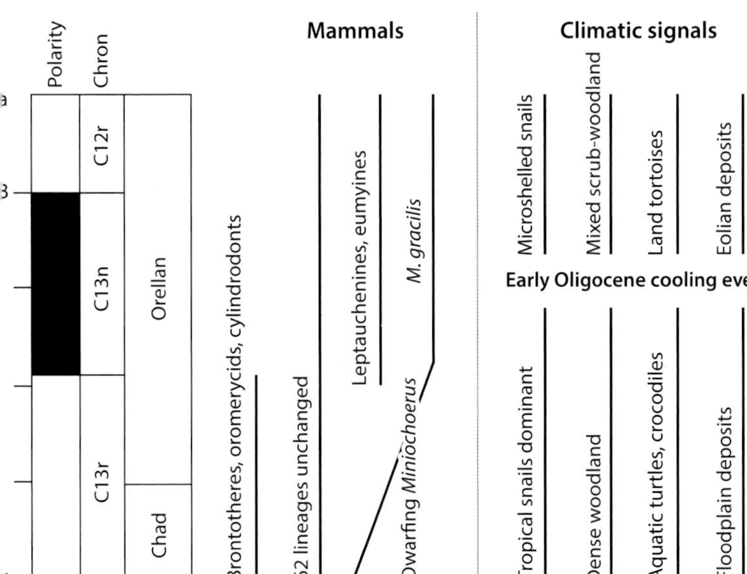

Figure 18.6. Detailed trend of the floral and faunal response to climatic change in the Eocene-Oligocene transition in the Big Badlands of South Dakota and correlated areas in Wyoming, Nebraska, and Colorado. There was a big change in the climatically sensitive ancient soils, land plants, land snails, and reptiles, indicating a change in vegetation from dense forests to open scrubland. Yet the land mammals show almost no response to the biggest climatic change of the past 50 m.y.

Such a dramatic change in climate and vegetation would be expected to have a significant effect on the land mammals. Instead, most of the mammals in this time interval passed right through the climate change with no apparent change whatsoever. Like Rhett Butler in *Gone with the Wind*, they demonstrate a "Frankly, my dear, I don't give a damn" reaction to climatic or vegetational change. Just a few lineages of archaic mammals, such as the multituberculates, oromerycids, cylindrodont rodents, and brontotheres vanished, while over 90% of the lineages remained unchanged (other than some dwarfing in one lineage of oreodonts). Only a few new lineages with higher-crowned teeth (such as eumyine cricetid rodents and leptauchenine oreodonts) appeared in response to the very different vegetation, and they were relatively rare until much later in the Oligocene.

Detailed examination of other intervals of rapid climatic change in the Cenozoic produces similar results. For example, my former students and I have spent a decade measuring the bones of mammals from nearly every well-dated pit at Rancho La Brea. These fossil samples span the interval from 35,000 years ago to only 9,000 years ago, and the complete transition from peak glacial coverage at 20,000 years ago to the interglacial period that began about 10,000 years ago. Pollen evidence shows that the area of Los Angeles around Rancho La Brea went from ponderosa-piñon pine forests such as are found in the nearby mountains to the modern sage scrub that grows today in the lower elevations. Yet there are no significant changes in the size or limb robustness of any of the mammals or birds we examined, including dire wolves, saber-toothed cats, Ice Age lions, bison, horses, camels, ground sloths, and nearly every other mamm for which there are enough samples from enough well-dated pi The same is true of the most abundant birds from Rancho Brea, including the bald eagle and golden eagle, the turkeys, t caracaras, the condors, and the great horned owls.

This paradox can be resolved if we look at evolution the w modern paleobiology has done in the last 40 years. The pr posal of the "punctuated equilibrium" model of evolution Niles Eldredge and Stephen Jay Gould in 1972 has transform the way paleontologists think about species and speciation. Pa ontologists have long known that individual species are stab over millions of years, despite many selective pressures su as climatic events. Species do not respond to climatic chang or other stresses by gradually evolving through time. Instea they respond by splitting into new species, or by going extin This response is known as "species sorting." When there is e treme selection pressure, species do not show gradual chan over time, but instead form new species as old ones vanish. T analyses shown in Figures 18.2, 18.3, and 18.5 are only looki at climatic change on the scale of genera and species as discre units, which apparently are responsive to these external en ronmental factors. But the fine-scale examination of hundre of specimens of species experiencing climatic change throu a short time interval, such as in my work in the Big Badlan or at Rancho La Brea, shows no response at all *within* lineag Instead, their only response when the environmental pressur got to be too great is extinction (see the Pleistocene extincti discussion earlier in the chapter).

WHAT ABOUT MASS EXTINCTIONS?

The rise of the dominance of mammals after the Age of Dinosaurs was caused by some sort of environmental catastrophe that wiped out the dinosaurian overlords of the Late Cretaceous. Some scientists think it was mostly due to an asteroid impact that hit Yucatan about 66 Ma. Other geologists and paleontologists point out that the second largest volcanic eruptions in earth history, the Deccan volcanoes of India and Pakistan, were erupting violently just before the end of the Cretaceous, changing the atmosphere and the climate in many ways. Although the media only talk about the impact model, currently most paleontologists think that the volcanic eruptions were more important than the impact, which may have been only the final *coup de grâce*.

A second mass extinction to affect Cenozoic mammals was the megamammal extinction at the end of the Pleistocene, which we have already discussed. The causes may have been climate or human overhunting, but the impact suggestion has been debunked. The third biggest wave of extinctions in the Cenozoic occurred through the late Eocene and Oligocene, which was a complex extinction that occurred in pulses at 37 Ma and 33 Ma, apparently due to climatic cooling. If the analyses in Figure 18 are correct, most of the extinctions and originations in Cenozc land mammals were due to warming and cooling events, ar their effects on the vegetation, as well.

Nevertheless, there have been many attempts to claim that e tinctions in Cenozoic mammals might be due to more dramat causes, such as impact events or major volcanic eruptions. Su causes have been suggested many times, usually by scientists wi no firsthand knowledge of the fossil record.

When the data are examined closely, however, it is cle that no impact in the Cenozoic caused significant extinctior Both my own analysis and another by John Alroy of Macquar University in Australia demonstrated no correlation (Fig. 18. between extinctions and impact craters. If we examine t database of well-dated impact craters, it turns out that the are many of them, and almost none occur in an interval mass extinction. The huge Ries impact that hit Germany in t early Miocene and the Montagnais impact off Nova Scotia the late Paleocene show absolutely no effect on anything fro plankton to mammals.

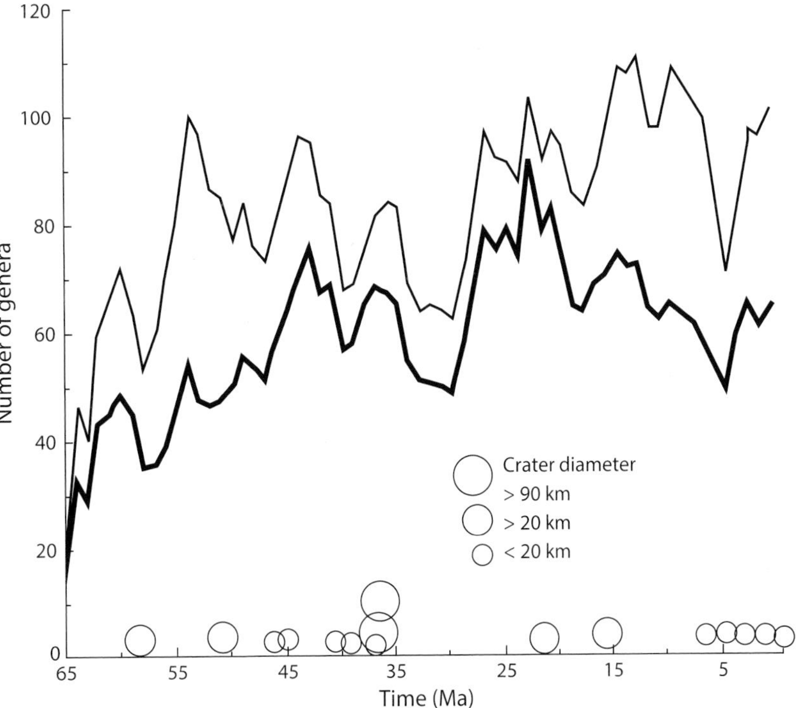

ure 18.7. Diversity in North American mammals mpared to the times of major extraterrestrial impacts rcles) in the Cenozoic. The size of the circle licates the size of the impact crater. There is no sociation between any Cenozoic impact events and ss extinctions in land mammals.

In fact, the largest extraterrestrial incursions since the end of e Cretaceous were a pair of impacts that formed Chesapeake ay and a meteorite that slammed into Popigai in northeastern beria, at 35.5 Ma. Close study of the fossil record shows that ere were *no* extinctions caused by these huge impacts, especially nong the mammals of the late Eocene. This casts doubt on the nportance of extraterrestrial impacts for extinction, since even steroid collisions just smaller than the end-Cretaceous impact pparently have had no effect. Since the days when the end-Cre-ceous impact was first proposed in 1980, there has been a sci-ntific bandwagon to blame all mass extinctions on impact. Now, ore than 36 years later, no other mass extinction has been clearly ssociated with an impact event, and this fad has died completely the scientific community (even if the media still perpetuate it).

What about volcanic eruptions? There were some large eruptions in Ethiopia near the end of the Eocene, but there is no clear association of them with the detailed pattern of Eocene-Oligocene extinctions in mammals. No other extinction horizons in the Cenozoic can definitively be associated with a major volcanic event.

Thus, the causes of extinction and diversification remain complex. There are still a lot of questions to be answered, and a lot of interesting ideas to research. We know a lot more than we did 36 years ago, when the mass extinction debates first began, or 45 years ago, when the debates about evolutionary mechanisms were escalating, but there is still much to learn.

THE FUTURE OF MAMMALS

aleontologists have documented five huge mass extinctions in e history of life, nicknamed the "Big Five." These include the reat end-Permian extinction 250 Ma, which wiped out 95% f marine life and nearly wiped out most of the protomam-als (see Chapter 2). Another was the end-Cretaceous extinc-on at 66 Ma, which killed off the non-bird dinosaurs (but not e crocodilians, turtles, amphibians, or most other land life) nd opened the landscape for mammals to diversify. The three ther extinctions of the Big Five were at the end of the Triassic 200 Ma), in the Late Ordovician (455 Ma), and in the Late Devonian (370 Ma).

In recent years, biologists and paleontologists have documented a "Sixth Extinction," referring to the large percentage of the life on this planet that has vanished in the past few centuries due to humans, and their associated animals, and their devastation of the earth's habitats. First, humans exterminated nearly all the native species on exotic islands like Madagascar, New Zealand, Hawaii, and many others. From the dodo to the passenger pigeon, wiped out in the previous centuries, to the many animals on the brink of extinction now, the Sixth Extinction is already occurring faster than any natural extinction in the geologic past. The rate of extinction is thought to be 100 times to 1,000 times that of any previous mass extinction, much faster and bigger than the extinction that ended the reign of the dinosaurs or the "Mother of all Mass Extinctions" in the late Permian. Some scientists estimate that about 140,000 species are vanishing each year, and most think that about 50% of all species on the planet will be gone by 2100, within the lifetime of people reading this book.

In some cases, the stories are heartrending. Only a few decades ago, there were hundreds of thousands of rhinoceroses in Africa, but brutal poaching has nearly wiped them out across the continent. Due to silly superstitions in Asian folk "medicine," rhino horn is thought to have medicinal value (even though it is just made of hair, with exactly the same kind of keratin you have in your hair or fingernails). Consequently, it is worth more per ounce than gold or cocaine, and as long as there is demand in China and Vietnam (where wealth is blossoming), the wild rhinos of Africa are doomed. No amount of bans on rhino horn or poaching can stop such huge market pressures, and smugglers just raise the price higher and make it more valuable and desirable. In many places, the last few wild rhinos have armed guards around the clock to protect them, but soon rhinos will occur only in zoos. The Sumatran rhino is down to at most a few hundred individuals, and the Javan rhino may have only about 20 to 30 individuals left in two reserves in western Java and in the remote jungles of Vietnam.

Similarly elephants have been hunted to near extinction due to the Asian demand for ivory, and no number of international bans on ivory stops this voracious market or the poachers and smugglers that feed it. These losses are particularly tragic, because the oldest and wisest females that lead the elephant herds are killed first, leaving juveniles to die or grow up not knowing how to survive and form healthy herds. As we mentioned earlier, elephants are "keystone species" essential to maintaining the openness of the savannas, and other species will vanish when elephants no longer break down the trees and keep the forests at bay.

Many other mammals are similarly threatened. Tigers are nearly extinct in the wild due to excessive poaching, again mostly for Asian "medicine" that has no real curative value. The rare pangolins are being ruthlessly hunted both for Asian "medicine" and also to supply Chinese restaurants. The list goes on and on, not only with sympathetic and popular animals like elephants, rhinos, tigers, and pandas, but across the entire spectrum of living mammals. Recent estimates have listed at least 80 species of mammals that have gone extinct since 1500. These include the Tasmanian "tiger" or thylacine, the Steller's sea cow, the South African zebra known as the quagga, the bluebuck, the Pyrenean ibex, the Caribbean monk seal, the Atlas bear, the Yangtze dolphin, and many others. About 1,100 to 1,600 spec of living mammals are now listed as threatened or endangere about 25% of all the mammals on the planet. You need o Google "endangered mammals" to see the list of over a thousa species that are on the brink of vanishing forever. Some will n be around by 2050, and most will be gone by 2100.

The culprit, of course, is well known—it is us. Because our direct hunting of these beasts, the introduction of our mesticated fellow travelers (especially rats, cats, pigs, goats, a cattle), and our destruction of wild habitat to nurture our ev expanding population of more than 7.4 billion people (growi rapidly as I write this), the rest of our planetary fellow travel are doomed to vanish, except in zoos. It is a tragedy of unima nable proportions, yet only a few people are making much of effort to try to stop it. We are the most destructive species t planet has ever seen, wiping out more life than any asteroid fr space or monstrous volcanic eruption ever could.

Some would say, "We humans need to take care of huma first, and feed our own people, and not worry about nonhum creatures." But this ignores some important facts. For pur selfish reasons, it is important that we preserve as much wildl as we can. Not only is it good for our own souls to see w animals in their native habitats, but wild animals and plants a often the sources of important products (especially importa new drugs) that may help humans survive. If we wipe them out, we may be killing ourselves in the process.

On a more philosophical and less anthropocentric lev many people believe that all living things have just as much rig to live on this planet as we do. They argue that it is arroga and selfish for us to take all of the world for ourselves and de other species their equal rights to life. If humans drive all oth wild creatures extinct, we will inherit a barren planet with on ourselves, our domesticated animals and plants, and rats ar cockroaches—a much poorer world than we inherited.

The great biologist Paul Ehrlich posed an interesting analo to this problem. Suppose you were riding in an airplane ar looked out your window to see a rivet pop out of the wing. Th would not be too alarming. Then another rivet falls out, the another. You would begin to worry, and sooner or later yo know that if enough rivets fall out, the wing will break up, an you will crash and die. Similarly, we are watching the rivets th hold Spaceship Earth together fall out, one by one. We kn this is not good for the planet, but we do nothing to curb ou population growth, our excessive consumption, our polluti of the water and air, and other ways in which we foul our ow nest. How much more can Spaceship Earth take before it to falls apart and crashes?

ILLUSTRATION CREDITS

roy et al. 2000 (modified): Fig. 18.3
merican Museum of Natural History (Department of Library Services): Fig. 15.12
. Archer: Fig. 2.7D
Barnes: Fig. 12.16B
. Bernor: Fig. 6.7C
Bloch and D. Boyer: Fig. 9.1
. Boyer: Fig. 7.1
. Caulfield/Wikimedia Commons: Fig. 15.4D
. Colbert: Fig. 15.5B
onty/Wikimedia Commons: Figs. 7.9B, 7.11E
. Cuello/Wikimedia Commons: Fig. 16.6
Cummings/Wikimedia Commons: Figs. 1.4A, 11.2A, 11.2B
aderot/Wikimedia Commons: Figs. 5.5B, 6.15B, 11.5, 14.5A
. Descouens/Wikimedia Commons: Fig. 7.11A
iagram Lajard/Wikimedia Commons: Fig. 5.3C
. Dimitrov/Wikimedia Commons: Fig. 6.5C
. B. Fenton: Figs. 10.3, 10.4
igueirido et al. 2012, Fig. 1: Fig. 18.5
. Ewan Fordyce/Museum Victoria: Fig. 14.16B, 14.18A
L. Franzen, P. D. Gingerich, J. Habersetzer, J. H. Hurum, W. von Koenigswald, B. H. Smith/Wikimedia Commons: Fig. 7.5A
reeimages.com: Fig. 12.17
unkMunk/Wikimedia Commons: Figs. 2.4, 12.10B
utureman1199/Wikimedia Commons: Fig. 6.5B
. Galiano: Figs. 1.7, 12.10A
ally242/Wikimedia Commons: Fig. 12.4A
hedo/Wikimedia Commons: Figs. 1.4B, 2.9C, 3.4, 3.8, 11.4, 12.6A. 14.3C
hilip Gingerich: Figs. 7.4B and C
. Gould/Wikimedia Commons: Fig. 2.7A
regory 1951: Fig. 15.5A
. Guérin/Wikimedia Commons: Fig. 12.11B
. Harris/Wikimedia Commons: Fig. 3.6A
. Hermans/Wikimedia Commons: Fig. 6.5A
. Michael Hogan/Wikimedia Commons: Fig. 7.6 (*Propithecus diadema*)
uhu Uet/Wikimedia Commons: Fig. 6.4B
Christine Janis, Professor of Biology, Brown University: Fig. 14.7
Jernvall et al. 1996, Fig. 2: Fig. 18.4
Karora/Wikimedia Commons: Fig. 3.9
Elaine Kasmer (reconstruction): Fig. 13.2B
Lydia Kibiuk (reconstruction): Fig. 7.5B
Kippel/Wikimedia Commons: Fig. 7.9A
A. Kouprianov/Wikimedia Commons: Fig. 13.4A
D. Krentzel/Wikimedia Commons: Figs. 17.2B, 17.5A
O. Lambert: Figs. 14.14A and B
Fabrice Lihoreau: Fig. 14.9B
Norman Lim: Fig. 7.3
Spencer G. Lucas: Fig. 17.5C
Z. Luo: Figs. 2.5, 2.6, 2.8B, 2.9A, 2.10, 3.2A, 4.1, 4.2
P. Maas/Wikimedia Commons: Fig. 9.3
Masur/Wikimedia Commons: Fig. 6.14A
Matthew 1926: Fig. 15.2A
J. S. Mellett: Fig. 12.3B
Meng Jin: Fig. 8.3
Meyers Konversations-Lexikon, 4th ed. (1885–90): Fig. 14.17A
Mihlbachler 2008 (modified): Fig. 15.11
D. Mitchell/Wikimedia Commons: Fig. 6.4E
Momotaru2012/Wikimedia Commons: Figs. 6.4C, 8.7A, 12.9, 13.3
Nachbarnebenan/Wikimedia Commons: Fig. 11.1
Elias Neideck/Wikimedia Commons: Fig. 6.15A
H. A. Nicholson/Wikimedia Commons: Fig. 14.25C
J. Orcutt: Fig. 8.8A
Orliac et al. 2010: Fig. 14.10
Osborn 1910, p. 110, Fig. 30: Fig. 17.5B
Osborn 1924: Fig. 13.4B
Osborn 1929: Fig. 15.10
Osborn and Granger 1932: Fig. 17.3
Patterson 1949 (modified): Fig. 17.6
D. Patton: Fig. 14.18B
R. Pogson, Australian Museum: Fig. 2.7C
Prothero 2005 (modified): Fig. 18.7
Prothero 2007, p. 112, Fig. 4.2 (modified): Fig. 3.5
Prothero 2007, p. 274, Fig. 13.3 (modified): Fig. 2.2
Prothero 2007, p. 319, Fig. 14.16 (modified): Fig. 14.12
Prothero 2013, p. 67, Fig. 4.2 (modified): Fig. 1.5
Prothero 2013, p. 69, Fig. 4.3 (modified): Fig. 1.6
Prothero and Schoch 2002 (modified): Fig. 15.6
C. R. Prothero: Figs. 2.11A, 6.2, 14.1, 14.19B, 14.20A, 14.25A, 15.2B, 15.8
D. R. Prothero: Figs. 1.4C and D, 1.10B, 5.5C and D, 6.6B, 6.11B, 7.11C, 8.7B, 8.8B, 10.1, 10.2A, 12.3A, 12.7A, B, C, E, 12.11A, 12.13, 12.14, 12.16C, 14.19C, 14.20B, 16.5, 18.6
E. T. Prothero: Figs. 1.2, 1.3, 1.8, 1.9, 3.3, 6.1, 7.2, 8.5, 12.2, 14.16, 15.2B, 15.8
L. Pycock/Wikimedia Commons: Fig. 15.3B
Ra'ike/Wikimedia Commons: Fig. 14.24B and C
Raul654/Wikimedia Commons: Fig. 14.25B
D. T. Rasmussen: Fig. 6.13
Rept0n1x/Wikimedia Commons: Fig. 7.11D
Kenneth D. Rose: Figs. 7.5B, 11.7, 13.2B, 14.3A and B
Russell 1964 (modified): Fig. 13.2A
Scott 1913 (modified): Fig. 14.19A
Scott and Jepsen 1936: Fig. 9.5C
Sevenstar/Wikimedia Commons: Fig. 8.10
D. Sifry/Wikimedia Commons: Fig. 15.4C
Simeon Stoilov Studio/Wikimedia Commons: Fig. 6.5D
Simon from United Kingdom/Wikimedia Commons: Fig. 14.4
Simpson 1941 (modified): Fig. 8.6B
Smith et al. 2010, Fig. 3: Fig. 18.2
Smithsonian Institution: Fig. 14.13
Smithsonian Institution Archives/Wikimedia Commons: Fig. 3.6B
Smokeybjb/Wikimedia Commons: Fig. 7.6 (*Archaeoindris fontoynonti*)
Nikos Solounias: Fig. 14.25D
R. Somma/Wikimedia Commons: Figs. 5.3B, 9.4A, 14.23, 15.5C, 16.7
Sporst/Wikimedia Commons: Fig. 12.15A
G. Storch: Fig. 11.3
Brian Switek: Fig. 6.4A, 6.10
Nobumichi Tamura: Figs. 2.8A and C, 2.9B,

2.12B, 3.2B, 6.4D, 6.7B, 6.11C, 7.4A, 12.10C
Ian Tattersall: Fig. 7.10
Tommy from Arad/Wikimedia Commons: Figs. 10.2B, 15.3A
Peter Trusler: Fig. 2.7E
Z. "Jack" Tseng: Figs. 2.8D, 6.14B, 11.6, 12.4C, 12.18A
Shane Tucker, University of Nebraska State Museum: Fig. 1.1
University of Nebraska State Museum: Fig. 15.1
Nathan S. Upham: Fig. 9.2C
F. Vassen/Wikimedia Commons: Fig. 7.6 (*Lepilemur ruficaudatus*)
Vinithehat/Wikimedia Commons: Fig. 7.7A
W. von Koenigswald: Figs. 9.5A and B, 9.6
Wang and Tedford 2008, p. 141 (modified by M. P. Williams): Fig. 12.12
E. Welsh: Fig. 8.9A
WereSpielChequers/Wikimedia Commons: Fig. 16.3
T. D. White: Fig. 7.11B (used by permission of AAAS)
A. Whittome/Wikimedia Commons: Fig. 2.7B
Wikimedia Commons: Figs. 2.11B, 5.3A, 5.5A, 6.6A, 6.7A, 6.7D, 6.11A, 6.15A and C, 7.6, 7.7B, 9.2A–C, 9.4B, 12.5A, 12.6B, 12.11C, 12.15B, 12.16D, 14.5B, 14.11, 14.17B, 14.24A, 14.28, 14.29, 15.4B, 16. 17.2A
M. P. Williams: Figs. 2.1, 2.9D, 2.12A, 3.1, 3.7, 5.1, 5.2, 5.4, 5.6, 6.3, 6.8, 6.9, 6.11C 6.12, 7.8, 8.1, 8.2, 8.9B, 9.4C, 12.4B, 12.5B, 12.7D, 12.16A, 12.18B, 13.1, 14.2 14.6, 14.8, 14.9A, 14.15C, 14.21, 14.22, 14.26, 14.27, 14.31, 15.7, 15.9, 16.1, 16.2 16.8, 17.1, 17.4, 17.8, 17.9, 18.1
Wood 1962: Fig. 8.6A
F. Wouters/Wikimedia Commons: Fig. 12.
M. Zanferrari/Wikimedia Commons: Fig. 15.4A

FURTHER READING

Agusti, J., and M. Anton. 2002. *Mammoths, Sabertooths, and Hominids: 65 Million Years of Mammalian Evolution in Europe*. New York: Columbia University Press.

Alroy, J., et al. 2000. Global climate change and North American mammalian evolution. *Paleobiology* 26 (4):259–88.

Anton, M., and M. Turner. 2000. *The Big Cats and Their Fossil Relatives*. New York: Columbia University Press.

Asher, R. J., J. H. Geisler, and M. R. Sánchez-Villagra. 2008. Morphology, paleontology, and placental mammal phylogeny. *Systematic Biology* 57:311–17.

Figueirido, B., et al. 2012. Cenozoic climate change influences mammalian evolutionary dynamics. *Proceedings of the National Academy of Sciences* 109:723–27.

Foley, N. M., M. S. Springer, and E. C. Teeling. 2016. Mammal madness: Is the mammal tree of life not yet resolved? *Philosophical Transactions of the Royal Society B* 371:21050140.

Gregory, W. K. 1910. The orders of mammals. *Bulletin of the American Museum of Natural History* 27:1–524.

Gregory, W. K. 1951. *Evolution Emerging*. New York: Macmillan.

Janis, C. 1993. Tertiary mammal evolution in the context of changing climates, vegetation, and tectonic events. *Annual Reviews of Ecology and Systematics* 24:467–500.

Janis, C., K. M. Scott, and L. L. Jacobs, eds. 1998. *Evolution of Tertiary Mammals of North America*. Vol. 1, *Terrestrial Carnivores, Ungulates, and Ungulate-like Mammals*. Cambridge: Cambridge University Press.

Janis, C., G. F. Gunnell, and M. D. Uhen, eds. 2008. *Evolution of Tertiary Mammals of North America*. Vol. 2, *Small Mammals, Xenarthrans, and Marine Mammals*. Cambridge: Cambridge University Press.

Jernvall, J., et al. 1996. Molar tooth diversity, disparity, and ecology in Cenozoic ungulate radiations. *Science* 274:1489–92.

Kemp, T. S. 2005. *The Origin and Evolution of Mammals*. Oxford: Oxford University Press.

Kielan-Jaworowska, Z., Z. Luo, and R. L. Cifelli. 2004. *Mammals from the Age of Dinosaurs: Origins, Evolution, and Structure*. New York: Columbia University Press.

Kurtén, B. 1968. *Pleistocene Mammals of Europe*. New York: Columbia University Press.

Kurtén, B. 1988. *Before the Indians*. New York: Columbia University Press.

Kurtén, B., and E. Anderson. 1980. *Pleistocene Mammals of North America*. New York: Columbia University Press.

Li, C. K., R. W. Wilson, and M. R. Dawson. 1987. The origin of rodents and lagomorphs. *Current Mammalogy* 1:97–108.

Lillegraven, J. A. 1974. Biological considerations of the marsupial-placental dichotomy. *Evolution* 29:707–22.

Lillegraven, J. A., Z. Kielan-Jaworowska, and W. A. Clemens, eds. 1979. *Mesozoic Mammals: The First Two-Thirds of Mammalian History*. Berkeley: University of California Press.

MacFadden, B. J. 1992. *Fossil Horses: Systematics, Paleobiology, and Evolution of the Family Equidae*. Cambridge: Cambridge University Press.

Madsen, O., M. Scally, C. J. Douady, D. J. Kao, W. DeBry, R. Adkins H. Amrine, M. J. Stanhope, W. W. de Jong, and M. S. Springer. 2001. Parallel adaptive radiations in two major clades of placental mammals. *Nature* 409:610–14.

Matthew, W. D. 1926. The evolution of the horse: A record and its interpretation. *Quarterly Review of Biology* 1 (2):139–85.

McKenna, M. C. 1975. Toward a phylogenetic classification of the Mammalia. In *Phylogeny of the Primates*, ed. W. P. Luckett and F. S. Szalay, 21–46. New York: Plenum Press.

McKenna, M.C., and S. K. Bell. 1997. *Classification of Mammals above the Species Level*. New York: Columbia University Press.

Mihlbachler, M. C. 2008. Species taxonomy, phylogeny, and biogeography of the Brontotheriidae [Mammalia: Perissodactyla].

Bulletin of the American Museum of Natural History, no. 311:1–475.
rphy W. J., E. Eizirik, S. J. O'Brien, O. Madsen, M. Scally, C. J. Douady, E. Teeling, O. A. Ryder, M. J. Stanhope, W. W. de Jong, and M. S. Springer. 2001. Resolution of the early placental mammal radiation using Bayesian phylogenetics. *Science* 294:2348–51.
rphy W. J., P. A. Pevzner, and S. J. O'Brien. 2004. Mammalian phylogenomics comes of age. *Trends in Genetics* 20:631–39.
vacek, M. J. 1992. Mammalian phylogeny: Shaking the tree. *Nature* 356:121–25.
vacek, M. J. 1994. The radiation of placental mammals. In *Major Features of Vertebrate Evolution*, ed. D. R. Prothero and R. M. Schoch, 220–37. Paleontological Society Short Course 7.
vacek, M. J. and A. R. Wyss. 1986. Higher-level relationships of the recent eutherian orders: morphological evidence. *Cladistics* 2:257–287.
vacek, M. J., A. R. Wyss, and M. C. McKenna. 1988. The major groups of eutherian mammals. In *The Phylogeny and Classification of the Tetrapods*, vol. 2, *Mammals*, ed. M. J. Benton, 31–73. Oxford: Clarendon Press.
liac, M., J.-R. Boisserie, L. MacLatchy, and F. Lihoreau. 2010. Early Miocene hippopotamids [Cetartiodactyla] constrain the phylogenetic and spatiotemporal setting of hippopotamid origin. *Proceedings of the National Academy of Sciences* 107:11871–76.
born, H. F. 1910. *The Age of Mammals in Europe, Asia, and North America*. New York: Macmillan.
—. 1924. *Andrewsarchus*, a giant mesonychid of Mongolia. *American Museum Novitates* 146:1–5.
—. 1929. *The Titanotheres of Ancient Wyoming, Dakota and Nebraska*, United States Geological Survey Monograph 55. Washington, DC: Government Printing Office.
born, H. F., and W. Granger. 1932. Coryphodonts and uintatheres from the Mongolian expedition of 1930. *American Museum Novitates* 552:1–16.
tterson, B. 1949. Rates of evolution in Taeniodonts. In *Genetics, Paleontology and Evolution*, ed. G. L. Jepsen, G. G. Simpson, and E. Mayr, 243–78. Princeton, NJ: Princeton University Press.
othero, D. R. 1981. New Jurassic mammals from Como Bluff, Wyoming, and the interrelations of the non-tribosphenic Theria. *Bulletin of the American Museum of Natural History* 167 (5):277–326.
—. 1999. Does climatic change drive mammalian evolution? *GSA Today* 9 (9):1–5.
—. 2005. Did impacts, volcanic eruptions, or climatic change affect mammalian evolution? *Palaeogeography, Palaeoclimatology, Palaeoecology* 214:283–94.
—. 2005. *The Evolution of North American Rhinoceroses*. Cambridge: Cambridge University Press.
—. 2006. *After the Dinosaurs: The Age of Mammals*. Bloomington: Indiana University Press.
—. 2007. *Evolution: What the Fossils Say and Why It Matters*. New York: Columbia University Press.
—. 2013. *Bringing Fossils to Life: An Introduction to Paleobiology*. 3rd ed. New York: Columbia University Press.
othero, D. R., and S. Foss, eds. 2007. *The Evolution of Artiodactyls*. Baltimore: Johns Hopkins University Press.
othero, D. R. and T. H. Heaton. 1996. Faunal stability during the early Oligocene climatic crash. *Palaeogeography, Palaeoclimatology, Palaeoecology* 127:239–56.
othero, D. R., and R. M. Schoch. 2002. *Horns, Tusks, and Flippers: The Evolution of Hoofed Mammals*. Baltimore: Johns Hopkins University Press.
Prothero, D. R., E. M. Manning, and M. Fischer. 1988. The phylogeny of the ungulates. In *The Phylogeny and Classification of the Tetrapods*, vol. 2, *Mammals*, ed. M. J. Benton, 201–235. Oxford: Clarendon Press.
Rose, K. D. 2006. *The Beginning of the Age of Mammals*. Baltimore: Johns Hopkins University Press.
Rose, K. D., and J. D. Archibald, eds. 2005. *The Rise of Placental Mammals: The Origin and Relationships of the Major Extant Clades*. Baltimore: Johns Hopkins University Press.
Russell, D. E. 1964. Les Mammiferes paleocenes d'Europe. *Memoires du Museum National d'Histoire Naturelle*, ser. 3, *Sciences de la Terre* 2 (16):1–99.
Savage, D. E., and D. E. Russell. 1983. *Mammalian Paleofaunas of the World*. Reading, MA: Addison Wesley.
Savage, R.J.G., and M. R. Long. 1986. *Mammal Evolution: An Illustrated Guide*. New York: Facts-on-File Publications.
Scott, W. B. 1913. *The History of Land Mammals in the Western Hemisphere*. New York: Macmillan.
Scott, W. B., and G. L. Jepsen. 1936. The mammalian fauna of the White River Oligocene. *Transactions of the American Philosophical Society* 28, pt. 1.
Simpson, G. G. 1941. A giant rodent from the Oligocene of South Dakota. *American Museum Novitates* 1149:1–16.
Smith, F. A., et al. 2010. The evolution of maximum body size of terrestrial mammals. *Science* 330:1216–19.
Springer, M. S., M. J. Stanhope, O. Madsen, and W. W. de Jong. 2004. Molecules consolidate the placental mammal tree. *Trends in Ecology and Evolution* 19:430–38.
Springer, M. S., A. Burk-Herrick, R. Meredith, E. Eizirik, E. Teeling, S. J. O'Brien, and W. J. Murphy. 2007a. The adequacy of morphology for reconstructing the early history of placental mammals. *Systematic Biology* 56:673–84.
Springer, M. S., R. W. Meredith, E. Eizrik, E. Teeling, and W. J. Murphy. 2007b. Morphology and placental mammal phylogeny. *Systematic Biology* 57:499–503.
Springer, M. S., R. W. Meredith, J. E. Janecka, and W. J. Murphy. 2011. A historical biogeography of Mammalia. *Proceedings of the Royal Society B*, 366:2478–502.
Stucky, R. K. 1990. Evolution of land mammal diversity in North America during the Cenozoic. *Current Mammalogy* 2:375–432.
Szalay, F. S., M. J. Novacek, and M. C. McKenna, eds. 1993. *Mammal Phylogeny*. Berlin: Springer-Verlag.
Tassy, P., and J. Shoshani. 1988. The Tethytheria: elephants and their relatives. In *The Phylogeny and Classification of the Tetrapods*, vol. 2, *Mammals*, ed. M. J. Benton, 283–316. Oxford: Clarendon Press.
Turner, A., and M. Anton. 2004. *National Geographic Prehistoric Mammals*. Washington, DC: National Geographic Society.
Turner, A., and M. Anton. 2004. *Evolving Eden: An Illustrated Guide to the Evolution of the African Large-Mammal Fauna*. New York: Columbia University Press.
Wang, X., and R. H. Tedford. 2008. *Dogs: Their Fossil Relatives and Evolutionary History*. New York: Columbia University Press.
Werdelin, L., and W. L. Sanders, eds. 2010. *Cenozoic Mammals of Africa*. Berkeley: University of California Press.
Wood, A. E. 1962. The early tertiary rodents of the family Paramyidae. *Transactions of the American Philosophical Society* 52 (1):3–261.

INDEX

For many scientific names in the index, a pronunciation guide [in brackets] is included. This represents the most common pronunciation by vertebrate paleontologists in the United States, although there are some that have no standard pronunciation.

INDEX

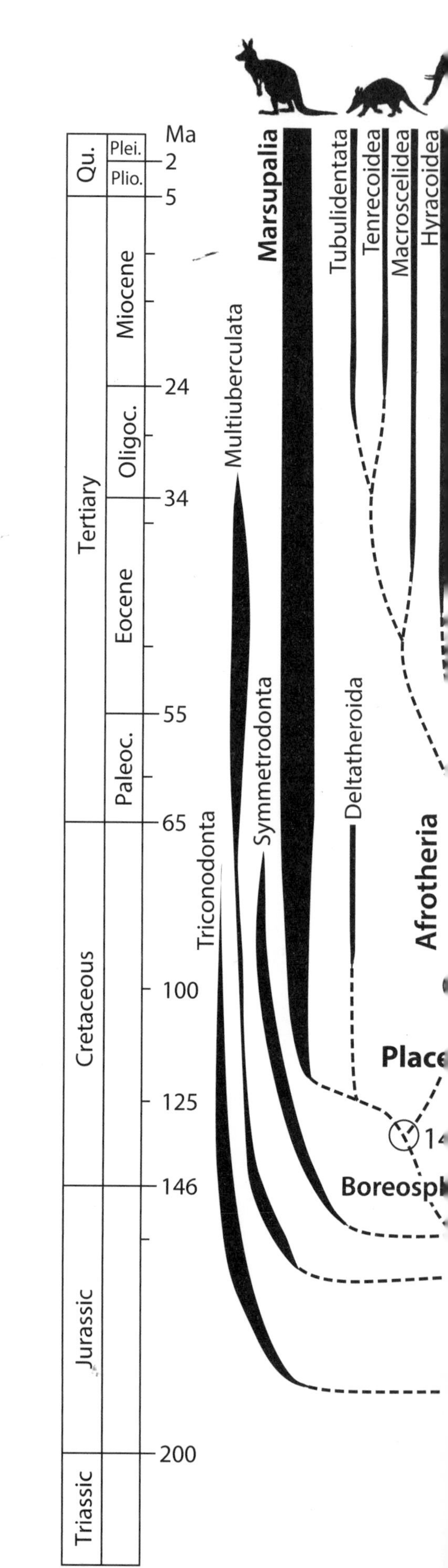

Ma
2
5
24
34
55
65
100
125
146
200
Qu.
Plei.
Plio.
Miocene
Oligoc.
Eocene
Paleoc.
Tertiary
Cretaceous
Jurassic
Triassic
Marsupalia
Multiuberculata
Symmetrodonta
Triconodonta
Deltatheroida
Tubulidentata
Tenrecoidea
Macroscelidea
Hyracoidea
Afrotheria